CHAINS
FOR
POWER TRANSMISSION
AND
MATERIAL HANDLING

MECHANICAL ENGINEERING

A Series of Textbooks and Reference Books

1. Spring Designer's Handbook, *by Harold Carlson*
2. Computer-Aided Graphics and Design, *by Daniel L. Ryan*
3. Lubrication Fundamentals, *by J. George Wills*
4. Solar Engineering for Domestic Buildings, *by William A. Himmelman*
5. Applied Engineering Mechanics: Statics and Dynamics, *by G. Boothroyd and C. Poli*
6. Centrifugal Pump Clinic, *by Igor J. Karassik*
7. Computer-Aided Kinetics for Machine Design, *by Daniel L. Ryan*
8. Plastics Products Design Handbook, Part A: Materials and Components, *edited by Edward Miller*
9. Turbomachinery: Basic Theory and Applications, *by Earl Logan, Jr.*
10. Vibrations of Shells and Plates, *by Werner Soedel*
11. Flat and Corrugated Diaphragm Design Handbook, *by Mario Di Giovanni*
12. Practical Stress Analysis in Engineering Design, *by Alexander Blake*
13. An Introduction to the Design and Behavior of Bolted Joints, *by John H. Bickford*
14. Optimal Engineering Design: Principles and Applications, *by James N. Siddall*
15. Spring Manufacturing Handbook, *by Harold Carlson*

16. Industrial Noise Control: Fundamentals and Applications, *by Lewis H. Bell*
17. Gears and Their Vibration: A Basic Approach to Understanding Gear Noise, *by J. Derek Smith* (in press)
18. Chains for Power Transmission and Material Handling: Design and Applications Handbook, *by the American Chain Association*

OTHER VOLUMES IN PREPARATION

CHAINS FOR POWER TRANSMISSION AND MATERIAL HANDLING

Design and Applications Handbook

AMERICAN CHAIN ASSOCIATION

MARCEL DEKKER, INC. New York and Basel

Library of Congress Cataloging in Publication Data
Main entry under title:

Chains for power transmission and material handling.

(Mechanical engineering ; 18)
Based, in part, on 2 earlier ACA manuals.
Includes index.
1. Chain drives—Handbooks, manuals, etc.
2. Chain conveyors—Handbooks, manuals, etc.
I. American Chain Association. II. Series.
TJ1051.C47 1982 621.8'58 82-17971
ISBN 0-8247-1701-5

MARCEL DEKKER, INC.
270 Madison Avenue, New York, New York 10016

Current printing (last digit):
10 9 8 7 6 5 4 3 2 1

PRINTED IN THE UNITED STATES OF AMERICA

Foreword

The predecessor organization of today's American Chain Association was formed in the 1930s. While a routine function of collecting and correlating marketing data for distribution was an important aspect, the primary purpose was to act for the continued development of the American chain industry and the benefit of its customers as a whole. This objective "to develop and promote standards of sound manufacturing and engineering practice, and through well-guided research, to foster improvements in the quality and utility of the industry's products" was spelled out in the first pages of one of its early publications.

This credo, as promulgated by the then Association of Roller and Silent Chain Manufacturers (ARSCM) soon led to a series of projects to build and operate chain test equipment on a laboratory basis. From those projects were developed the horsepower curves covering all basic sizes of standard roller drive chains. These were included in a hardbound chain design manual, published by ARSCM in 1955, and entitled *Design Manual, Roller and Silent Chain Drives.* That edition and its softbound successors, published in 1968 and 1975, have seen nearly 30,000 copies distributed by the Association.

The Association soon expanded to include manufacturers of Engineering Class Steel Chains and Malleable Chains, and the Association name was changed to American Sprocket Chain Manufacturers Association (ASCMA) and then to the present American Chain Association (ACA).

The laboratory projects were extended to Engineering Steel drive chains and horsepower ratings were developed for those chains. These tables and other data developed by the Association were adopted by the then American Standards Association. This latter organization also went through name changes and is known today as the American National Standards Institute (ANSI). The standards and tables developed by the American Chain Association are revised and updated by ACA periodically and resubmitted to ANSI for adoption.

The inclusion of other types of chain in the ACA program and the development of horsepower curves for the Engineering Steel drive chains also entailed considerable standardization of sizes. It also led to the publication in 1971 of another new manual, *Engineering Steel Chains, ACA Applications Handbook* of which a slightly revised version appeared in 1973.

The Roller and Silent Chain and the Engineering Steel Chain manuals, published as separate handbooks, contain some duplicate information. The text of the combined manual has been formed by an editorial blending of the various sections of the two original manuals, plus some additional material not appearing in those publications.

The publication of the combined manual also represents a continuation of the objectives of the American Chain Association to enhance the quality and utility of its products. The manual presents the information developed by the Association and all its member companies.

The authorship of the combined manual is a combined effort of the technical committee and other representatives of the member organizations of the ACA over a period of 25 years.

ACA member organizations participating in the combined manual are as follows:

Acme Chain, Incom International, Inc., Holyoke, Massachusetts
Diamond Chain Company, Indianapolis, Indiana
Drives, Inc., Fulton, Illinois
Jeffrey Chain Operations, Dresser Industries, Inc., Morristown, Tennessee
Moline Corporation, St. Charles, Illinois
Morse Industrial Products, Borg-Warner Corporation, Ithaca, New York
Portland Chain Mfg. Co., Div. Webster Industries, Portland, Oregon
PT Components, Inc., Link Belt Chain Division, Indianapolis, Indiana
Ramsey Products Corporation, Charlotte, North Carolina
Renold Power Transmission Corporation, West Pittston, Pennsylvania
Rexnord, Inc., Milwaukee, Wisconsin
Webster Mfg. Co., Div. Webster Industries, Tiffin, Ohio
Whitney Chain Operations, Dresser Industries, Inc., Hartford, Connecticut

American Chain Association

Preface

Just about anyone recognizes a chain when it is seen. It is also taken for granted that many different kinds of chain exist, although identification of many of the types and styles by name is not an offhand matter to most viewers. Most of us have not inquired beyond "log" and "tow" chains, the bicycle and motorcycle chains we saw at an early age, and, perhaps, the bead chain used to operate the overhead light in the attic.

Thus, assuming a basic, if somewhat casual knowledge of chain upon the part of readers unfamiliar with the subject, the early chapters of this book attempt to better organize and expand that knowledge. This is done by an overview of how chain was developed, a summary of the types, and how they are made and used.

For the designer, unfamiliar with chain and attempting to select a chain from a manufacturer's catalog for the first time, this first attempt can be rather perplexing. Not only are there a mass of types and data available but there exists what at first glance appears to be an inconsistent terminology throughout the field. (Actually, the terminology is quite logical, differing only for different styles of chain, as will be seen.)

It is important for the designer to select the right kind of chain for a given application and this manual will help clarify the data shown in the catalogs as well as explain the backgrounds of the various chain types. While the multiplicity of these types may appear confusing at first glance, they and their terminology are logical products springing from a constantly growing spectra of industrial needs.

Chapter 1 traces the development of chain as it exists today, showing it to be a parallel development in competitive paths of today's classes of chains. Each individual group developed its own terminology to a degree to differentiate its products from other types of chain available.

The early chapters of this book, which constitute a "primer" on chain, trace the development of chain with the purpose of allowing the reader to comprehend the terminology by examining it as a logical evolution. As the industry grew in parallel paths, so did the terminology until the advent of the American Chain Association with its published standards. This has caused a gradual change to more nearly uniform terms throughout the industry except where a real need for differences exists.

The manual is organized as a whole in sections aimed at providing an understanding of chain from the standpoint of applications. As has been already stated, the early chapters are a chain primer; broad areas of usage as correlated with product design are discussed, and the section concludes with a chapter about sprockets.

The next four chapters concern themselves with selection and application of chains for power transmission. This includes not only power transmission drives but tension linkages. Drive chains transmit power between parallel rotating shafts. Tension linkages simply serve as flexible tension members to conduct a force between points; such chains are often not "endless" in that they are not connected as a loop.

Selection of power transmission chains is covered in chapters on Roller, Engineering Steel, and Silent chains. Although there is overlap in the sense that any of the three types could be used for some applications, the type of need plays a very large part in the style of chain selected.

Silent Chains can be used for some very high levels of chain speed; Roller Chains are also often used in high speed drive applications. Engineering Steel offset heavy duty drive chains is normally used where heavy loads and shocks, lower speed ranges, and "unfriendly" application environments are involved.

The process of selection of chains for power transmission is relatively similar regardless of the chain considered. The same information in terms of speed, load, and details of operation is needed.

Chapters 9 through 11 deal with chains for material handling, and include chains for heavy and light conveyors and for bucket elevators.

Chain conveyors of an endless number of styles are used for handling parts, packages, and bulk materials weighing from a few ounces to several tons. They are used in every conceivable basic industry: mining, machine shops, construction equipment, packaging machines, agricultural, food production and processing, cement, automated assembly lines, aggregates, steel, lumber, foundries, and sewage treatment to name a few.

It is probable that every chain covered in this manual has at one time or another been used for material handling. Even Silent Chain, considered by most to be purely for power transmission, is extensively used to convey hot glass and metal components.

The last chapter of the manual is about the care and servicing of chain installations and concludes with a check-list of safety activities to be considered in dealing with chains. A glossary follows, intended to supplement the text by including terms not totally defined in the text, or which may be encountered by the reader but are not otherwise included in the manual.

Operational chain problems are often traceable to misapplication. The use of this manual, leading to a better understanding of manufacturers' catalogs and chain in general, should serve to greatly reduce if not eliminate this problem. As an engineering reference, the manual will provide a sound basis for correct selection of the best class of chain for a proposed application. The engineering school student seeking a subject for thesis can find many challenging problems in the field of chain. This manual should serve as a basic reference source for such projects, as well as indicate potential fields of study.

Some of the aspects of design of chain and allied components can be technically rather complex. A professor of machine design at a major university once informed the writer that the chain-sprocket relationship was the "classic problem in kinematics and design." Professor Binder at Purdue University devoted an entire book to this relationship relative to Roller Chains. Actually, these types of studies need not be the concern of those selecting and using chains. The chain-sprocket relationships are the responsibility of the manufacturers and have been duly considered in establishing chain ratings and recommendations for use.

The information contained in the manufacturers' catalogs in this manual is a distillation of over a hundred years of chain engineering effort. It may be used with confidence.

Richard B. Graham
Columbus, Ohio

Contents

CHAINS
FOR
POWER TRANSMISSION
AND
MATERIAL HANDLING

1

A Brief History of the Development of Chain

Chains have been used to transmit power to drive machinery and to transport materials along conveyors and up elevators for centuries. In 225 B. C. per Philo's treatise, a chain-driven water lift was used, as shown in Figure 1-1. This was really a form of bucket elevator, and it has a more modern counterpart in a form of pumps developed in the nineteenth century to raise water from shallow wells in which disks were substituted for the buckets.

In the 1500s, Leonardo daVinci sketched a group of chain designs (Fig. 1-2), some of which have modern counterparts of remarkable similarity. Ancestors of modern bar link, leaf, and perhaps Silent Chains may be noted among his designs.

Figure 1-3 pictures Ramelli's waterwheel-driven rotary pump, in which he used a chain drive to transmit power in the sixteenth century.

Undoubtedly, many other examples may be found of uses of chain for power transmission and conveying in pre-nineteenth-century times. However, it was not until the 1800s that the history of modern chain development really began.

In the early 1800s, the cog chain was developed to transmit power or motion between the shafts of treadmills to water elevators, weaving looms, and harvesting machinery. This chain, shown in Figure 1-4, consisted of rectangular cast links connected by looped and riveted iron bands. Because the links contacted the sprocket tooth, or "cog" as it was then called, it became known as *cog chain.*

Although this chain made a strong contribution to the mechanization of farm implements, it broke easily and was troublesome to repair in the field. These problems were overcome in 1873 with the development of cast detachable chain, as shown in Figure 1-5. This was a simple cast metal chain composed of identical links which could easily be coupled or uncoupled by hand. It also greatly improved the performance of power takeoffs from cleated bull-wheels in contact with the ground under horse-drawn farm implements. Thus the mechanization of agricultural equipment soon became a practical reality, and the use of chain for power transmission drives rapidly spread to other major industries.

Figure 1-1 Philo's chain drive to a bucket elevator (225 B.C.).

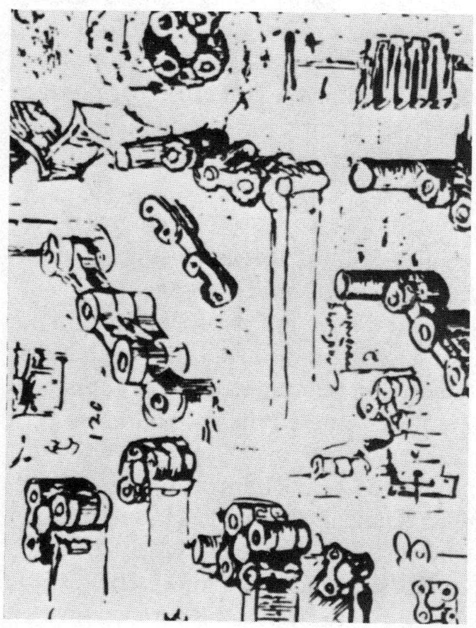

Figure 1-2 Leonardo daVinci's sketch of chains.

Figure 1-3 Chain drive from undershot waterwheel to rotary pump (Ramelli, 1588). (W. A. Williams, *Mechanical Power Transmission Manual,* Conover-Mast, New York, 1953.)

Figure 1-4 An early type of chain known as cog chain.

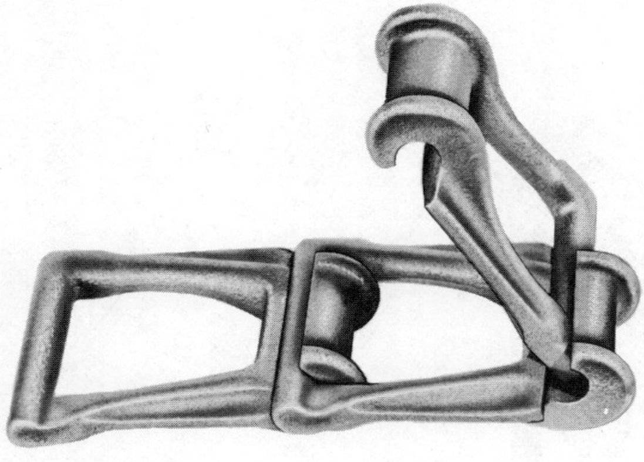

Figure 1-5 Cast detachable chain.

This basic detachable chain design is the earliest of the industrial chain concepts that have come down to us today virtually unchanged. Malleable iron detachable chain and its fabricated steel counterparts, made with highly sophisticated and automated equipment, is still in use today and is available from manufacturers and their distributors.

By the 1800s, the cast basic detachable links had been complemented with integral cast *appendage* or *attachment* links that could be installed at intervals in the basic chains. Such a detachable chain attachment link is shown in Figure 1-6. To attachment links somewhat like these malleable iron *buckets* could be bolted. Used vertically, in an arrangement such as that of Philo's water elevator of Figure 1-1, this became a bucket elevator for loose bulk materials. By the 1890s, designs of bucket elevators such as the one shown in Figure 1-7 were being offered as a standardized machine by

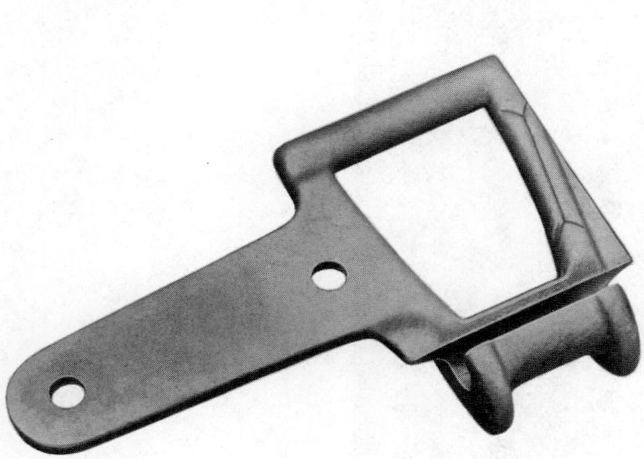

Figure 1-6 A-113 attachment link for detachable chain.

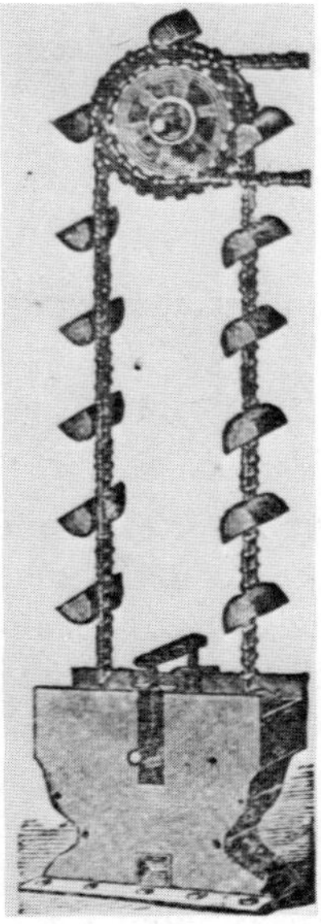

Figure 1-7 Early bucket elevator using cast chain with malleable iron buckets bolted to attachments.

many manufacturers. As will be seen later in this book, there are many modern kinds of bucket elevators in use today.

For conveying of bulk materials horizontally or up mild inclines, slats or *flights* rather than buckets were bolted to the cast attachments, this becoming the modern apron conveyor. To convey bulk materials up fairly steep inclines as well as horizontally, *drag* conveyors using different styles of attachments and scraper flights in a trough were developed. An early version, as illustrated in a manufacturer's catalog, is shown in Figure 1-8.

In the same span of time, the cog chain of the early nineteenth century was adapted for conveying service by stretching out the links and providing either cast integral attachment appendages to the plain link form, or, in the blacksmith shop, hammering and welding all-steel links to which attachment sections could be welded. An early chain of this nature with scraper attachments is shown in Figure 1-9.

This type of chain, used primarily for scraper conveyors handling wet or abrasive materials, is to a degree the direct ancestor of the steel forged open coil and rivetless chains used for similar materials handling applications today.

Figure 1-8 Scraper conveyor of the 1890s.

Figure 1-9 Extended-pitch (flattened) cog-style chain of the late 1800s with scraper flights.

Pintle Chain

The modern cast descendants of the types of chains discussed previously are outside the scope of this book. The next important development after the cast detachable chain was the cast *pintle* chain, shown in Figure 1-10. This type of chain, the direct ancestor of both standard roller chain and engineering class steel chains, had a "closed barrel" design, heavier link sections, and steel pins or *pintles*. It evolved from the cast detachable chain link design to meet the requirements of heavier loads, higher speeds, and more severe operating conditions in both drives and conveying applications. Modifications were soon made to the original pintle chain design to meet specialized needs. For example, "shoes" were added to provide greater resistance to sliding wear, as illustrated by the links shown in Figure 1-11. This design is still widely used today. Not all the innovations, however, were so successful; the patented design shown in Figure 1-12, with ribs on the sides carefully interlocked to present a smooth and continuous surface against potential hang-ups, is long gone. Although the elaborate ribs probably did serve a useful purpose for some long-forgotten application, lateral control of the chain path has since been provided by other methods of a great deal less elaborate nature. The design is, however, typical of the rather "gingerbready" look apparently deemed to be a sign of great competence in manufacturing in the 1890s. A great deal of effort was spent on features with more value in appearance than function by manufacturers to obtain a larger share of what was becoming a booming market.

The short-pitch pintle chains were used for both drive and conveying applications, but for the latter, new types of chains of longer pitches were soon developed. Not only did these chains have better economics because of having fewer joints, but the longer pitches provided more space for attachment wings or more complex designs, such as that of Figure 1-13. Such chains were used for conveyors of increased length, and soon, exposed hardened steel bushings were installed in the barrels of the

Figure 1-10 Cast pintle chain.

Figure 1-11 Cast pintle chain with wear shoes.

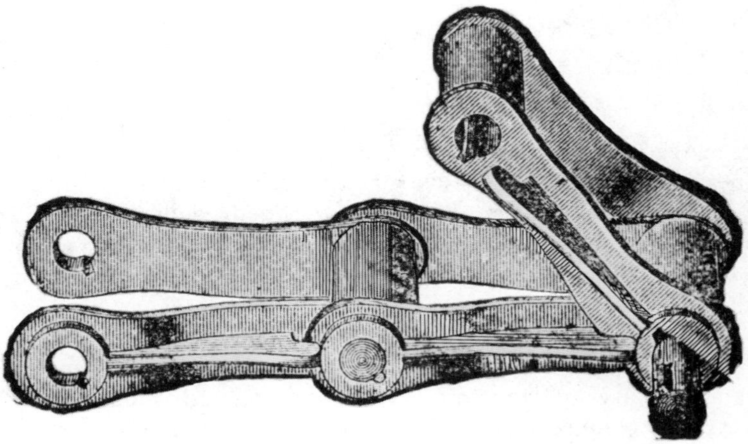

Figure 1-12 Pintle chain with interlocking side ribs presenting a smooth surface to prevent hang-ups.

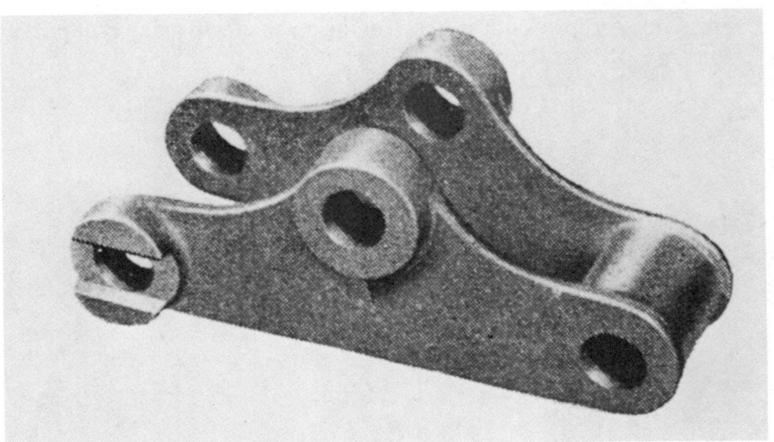

Figure 1-13 Long-pitch pintle chain style M-3 attachment link.

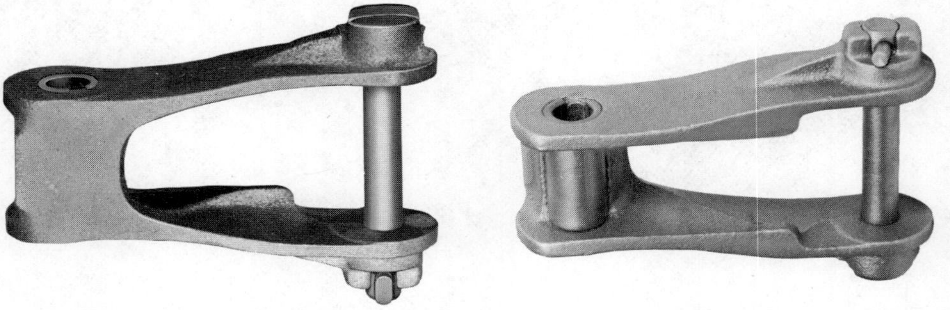

Figure 1-14 Pintle chain with hardened steel bushings. Top and bottom are shown. The sprocket makes contact with the exposed portion of the steel bushing (right).

cast links to provide increased resistance to wear. Links of this type, shown in Figure 1-14, were found to work quite well in abrasive materials and became the standard bucket elevator chain used during the first third of the twentieth century.

Both pitch and width of cast pintle links were expanded to provide *drag* chains of high efficiency and dependability, as shown in Figure 1-15. This design is still considered an extremely effective product for conveying of abrasive or wet and sticky bulk materials.

Rollers revolving on the link barrels, the links being split through the chain centerline perpendicular to the axes of the joints, were found to greatly reduce power needs on long conveyors by eliminating the sliding friction of the links on conveyor ways or tracks. This design is shown in Figure 1-16.

Thus by the early 1900s, the cast pintle chain had been developed to the point that industry, in the United States and abroad, had an efficient and dependable product available for power transmission, conveying, and elevating. Although cast chains are not discussed in this book, they occupy an important spot in today's marketplace. Figure 1-17 shows a typical modern application of cast malleable iron pintle chain used for power transmission. A representative conveying application using cast pintle chain is depicted in Figure 1-18.

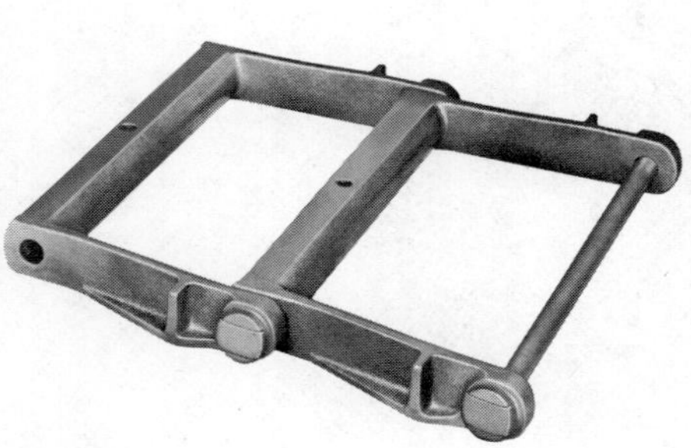

Figure 1-15 Long-pitch expanded-width pintle chain, known as drag chain.

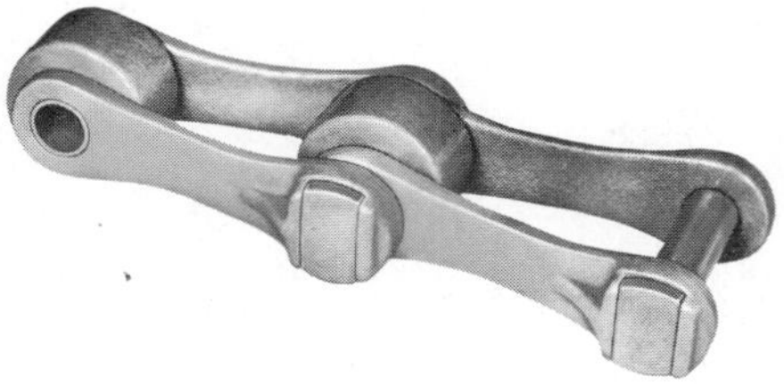

Figure 1-16 Pintle chain with cast rollers.

Figure 1-17 Malleable iron "mill" class chain in power transmission service on a traveling water screen used to prevent debris in the water from entering the intake.

Figure 1-18 Malleable iron "pulpwood"-style combination class chain used to lift logs from a flume on a "jack ladder" in a forest products plant.

Figure 1-19 Chain drive on safety bicycle (1895).

Precision Roller Chain

About a decade after cast detachable chain was introduced, a chain made entirely of steel components was introduced for use in driving bicycles. This was the beginning of the Roller Chain market. By the early 1900s, Roller Chain was in use to drive the wheels of the safety bicycle (Fig. 1-19), the newly invented automobile, and the propellers of the Wright Brothers' airplane flown at Kitty Hawk (Fig. 1-20).

The chains developed in the nineteenth century were a key component and thus a major factor in the development of the vehicles and other industrial equipment upon which they were used. They either met or exceeded the needs of their era in terms of load-carrying capacity, speed of operation, and wear life. However, so fast was the development of the mechanization of which they were a part that the early designs would have fallen far short of being adequate within a decade had they not been upgraded constantly. The public demand for large quantities of chain to be installed on bicycles and soon, on motorcycles, automobiles, and trucks, sparked organized methods of manufacturing that led to the size and success of the modern chain industry.

Two major factors, both being innovative and historically important as examples of the early industrial and technological development of the country, were largely responsible for the development of the Roller Chain industry as it exists today. These, in today's terminology and stated briefly, are automation and standardization.

The former was needed to meet the demands for large volumes of chains of uniformly high quality to be used on machinery manufactured by other industries. Man-

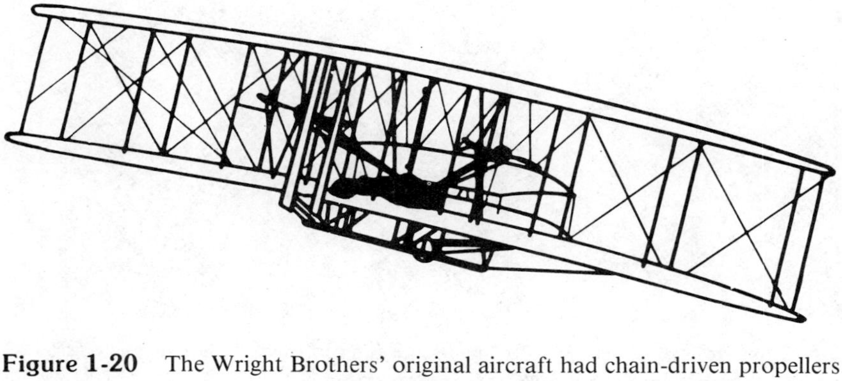

Figure 1-20 The Wright Brothers' original aircraft had chain-driven propellers.

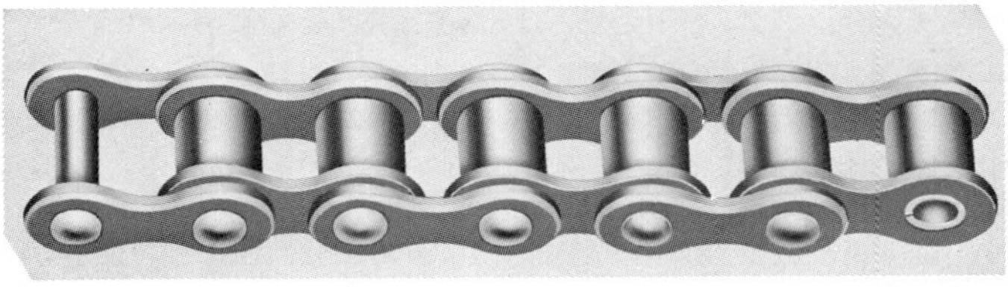

Figure 1-21 Typical Roller Chain.

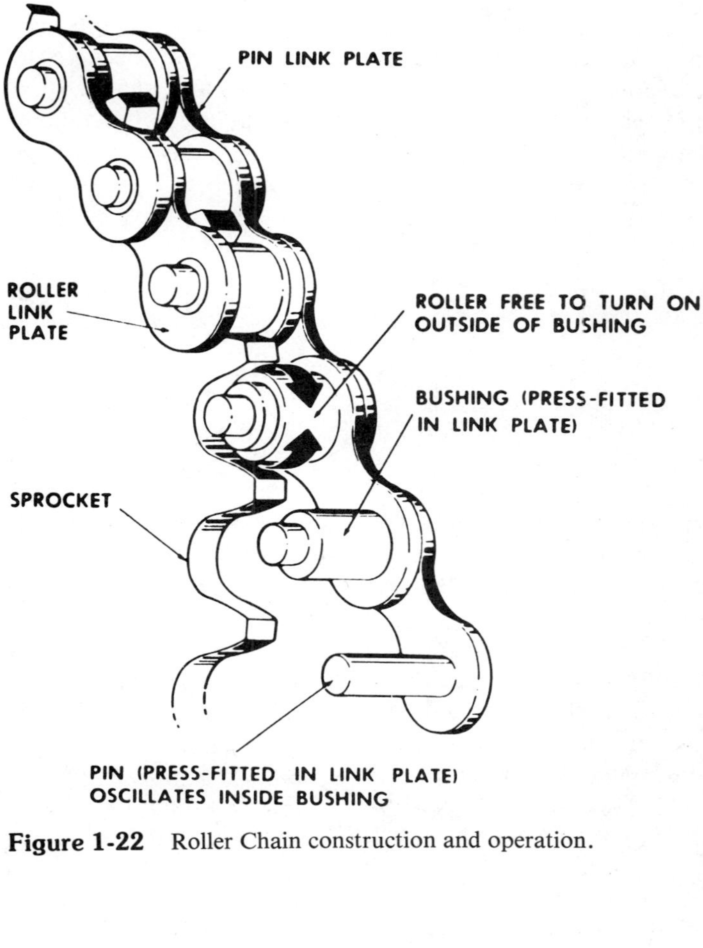

Figure 1-22 Roller Chain construction and operation.

ufacturing equipment such as punch presses and automatic screw machines had to run long hours at high speeds to meet production needs. Mechanical material feed equipment on such machines required high-quality raw materials in terms of both dimensional and metallurgical properties, and great uniformity of parts' details was required to permit high-volume assembly machines to run without interruption.

Since many manufacturers of roller chains had been organized to participate in what was obviously a booming market with long-term potential, the need for uniformity of details such as product dimensions and capacity was soon apparent. This need was in terms of those details required to fit and operate with mating parts of the machines upon which the chains were to be installed, as determined by the designers of the machines. Standardized dimensions among the manufacturers were also necessary in terms of replacement links and chain being available from stock everywhere.

The outcome of all this was that the industry soon became identified as the *precision Roller Chain industry,* and efforts aimed at standardization, beginning as early as 1913, led to the precision Roller Chain industry being one of the first groups in the world to have published standards available to designers and others concerned with chains.

Figure 1-21 shows a section of standard Roller Chain, and Figure 1-22 is a cutaway view showing the component parts of a Roller Chain as it engages a sprocket. The effect of the Roller Chain industry standard is that chain made by any manufacturer will fit the sprockets furnished by any other manufacturer complying with the standards. In addition, the terms "Roller Chain" and "Roller Chain industry" have become accepted generic terms and refer to the class of chain being discussed quite specifically. More precisely defined, *Roller Chain* is that chain covered by American National Standards Institute (ANSI) Standard B29.1 and related published standards.

Terminology

In this book, which discusses many chains of various types that have rollers, the term "Roller Chain," always capitalized, will refer to standard Roller Chain made to conform with ANSI Standard B29.1. When the words "roller chain" are not capitalized, the usage will be a more general reference to chain equipped with rollers, regardless of standards.

Certain terms or phrases commonly used in discussions of chain are well known and accepted. "Chain pitch," and "pitch diameter," and "link" and "sprocket" are the most common. These are defined in the glossary of this book and also, implicitly at least, at many places in the text.

As noted in the Editor's Preface, varying terminology is applied by chain catalogs and other published works to what are functionally similar components of various types and classes of chains. This divergent terminology stems from the widely separated, although parallel development of Roller Chain and Engineering Steel Chains over a span of nearly 100 years. Each group developed its own terminology, the usages being reflected not only in the catalogs and advertising of manufacturers, but also in technical papers, ANSI standards, and other published works referring to chain throughout the world.

At first glance, the differences in terminology would seem to add confusion to what appears to those not involved more than casually with chain, to be a rather complex subject. Actually, the variations in terms are quite useful, since the differing names are generally applied to components only functionally similar, but differing widely in modes of manufacture, shape, finish, and in many other ways.

For example, the tension members between joints of Roller Chain are called *link plates*; the functionally similar member of an Engineering Steel Chain is usually termed a *sidebar,* but in some types of chains used for tension linkage applications, these parts are called *blocks and bars.*

Link plates for Roller Chain (and Silent Chain) are usually made in very large quantities using high-speed machinery and tooling having a high degree of sophistication. The characteristic shapes of the link plates are completely derived from the tooling. Additional operations, often done with progressive die tooling and automatic materials' feed and transfer devices, have been performed to provide improved quality of holes for pins and bushings and sheared edges of link plates. The link plates are normally nicely finished, surface conversion treatments often being provided and rust preventive treatments or coatings being routine.

Figure 1-23 shows a Roller Chain link plate and an Engineering Steel Chain typical sidebar. Obviously, although there are similarities, the two parts differ greatly in appearance.

The sidebars produced for Engineering Steel Chains are normally manufactured in relatively small numbers compared to the quantities of link plates for Roller Chain that would be made on a single production run. Conversely, a single sidebar of, for example, a long-pitch conveyor chain may weigh as much as 100 link plates of intermediate size. Such a sidebar would be made from hot-rolled steel bar, and the cross-sectional shape and side and edge finishes would be that of the raw stock with scale removed. Holes for pins and bushings will have been punched with possibly secondary operations to improve hole quality and control pitch. Rust preventive treatment will probably be in the form of a dip applied to coils of the finished chain.

The link plate usually will be placed in a chain running in a fairly friendly environment—but much will be expected of it. A typical application could involve running a drive in an oil bath, under relatively heavy loads (i.e., a high percentage of the rated chain capacity) at high speeds for months at a time. On the other hand, the long-pitch conveyor chain sidebar might be installed in a chain used in a woodyard. Subject to moisture from both weather and the materials passing over the conveyor, joints of the chain would have to flex in an environment of water and sand, in which the chain may also slide. Loading would be quite irregular, as would operation. The relatively slow conveyor speeds and the longer pitch and greatly longer amount of chain used in the

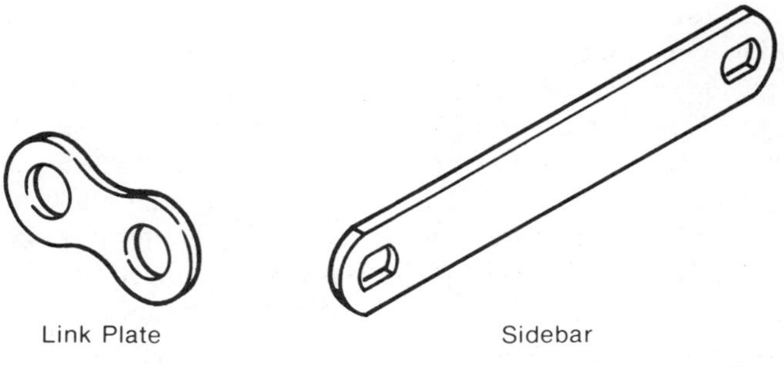

Figure 1-23 Roller Chain link plate and Engineering Steel Chain sidebar.

conveyor than in the drive produce an entirely different kind of life cycle for each. The sidebar will have had relatively few cycles of operation compared with the enormous number of times the link plate in the drive chain would have passed over its sprockets. Elongation from wear or link fatigue failure may eventually cause a drive chain to be replaced, but a conveyor chain is usually the victim of corrosion and abrasive wear.

Each part, link plate, or sidebar is produced by processes calculated to generate components appropriate for the end use of the kind of chain in which it is installed. Once the beholder understands that the parts, although functionally similar, represent substantial differences in size, shape, speed of operation, and end use or application, the terminology becomes an easy key to a great deal of information.

Figure 1-24 illustrates a typical application of Roller Chain on a drive—if any such application can be called typical. Drives are the predominant use of Roller Chain, although a great deal of it is used in extended-pitch and large-diameter roller versions for conveyor service. A typical conveyor using Roller Chain is shown in Figure 1-25.

Engineering Steel Chains

Engineering Steel Chains were first developed in the 1880s for about the same reasons that steel Roller Chain was developed—to meet the needs for higher strength, speeds, and shock resistance and for better dimensional control than was obtainable from the materials and methods of manufacturing cast links.

The major difference was that whereas Roller Chain was developed primarily to meet drive chain needs, the early Engineering Steel Chains were designed for difficult conveying applications. Just as with Roller Chains, Engineering Steel Chains were developed as all-steel products fabricated from rolled shapes. An occasional exception was that rollers, particularly those of the flanged variety, were made of cast iron, and this exception has continued into more recent years.

The early Engineering Steel Chains were a direct takeoff from cast pintle chains in that they were designed with equivalent pitches and similar dimensions to those of cast chains so that the steel chains could be fitted to identical sprockets. This procedure meant that higher-strength chains could be substituted for cast chains when necessary without needing to change to sprockets of a different design.

Larger sizes of Engineering Steel Chains were soon developed, in terms of pitch, strength, wear life, and carrying capacity as required to meet the heavy duty needs of industry. For convenient and practical attachment spacing, many Engineering Steel Chains of the conveyor class were made in pitch sizes of even increments of a foot. Typical sizes, as manufactured for many years, are 4-, 6-, 9-, 12-, and 18-in. pitches. Many of these designs have identical joint details and are available in three or more of the standard pitches, with a wide range of attachment types available.

Regardless of size and type, Engineering Steel Chains have been developed specifically for dependable operation under the most demanding conditions required by industry. Whereas Roller Chains are to a considerable extent sold in consumer markets, Engineering Steel Chains are used almost entirely for industrial applications. ANSI standards exist or are being considered for nearly all classes of Engineering Steel Chains, and the standard chains are, to a large extent, available from manufacturers' stocks. An important segment of the Engineering Steel Chain market is in the form of

Figure 1-24 Typical Roller Chain power transmission and speed reduction application.

Figure 1-25 This packaging machine is typical of Roller Chain applications.

special chains for unique installations or specialized duplicate machinery where unusual designs are needed.

Types of Engineering Steel Chains

There are many general types of Engineering Steel Chains, of which some of those covered in this book are shown in Figures 1-26 to 1-29. Those with steel rollers are perhaps the most widely used, and are applied for both drives and conveyors. A bushed, rollerless style meets the needs of many conveyor and bucket elevator applications. Welded-steel versions of the basic cast chains are becoming increasingly popular, and a simple bar-link variety is used for slow-moving conveyors and tension linkages. Each of these types is illustrated and described in greater detail in a subsequent chapter.

It is only in the last few decades that certain of the Engineering Steel Chains have been standardized so that recognized standards on them could be issued. Heavy duty roller-type conveyor chains are covered in ANSI Standard B29.15, steel bushed rollerless chain in B29.12, and welded steel mill and drag chains in B29.16 and B29.18, respectively. Heavy duty offset sidebar power transmission roller chains were covered in the first of the Engineering Steel Chain standards to be issued, ANSI B29.10. No standards are currently available on steel bar-link (or block-and-bar) chains. However, standards are constantly being considered for such types as open barrel steel pintle chain, where large quantities of nearly identical chain are being supplied by manufacturers.

Figures 1-30 to 1-32 illustrate typical applications of Engineering Steel Chains on a conveyor, a bucket elevator, and a drive, respectively.

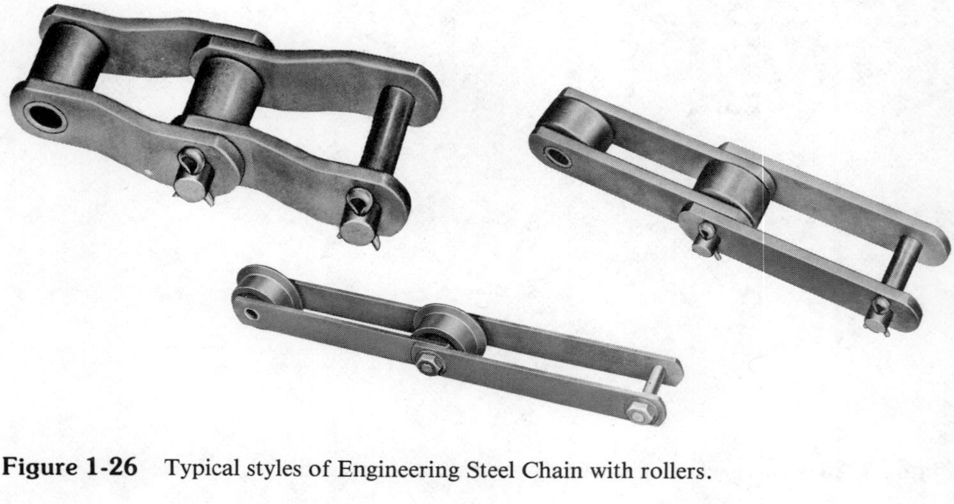

Figure 1-26 Typical styles of Engineering Steel Chain with rollers.

Figure 1-27 Bushed rollerless Engineering Steel Chain.

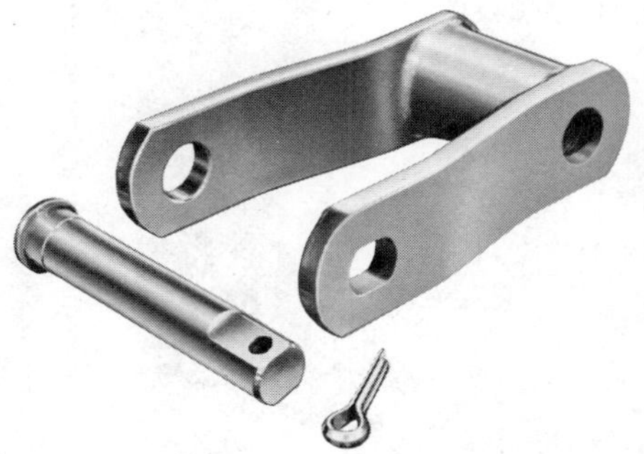

Figure 1-28 Welded steel version of basic cast pintle chain.

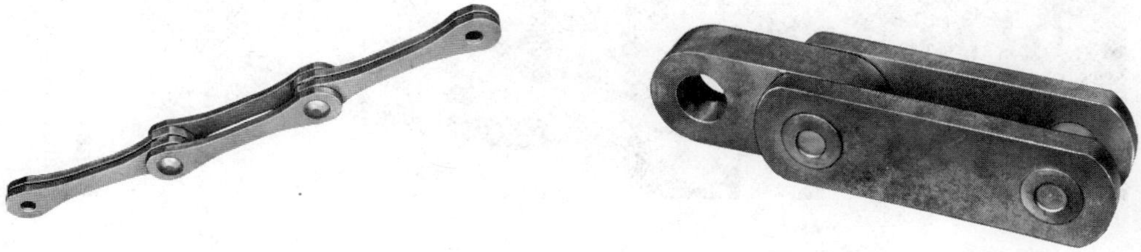

Figure 1-29 Engineering Steel Chains of the bar-link, often called block-and-bar style.

Figure 1-30 Engineering Steel Chain used on a scraper conveyor with the scraper flights between the strands—a typical application for this class of chain.

Figure 1-31 Typical bucket elevator installation. The cutaway view shows how chains, buckets, and sprockets are arranged in a "positive discharge" bucket elevator.

Figure 1-32 Engineering Steel heavy duty offset drive chain (left) may be used for power transmission as a "cradle"-style drum drive, as shown here for rotary kilns or grinding mills.

Silent Chains

Silent Chain was developed in the early part of the nineteenth century. Its advent was spectacular; very few products have ever appeared in such a large version in a highly publicized application and performed with an effectiveness of products manufactured over 100 years later.

The S.S. *Great Britain,* built in the early 1840s, was truly a landmark vessel in ship construction (see Fig. 1-33). She was the first seagoing iron steamer; her hull, with double-bottom construction and excellent lines, was not unlike that of a modern steel ship. In addition, she was the first transatlantic screw steamer ever built. Since the really large steam engines of her day were slow-speed, a "step-up" drive was needed to attain the necessary propeller shaft revolutions required. Powered by two steam engines producing a total of nearly 2000 hp, engine speed was limited to 20 rpm, but a shaft speed of 53 rpm was needed.

The ship's developer, Sir Isambard Kingdom Brunel, is well known in history for his development not only of the *Great Britain,* but for his later vessel, the "great iron ship," the *Great Eastern.* It is doubtful that any development of Brunel's had more technical impact with less fanfare than his method of transmitting power from engines to the screw shaft of the *Great Britain,* for he apparently developed the multistrand inverted tooth or "silent" chain for the purpose. The drive, from an 18-ft teakwood drum between the engines to a 6-ft lignum vitae wheel on the propeller or screw shaft, is shown in Figure 1-34. It is reported that the drive ran very quietly. Further discussion of this drive is included in Chapter 7.

Silent Chain did not see much growth in the United States, in spite of its auspici-

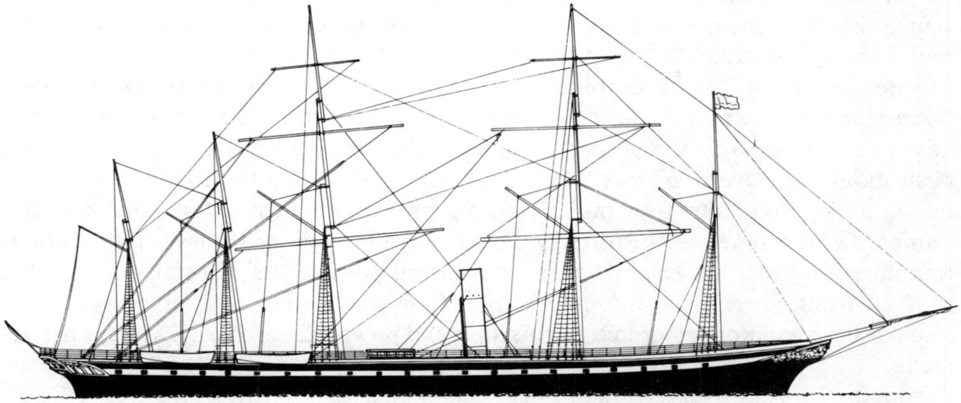

Figure 1-33 The *Great Britain* of 1845. The first oceangoing iron steamer, the first transatlantic screw steamship, and the first Silent Chain drive of importance.

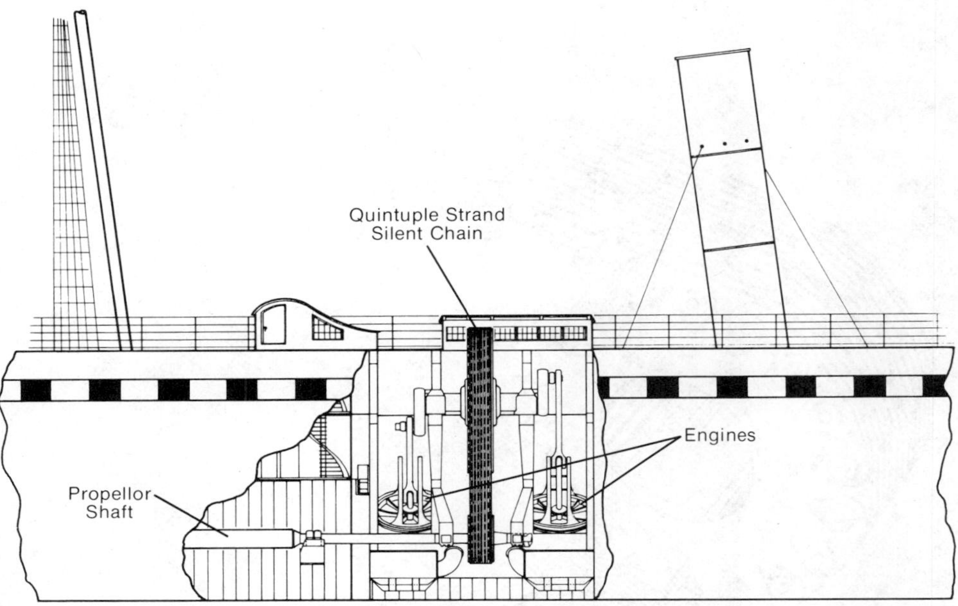

Figure 1-34 The drive arrangement of the *Great Britain.*

ous European beginning, until the 1920s. By the 1930s Silent Chain was used as a high-speed, quiet alternative to precision Roller Chain.

Silent Chains are made up of a series of leaf links with gear-type contours so designed that they engage the sprocket teeth in a manner similar to the way in which a rack engages a gear.

Because Silent Chains combine the flexibility and quietness of a belt and the positive action and durability of gears with the convenience and efficiency of a chain, they have been found to be the proper selection in the developments of drives for many basic industries. Over the years Silent Chains have been used for drives in paper and textile mills, flour and feed mills, printing presses, industrial fans and blowers, pumps, and machine tools. Probably the application receiving the most public attention in recent years has been their use to transmit power from transmission to drive shaft in front-wheel-drive automobiles and four-wheel-drive vehicles. Silent Chain also has been a major factor in transmissions and power trains of trucks and other vehicles.

Figure 1-35 shows a section of Silent Chain on a sprocket. Like the Roller Chain and Engineering Steel Chain industries, Silent Chain has components and terminology peculiar to itself. Tension members are called link plates, as with Roller Chain, but pin and joint designs vary widely within the industry.

Figure 1-36 shows a Silent Chain used in an automotive-type transmission design.

Figure 1-35 Typical Silent Chain.

Figure 1-36 Silent Chain used in a modern automotive drive.

2
A Chain Overview: Uses and Advantages

This chapter is intended to give the reader an overview of the various standardized types of chain and their uses, as covered in this book.

Chains may be classified in many different ways. From one rather abstract theoretical viewpoint, chain is a continuous flexible rack engaging a gear or pinion or a pair of gears. Certainly, a sprocket, being a toothed wheel whose teeth are shaped to mesh with a chain, is a form of gear. Another viewpoint, based on the history and development of the product, terms chain a mechanical belt, running over wheels, which can be used either to transmit power or to convey materials. Still another concept stems from the anchor chains of a ship or an auto wrecker's tow chain—a form of flexible cable used to tether or pull. None of these aspects relate to the practical standpoints of either the producer or the user of the product.

To the manufacturer and his engineers and shop people, chain is probably best defined by how it is made and assembled. To the salesperson it is defined in terms of cost and how it will be used. To the user, classification is unimportant aside from being an aid to secure the right product for the function wanted.

Many kinds of chains are described in this chapter, but major emphasis is on the shapes and forms of chains together with how they are to be used.

Types of Chain as Manufactured

In Chapter 1 the evolution of the chain industry was outlined from the first cast chains to silent chains. This book outlines three major kinds of chain from an industry standpoint:

1. Roller Chains
2. Silent Chains
3. Engineering Steel Chains

Uses of Steel Chain

In terms of applications, the major uses for these chains are as

1. Drives (power transmission)
2. Conveyors
3. Bucket elevators
4. Tension linkages

Each of these applications is discussed in detail in subsequent chapters. Many standard chains are intended to be used for only one of these four areas of application. However, some chains are designed so that they can be adapted to more than one use.

Roller Chains and Engineering Steel Chains are used in all areas of application in varying degrees. Silent Chain is essentially for drive applications, although a few conveying applications exist.

Although Roller Chain is usually thought of as being more for drive service, and Engineering Steel Chains for conveyors and bucket elevators, exceptions are frequent, as will be seen. In fact, in this book the exceptions are so frequent that a chapter in the drive chain applications section is devoted to offset heavy duty Engineering Steel drive chains and another chapter in the conveyor section covers the application of Roller Chains to conveyor service. Both are the subject of ANSI standards.

The categories of use of drives and tension linkages may be logically grouped. A drive is a means of transmitting power, of which force is a component, with distance, speed, and time. A tension linkage transmits force, with or without the other components. In the same vein, both conveyors and bucket elevators handle materials. The former moves materials horizontally or at an angle with the horizontal. A bucket elevator moves materials vertically upward, although it may do so at an angle with the vertical.

The classification described above is recognized herein by grouping drive and tension linkage usages together and following with conveyors and bucket elevators in materials handling applications.

Scope of Chains Covered

The standard chains discussed herein are assembled from parts fabricated from steel. Although cast link chains are beyond the scope of the book, occasional cast carrier rollers are encountered in some conveyor chains. Assembly is usually by means of interference fits or by welding, or for pins, spinning, or riveting.

The basic designs of these chains are predicated on function and long life in service. Joints encountering constant flexure over sprocket wheels are designed with maximum bearing areas and, usually, with hardened pins and bushings to resist wear. Chain tension members such as link plates and sidebars are often heat treated for toughness and shock resistance as well as to augment yield strength.

Styles and Forms of Chains as Manufactured

Regardless of whether chains are Roller Chains, Silent Chains, or Engineering Steel Chains, many styles may be classified by two basic modes, as follows:

1. Straight link chains; having alternate "inside" and "outside" links
2. Offset link chains, with all links alike

a. Chains with rollers
b. Chains similar to chains with roller, but "rollerless"
c. Integral link chains, as bar link and welded steel, where internal rollers cannot be installed.

These differences in configuration and details are important to both manufacturers and users, not only because of cost but for other reasons of a technical nature.

That chains exist in about every possible combination of the modes listed above is obvious. All told, however, there is probably a good deal less offset chain furnished than straight link and also, more with rollers than without them.

Straight and Offset Link Chains

Figure 2-1 illustrates typical and similar straight and offset link chains. The configuration of the tension members of these chains (in this case, sidebars, as these are Engineering Steel drive chains) could be the only difference in the construction of the two chains, if pitch and joint details are assumed to be identical. Assuming this and also that the sidebar cross sections are identical, so that the only difference is straight or offset link construction, the chains could be operated over the same sprockets and for the same purpose. (But they should not be "mixed" by intercoupling links of one into the other unless the manufacturer specifies that this is permissible.)

The differences in usage for a given application stem from the fact that straight link chains consist of *inside* and *outside* links and that offset links are all alike. For this reason, strands of straight link chains must be used in even numbers of links, whereas offset chain strands can be used in either odd or even numbers. Straight link chains are symmetrical relative to direction of travel, but offset chains should be operated in a specific direction (referring to *closed* or *open* end forward) to obtain the best service. The direction of travel for offset chain strands varies somewhat with application details, and manufacturers' recommendations should be obtained. The details of travel direction are discussed in subsequent chapters.

Each style has advantages. Straight link chain is somewhat easier to manufacture, which may be a cost advantage to the user. Attachments are more easily provided in straight link designs, and may cause fewer problems in use. For a given strength, a slightly shorter pitch chain can be provided in straight link chain, since space for an offset need not be provided. Figure 2-1 demonstrates this factor; although sidebar thickness and cross section seem similar, the pitch of the straight link drive chain is

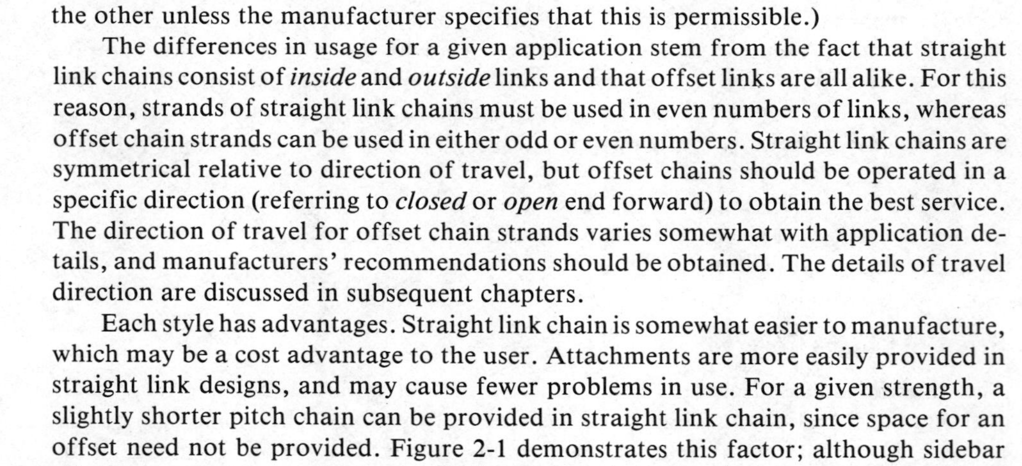

Figure 2-1 Otherwise similar offset and straight link steel drive chains.

considerably less. As may be noted from Figure 2-2, contoured offsets are often used to provide offset links of the shortest possible pitch. In drives, short pitch is reflected in terms of smaller sprockets and quieter operation.

An advantage of offset chain is that a worn drive strand can be easily shortened by removal of a single link. To shorten a straight link Roller Chain, a section consisting of five links—two inside and three outside links—must be removed, the central three links replaced with a two-pitch offset section (as shown in Fig. 2-2), and together with the two outside links previously removed, the section of four links reinstalled in the chain. With straight link chains having individually removable pins, however, the process is simply that a pair of links is removed and replaced with a single offset link. Care should be taken in such cases that the offset replacement link has offsets that will clear the ends of the adjacent straight links when the chain flexes.

There are certain types of applications, mostly with rollerless chains, where offset chains have the advantage of excellent wear life when operated open end forward on a conveyor or bucket elevator. Such applications usually stem from manufacturers' recommendations.

Chains With and Without Rollers

To a user, the presence or absence of rollers is probably of far greater interest than whether straight or offset links are present in a chain being considered for an application.

The major purpose of rollers is to reduce friction, but the rollers in chains are used in two separate functions to do this, the two functions usually being provided by the same roller. The functions are:

1. To engage the sprocket teeth and thus transfer any sliding action to the internal members of the chain, which are designed for that purpose. This is discussed further in subsequent chapters.
2. To serve as a guide or to support a chain and material carried on it on tracks or ways, as is characteristic of conveyors and some bucket elevators.

Rollers in chain in drive service, of which Roller Chain is the prime example, normally are smaller in diameter than the height of the link plates of the chain, as shown in Figure 2-3. Thus the link plates serve as guides when the chain engages the sprockets, and may also do so when the chain is riding on guides, as in a bucket elevator.

Rollers on conveyor chains normally have diameters considerably larger than the widths of their adjacent sidebars, as shown in Figure 2-4. This is done for two reasons. First, and most obvious, the large rollers, called *carrier rollers,* carry the sidebars well above conveyor tracks and thus prevent frictional dragging. Second, the larger roller has a definite mechanical advantage over smaller rollers relative to rotational friction, and thus help reduce chain pull. This is discussed more fully in Chapter 9.

The carrier rollers in most chains are also used to engage the sprockets, although carrier rollers, equipped with antifriction bearings, are sometimes used as outboard rollers on rollerless chains.

Rollerless chain is similar to chains with rollers in appearance, and is used on applications where rollers are not required. There are also other types of chain which do not have rollers but for which rollers would obviously not be appropriate. Examples

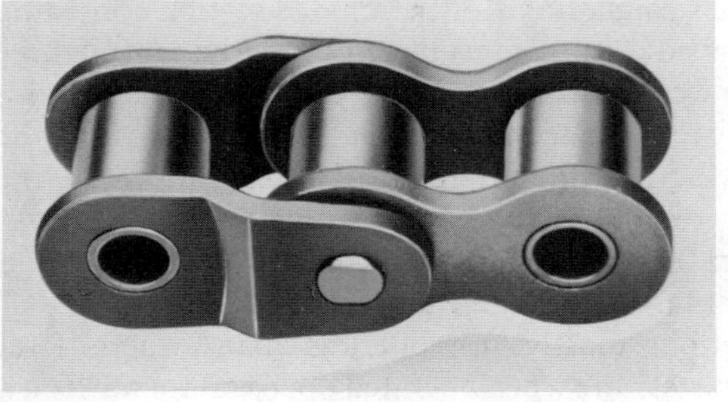

Figure 2-2 Roller Chain offset link section consisting of a roller link connected by a riveted pin to an offset link with contoured offset link plates.

Figure 2-3 Chain with standard rollers, the rollers being of a diameter less than the height or width of the adjacent link plates.

Figure 2-4 Chain with carrier rollers, the rollers being of larger diameter than the width of the adjacent link plates.

are Silent and bar-link chains, described in Figures 1-29 and 1-35. Most steel chains not having rollers are of the Engineering classes.

The following sections list the various types of chain with some details of construction and intended usage. However, for detailed information regarding the chains, such as dimensions, working loads, and similar data, appropriate ANSI standards and manufacturers' catalogs should be consulted.

Types and Styles of Standard Chains

Roller Chains

As outlined in Chapter 1, Roller Chain is the American standard series of roller chains covered by ANSI Standard B29.1 and several related standards.

Roller Chains are manufactured in several types, each designed for the particular service required. All Roller Chains are so constructed that the rollers are evenly spaced throughout the chain. The outstanding advantage of this type of chain is the ability of the rollers to rotate when contacting the teeth of the sprocket.

Two arrangements of Roller Chains are in common use—the single strand type and the multiple strand type. In the latter type, two or more chains are joined, side by side, by means of common pins that maintain the alignment of the rollers in the various strands.

Standard Roller Chain Number Designations. A standard Roller Chain number is merely a shorthand device that provides complete identification of an individual chain. The right-hand digit in the chain designation is 0 for Roller Chains of the usual proportions, 1 for a lightweight chain, and 5 for a rollerless bushed chain. The numbers to the left of the right-hand figure denote the number of 1/8-in. segments in the pitch. The hyphenated number 2 suffixed to the chain number denotes a double-strand, 3 a triple-strand, 4 a quadruple-strand chain, and so forth.

For example, the number 50 chain indicates a 5/8-in.-pitch chain of basic proportions; number 41 designates a narrow, lightweight 1/2-in.-pitch chain; number 35 indicates a 3/8-in.-pitch rollerless bushing chain; and number 100 indicates a chain of the basic proportions—10/8 in. or 1¼-in. pitch.

In multiple strand Roller Chains, for example (see Fig. 2-8), 60-2 designates two strands of a number 60 chain in parallel having common pins, 60-3 designates a triple strand, 60-4 a quadruple strand, and so on.

The letter H following the number of the chain denotes a chain of the "heavy" series having all link plates 1/32 in. thicker than the link plates of a basic chain of corresponding pitch.

These Roller Chains are defined as *pitch proportioned,* which makes them different from other types of chains with rollers. The ANSI standards' nominal proportions for these chains are based on a proportional relationship to the chain pitch. The pitch of a Roller Chain is the distance, in inches, between the centers of adjacent joint members. The three most important Roller Chain dimensions are pitch, roller diameter, and inside chain width, as indicated in Figure 2-5. These dimensions determine the fit of the sprockets.

Chain width (sometimes called "nominal chain width") is the minimum distance between the link plates of a roller link, and is approximately 5/8 of the pitch. Note that this is not the overall width of the chain.

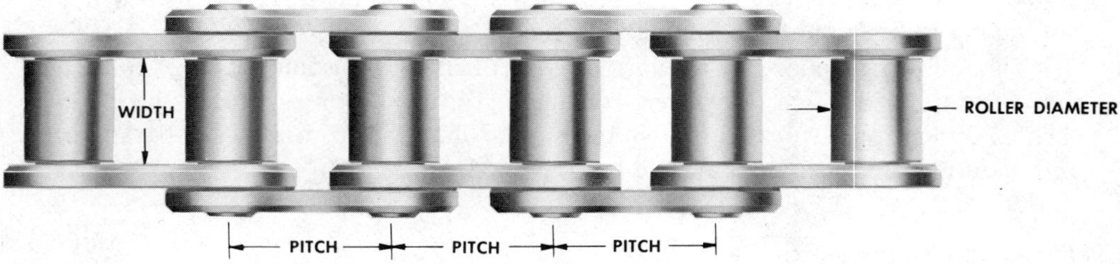

Figure 2-5 Roller Chain principal dimensions.

Roller diameter is the outside diameter of the roller, and is approximately 5/8 of the pitch.

For the various chain sizes the standard specifies that the pin diameters be approximately 5/16 of the pitch, and the link plate thicknesses be approximately 1/8 of the pitch.

In addition, the limiting dimensions required for interchangeability are given in the standard.

General Roller Chain dimensions, as specified in the standard, are given in Table 2-6. The purpose of the standard, as indicated by the dimensions given in Table 2-6, is that chains and sprockets be interchangeable. Identification is standardized; the chain of one manufacturer may be used to replace the chain of another. This does not mean that for any given type, all chains are identical; factors not affecting intercoupling or interchangeability on sprockets are left to the preference of the manufacturers.

Types of Roller Chains. Roller Chains are used for both drive and conveyor applications. There is a separate series of chains for each application area.

1. Drive series Roller Chains are of three strand types.
   a. Standard series: single strand, shown in Figure 2-7.
   b. Standard series: multiple strand, shown in Figure 2-8.
   c. Standard series: double-pitch Roller Drive chain, shown in Figure 2-9.

The most commonly used chain for drives is the single-strand standard series Roller Chain. The horsepower rating capacities of these chains cover a wide range of drive load requirements.

A multiple strand Roller Chain is essentially an assembly of two or more single-strand chains, placed side by side with common pins (Fig. 2-8). The roller links are identical with those of single-strand chains.

The pin link plates carry pins that extend through the entire width of the chain, and thus maintain uniform alignment of the different strands. The load-carrying capacity of multiple strand chain is less than the sum of the strands used and may be determined from factors in Table 5-5.

Connecting links, offset links, and offset sections, similar to those for single-strand chains, are used to join the free ends of multiple strand chain.

Multiple strand Roller Chains are used to provide increased power capacity without the need for increasing the chain pitch or its linear speed. For a given power load, a multiple strand chain with smaller pitch can be run at a higher speed than the single-strand Roller Chain of larger pitch required otherwise.

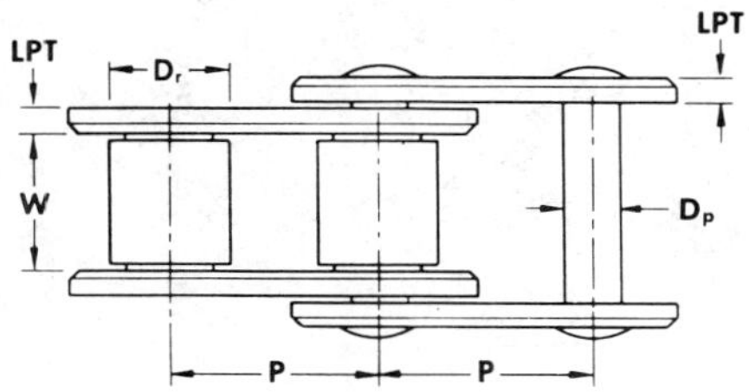

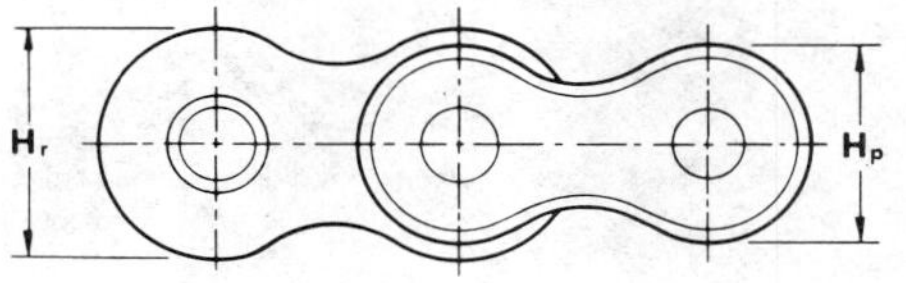

ANSI Standard Chain No.		Pitch P	Max Roller Diameter Dr	Width W	Pin Diameter Dp	Link Plate Thickness L P T		Measuring Load, Lb
Std.	Heavy					Std.	Heavy	
25*	—	1/4	0.130**	1/8	0.0905	0.030	—	18
35*	—	3/8	0.200**	3/16	0.141	0.050	—	18
41†	—	1/2	0.306	1/4	0.141	0.050	—	18
40	—	1/2	5/16	5/16	0.156	0.060	—	31
50	—	5/8	0.400	3/8	0.200	0.080	—	49
60	60H	3/4	15/32	1/2	0.234	0.094	.125	70
80	80H	1	5/8	5/8	0.312	0.125	.156	125
100	100H	1 1/4	3/4	3/4	0.375	0.156	.187	195
120	120H	1 1/2	7/8	1	0.437	0.187	.219	281
140	140H	1 3/4	1	1	0.500	0.219	.250	383
160	160H	2	1 1/8	1 1/4	0.562	0.250	.281	500
180	180H	2 1/4	1 13/32	1 13/32	0.687	0.281	.312	633
200	200H	2 1/2	1 9/16	1 1/2	0.781	0.312	.375	781
240	240H	3	1 7/8	1 7/8	0.937	0.375	.500	1125

*Without rollers.
** Bushing Diameter. Chain is rollerless.
†Light Machinery Series.

Table 2-6 General Dimensions (in.), ANSI Standard, Roller Chain

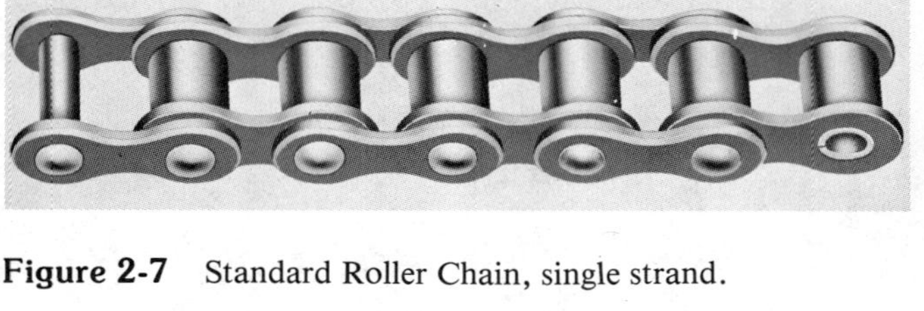

Figure 2-7 Standard Roller Chain, single strand.

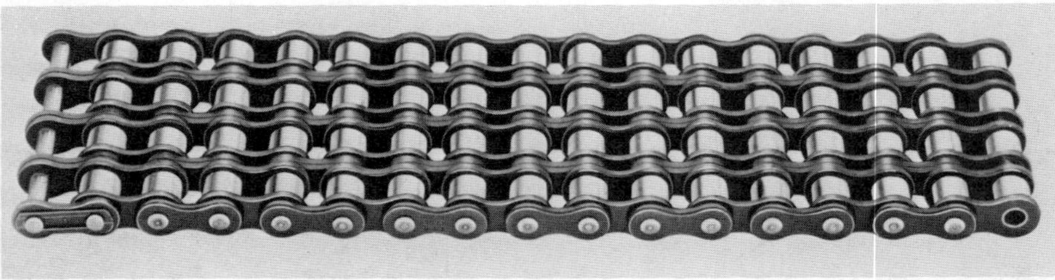

Figure 2-8 Standard Roller Chain, multiple strand.

The double-pitch series of drive chains is similar to the standard series (having the same pin, bushing, and roller diameters), except that the pitch is twice that of the standard chain. This form of double-pitch Roller Chain, which has figure-eight-shaped link plates, is described in ANSI Standard B29.3.

2. Conveyor series Roller Chains are of three standard types, as described in ANSI Standard B29.4.
   a. Standard series Roller Chain, shown with A-1 attachments in Figure 2-10. This chain has the standard figure-eight-shaped link plates.
   b. Double-pitch series Roller Chain, with standard Roller diameter, but with straight-sided link plates, shown with A-2 attachments in Figure 2-11.
   c. Double-pitch series Roller Chain, with large-diameter or carrier rollers and straight-sided link plates, shown with A-2 attachments in Figure 2-12.

The standard series Roller Chains used for conveyor applications are essentially the same as the drive chains, with the addition of attachments. The double-pitch conveyor series of Roller Chains with either the standard or large-diameter rollers have the same pin and bushing diameters as the standard series chain and a straight-sided rather than the contour-sided link plate.

A wide range of attachments are available for the conveyor Roller Chains in most of the same configurations as shown for Engineering Steel Chains in Figure 2-25.

Roller Chain Uses. In the power transmission field, standard Roller Chain applications range from fractional horsepower drives to those requiring in excess of 1000 hp. The ¼-in.-pitch chains, weighing less than 2 oz/ft, have been applied to such intricate machines as microfilm projectors. On the other hand, large-pitch chains of multiple strands, weighing over 50 lb/ft, meet the requirements of such heavy duty service as oil field equipment.

Double-pitch Roller Chains are particularly applicable to power drives where

Figure 2-9 Double-pitch Roller Chain, drive chain series.

Figure 2-10 Standard Roller Chain, with attachments for conveyor service.

Figure 2-11 Double-pitch Roller Chain, with standard rollers and attachments for conveying.

Figure 2-12 Double-pitch Roller Chain, with carrier rollers and attachments for conveying.

speeds are low, loads are moderate, or center distances are long. For such applications, the longer pitch results in a lighter and less expensive chain. They are therefore generally preferred for conveyor service. A further advantage is that the longer pitch provides space for the large-diameter carrier-type rollers shown in Figure 2-12.

The term *chain tension linkage* is applied to a chain installation in which linear motion is not continuous in direction—where the chain must move back and forth. Roller Chain is often specified where the positive action of a chain is required for certain hoists, marine steering equipment, and similar applications.

Leaf Chains

Leaf chain, sometimes called "cable chain" or "balance chain," is used for tension linkage applications.

It is constructed from interlaced heat-treated steel link plates similar in contour to Roller Chain link plates. Pins, used to connect the links, are usually made from case-hardened steels with the ends riveted. A typical section of leaf chain is shown in Figure 2-13 together with a drawing showing the lacing pattern as given in the ANSI standard for these chains.

Leaf chain *lacing patterns* are the mode by which the link plates are interlaced in alternate links. The pattern shown in Figure 2-13 consists of alternate links of three and four link plates. Those links with four link plates have the plates arranged in pairs interspersed between the three spaced link plates of adjacent links.

ANSI leaf chain number designations consist of a BL prefix followed by three or four digits. The last two digits indicate the lacing pattern and the first two indicate the chain pitch in eights of an inch. For example, a BL-534 leaf chain would be of 5/8-in. pitch with alternate links having three and four link plates, just as in the example of Figure 2-13.

Leaf chain is not usually applied in endless strands, as it has no provision for sprocket engagement, force being applied to the ends through clevises or other means of attachment through the chain pins. The applications where leaf chains are used are those where a strong, flexible linkage with very little elastic stretch (and a resulting positive action) is required. Changes of direction are usually made over sheaves or rollers.

Leaf chain is commonly used on lift trucks and for counterweights on heavy machine tools. It is equally appropriate for similar applications requiring continuous support of a load randomly movable in either direction between limits.

Silent Chains

Silent Chain (Fig. 2-14) consists of a series of toothed link plates assembled on joint components in such a fashion that articulation occurs between adjoining pitches. Detailed descriptions of the various types of joint components are discussed in Chapter 3 for component parts and design of the chains covered in this chapter.

The American National Standards Institute standard on Silent Chain drives is ANSI Standard B29.2, entitled "Inverted Tooth (Silent) Chains and Sprocket Teeth." This standard is primarily to provide for interchangeability between chains and sprockets of different manufacturers. It does not provide for the standardization of joint contours, so the available chain brands cannot be connected to one another even though all such chains will fit the same sprockets. There are many Silent Chain

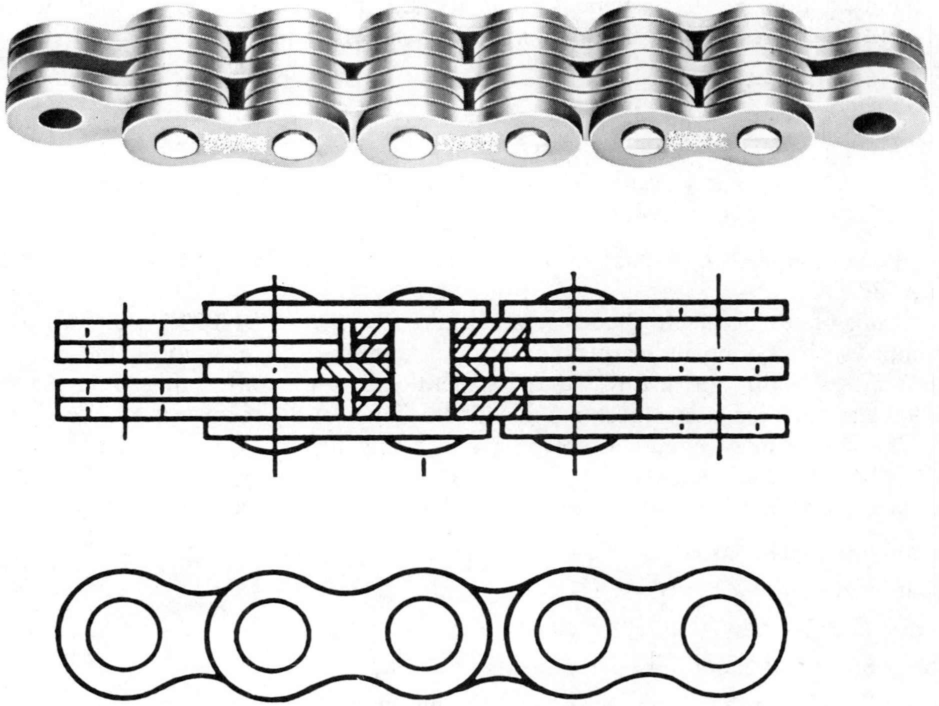

Figure 2-13 Leaf chain, and cross section showing construction.

styles not included in the standard. Most of these are made by more than one but not by all of the Silent Chain manufacturers.

Description

Silent Chain Designations. In the standard, Silent Chains are designated by a combined letter-and-number symbol as follows:

1. For chains of 3/8-in. pitch and larger:
   a. A two-letter symbol, SC (denotes conformance with ANSI Standard B29.2)
   b. One or two numerical digits indicating the chain pitch in eighths of an inch
   c. Two or three numerical digits indicating the chain width in quarter inches

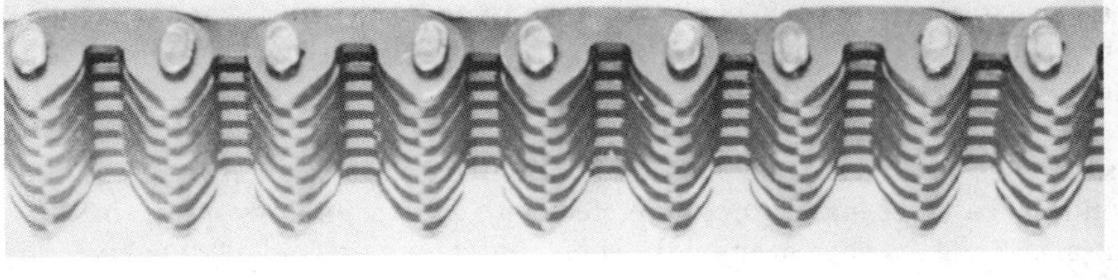

Figure 2-14 Typical Silent Chain.

For example, SC302 designates a Silent Chain of 3/8-in. pitch and 1/2-in. width; SC1012 designates a Silent Chain of 1¼-in. pitch, 3 in. wide.

2. For 3/16-in.-pitch chains
   a. A two-digit symbol, SC
   b. The digits 03, indicating 3/16-in. pitch
   c. Two digits indicating the total number of links wide and the approximate width in thirty-seconds of an inch

For example, SC0315 chain is 3/16-in.-pitch Silent Chain, approximately 15/32 in. wide.

Link plates of Silent Chains manufactured to the ANSI B29.2 standard are stamped with a symbol indicating the pitch; for example, SC6, or simply 6, indicates a plate with a 3/4-in. pitch. Link plates of Silent Chains not conforming to the ANSI B29.2 standard are generally stamped with a number of the manufacturer's choice.

Types of Silent Chain. There are a great many different types of Silent Chain. The ANSI standard chain is only one type. Generally speaking, in addition to pitch, Silent Chains can be identified and classified according to the following factors:

Chain link plate contour

Chain width

Methods of laterally guiding the chain on the sprockets

Shape or type of chain joint components

Direction in which the chain wraps around the sprocket (for some nonstandard designs)

The design considerations of Silent Chain are discussed in Chapter 3 and the application of this kind of chain is reviewed in Chapter 7.

Engineering Steel Chains

Engineering Steel Chains are made of fabricated and machined steel. They are designed to perform a wide variety of functions to satisfy known or anticipated operating conditions. Because of their broad scope of usage, there are many types available and ways to classify them.

Over the period of nearly a century that steel chains have been in existence in the United States, similar types of Engineering Steel Chains have been identified by widely divergent terms in manufacturers' catalogs. The same applies to the component parts of this class of chains.

In the last two decades, a concerted effort has been made by the American Chain Association to standardize not only the terminology for this class, but also the dimensions pertinent to fitting sprockets and other components of conveyors and bucket elevators. ANSI standards now exist for many styles of Engineering Steel Chains, and others are under consideration.

The classifications and terminology used in this manual are in accordance with current standards or current practice where standards do not exist. However, not all manufacturers' catalogs may agree completely with the terminology of this manual, and many such catalogs have been on the shelves of designers' and plant engineering departments for long periods of time. Some of the terminology used in those catalogs does not always seem to correlate entirely with current practice. Some of those terms, not current industry-wide practice but used by various manufacturers over long periods of time, are included in the Glossary. Reference should be made to the Glossary when terms not included in this manual are encountered.

Types of Engineering Steel Chain. Generally speaking, the types of Engineering Steel Chains as classified under ANSI standards will fall into the following categories or types:

1. Straight sidebar steel chains
   a. With rollers
   b. Rollerless
2. Offset sidebar steel chains
   a. With rollers
   b. Rollerless
3. Bar-link chains (often called "block and bar")
4. Welded steel chains
5. Open barrel steel pintle chains

Taking these up in the sequence listed, it will be found that there are many variations and subtypes.

Straight sidebar steel fabricated roller chain, such as that shown in Figure 2-16, is with or without rollers. Figure 2-15 shows a two-link section of straight sidebar rollerless chain. These chains have steel bushings and pins and are assembled using interference fits between those parts and the sidebars. The inside link, made up of two sidebars and two bushings, is called the *bushing* (or, sometimes, the "block") link. The outside link is the *pin* link.

The major use of this type of chain is on bucket elevators, although large amounts are also used in conveyor service. The chain is normally furnished with attachments specified for particular applications, most of which are in exposed locations with abrasive materials or other difficult-to-handle materials involved.

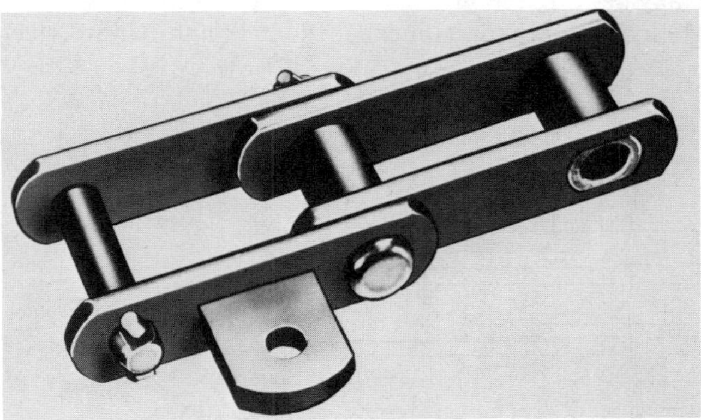

Figure 2-15 Straight sidebar rollerless chain with, typically, an attachment lug on the outside link.

ANSI Standard B29.12 currently lists seven straight link steel fabricated bushed rollerless chains ranging in pitch from 2.609 to 6.050 in. Manufacturers' catalogs list many more, most of which are usually available from stock unless unusual attachments or features are involved. These chains are built for heavy duty in severe service.

Straight sidebar steel fabricated roller chain, such as that shown in Figure 2-16, is used mainly for conveyor service. These chains are furnished with a wide variety of attachments, although the types designed for use with flights or slats are probably encountered most often. However, bucket elevators are often equipped with straight sidebar fabricated roller chain, as are scraper conveyors (see Chapter 9).

Straight sidebar steel fabricated roller chains are similar to the rollerless type. The rollers are installed to turn freely on the bushings between the inside sidebars, the inside links usually being termed roller links. Most chains of this type have rollers of the carrier type, their diameter being considerably larger than the width of the sidebars on each side.

ANSI Standard B29.15 presently lists 19 different chains of this type, with pitches ranging from 3 to 24 in. Although the chains listed as standard are probably the most frequently used, manufacturers list a great many with variations from the standard that are readily available. Many special chains use mostly standard components except, perhaps, for a special attachment or roller.

Offset Steel Fabricated Chains. Each link of an offset steel fabricated chain is identical. The link components consist of sidebars, a bushing, and a pin with a cotter. Figure 2-17 shows an example of this type of chain.

Offset steel fabricated bushed rollerless chain is used mostly on conveyors and bucket elevators. Such chains usually slide on ways or between guides. A rather specialized usage is on conveyors, where the load is carried on the ways with outboard rollers having antifriction bearings, the chain bushings being engaged with the sprocket teeth to drive the conveyor.

Current ANSI standards list no offset steel fabricated rollerless chain of this type.

Offset steel fabricated chain with rollers is listed in a heavy drive chain series under ANSI B29.10. This is probably the predominant use of this style of chain, an example of which is shown in Figure 2-18. Other types of offset steel fabricated roller chains are used in conveying and elevating service. None are listed in current ANSI standards, although listings may be found in manufacturers' catalogs.

Bar-Link Chains. Bar-link chains are sometimes called block-and-bar or "steel

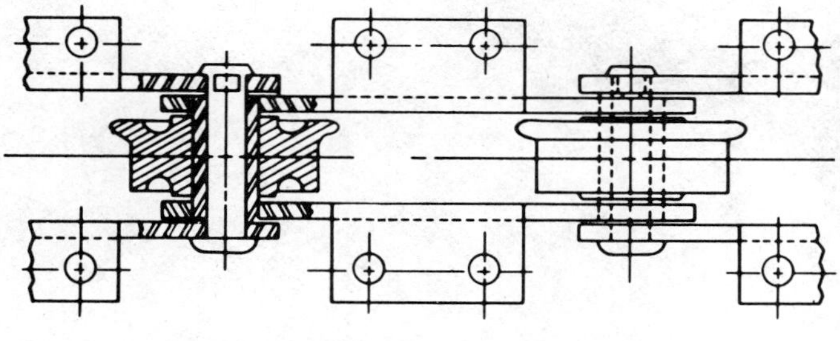

Figure 2-16 Straight sidebar Engineering Steel conveyor roller chain with flanged rollers and K-2 attachments, probably the most common combination for this class of chain.

Figure 2-17 Offset sidebar rollerless chain with K-2 attachments.

block'' chains. Figure 2-19 shows a typical example of a bar-link chain. This type of chain usually consists of two outer sidebars, one center bar, and two pins making up a two-link section. Bar-link chain usually does not have bushings, the center bar flexing directly on the pin. Sprocket contact is with the ends of the center bars.

These chains are also made in multilaced construction, as shown in Figure 2-20. Chains of this type of construction are often termed leaf chains.

Chains of the bar-link type are frequently used in tension linkage applications. A design of this type is so commonly used in steel mills for this application that it is known as *draw bench* chain. Steel bar-link chains are also used on slow-moving conveyors such as steel mill coil conveyors (with outboard rollers) and are made in very large versions to raise and lower river and canal lock and dam gates.

No ANSI standard presently exists on steel bar-link chains.

Welded Steel Chains. Welded steel chains are similar to fabricated steel chains except that links are made as integral weldments rather than being held together by means of tight fits and locking surfaces.

Welded steel chains can be classified as mill chains (Fig. 2-21) and drag chains (Fig. 2-22). With a few exceptions, mill-type welded steel chains are of offset construction. They are covered in ANSI Standard B29.16, some eight pitches now being covered. Welded steel mill chains are offered with a wide variety of attachments for general conveying service. They are also often used as plain chain for slow-moving conveyor drives.

Welded steel drag chains are covered in ANSI Standard B29.18. Pitches of 5, 6, and 8 in. with widths from 4 to over 12 in. are listed. These chains are of offset construction, and are used for drag conveyors handling bulk materials.

Open Barrel Steel Pintle Chains. Open barrel steel pintle chain is an economical lightweight conveyor chain. The basic chain consists of one-piece offset-formed steel links and pins. The pins are fixed against rotation by mechanical locks or interference fits. A section of the basic chain is illustrated in Figure 2-23.

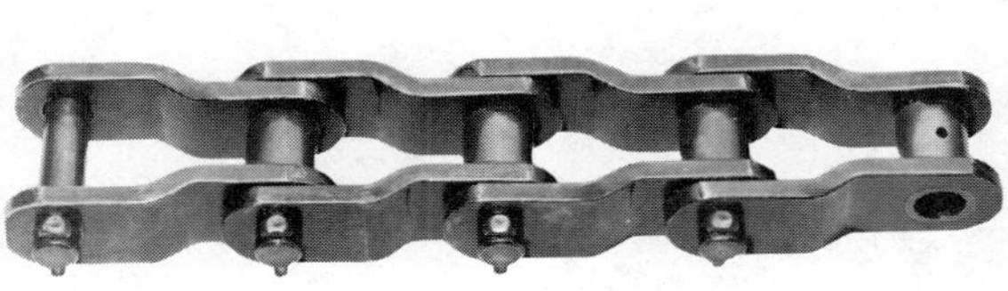

Figure 2-18 Offset sidebar, heavy duty roller chain for drive service.

Figure 2-19 Bar-link, often called block-and-bar chain.

Figure 2-20 Multilaced leaf-type bar-link chain.

Figure 2-21 Welded steel chain, mill chain type.

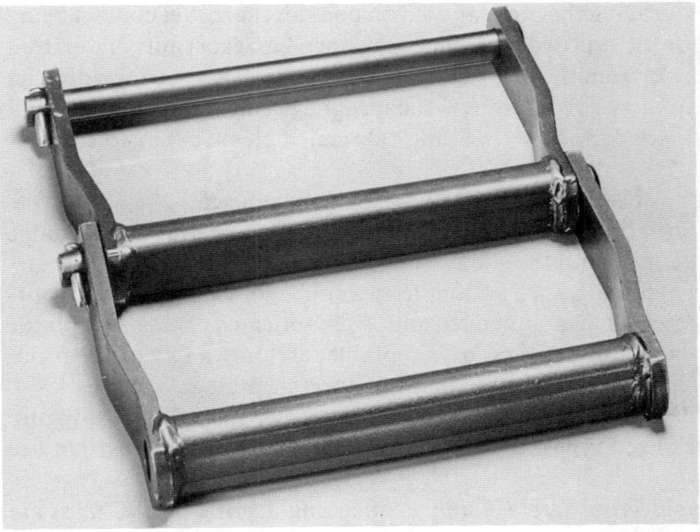

Figure 2-22 Welded steel chain, drag chain type.

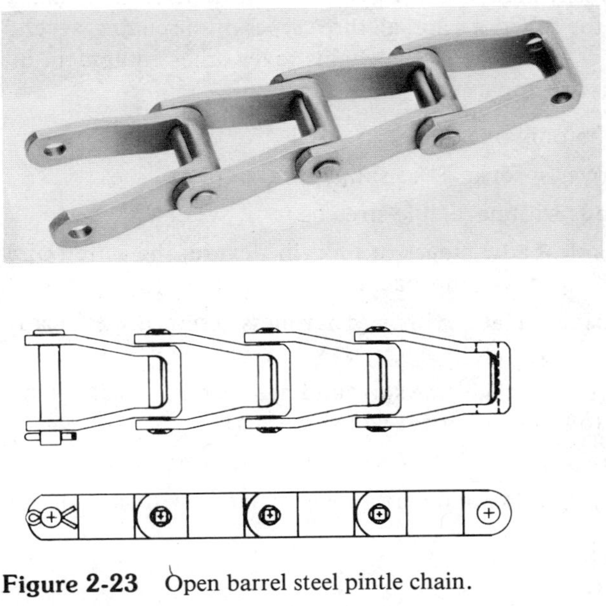

Figure 2-23 Open barrel steel pintle chain.

The barrels are open, leaving the pins exposed on one side, sprocket contact being against either the barrel or the exposed pins. Since the open barrel permits dissipation of foreign material and rust from the pin area, and makes corrosion seizing unlikely, the chain operates well in dirty surroundings, such as in agricultural equipment and conveyors. Attachments are formed by welding fabricated steel shapes to links or pins.

An ANSI standard has been formulated and is being proposed.

Other Engineering Steel Chains. Many other types of Engineering Steel Chains exist, although most of them are not the subject of a published standard. Special-purpose chains for handling materials, such as double-flex conveyor chains, welded roof-top and flat-top chains, steel detachable chains, and steel fabricated chains with plastic bushed rollers for use where lubrication is unacceptable, are only a few of the products available. All of these may be found in manufacturers' catalogs.

The types of chain described previously also exist in many special forms. Figure 2-24 shows various specialized types and standard types with attachments that are available.

Attachments for Engineering Steel Chains. Since Engineering Steel Chains are used in such a wide variety of attachments, a great variety is available. Figure 2-25 illustrates some of the more common attachments available for Engineering Steel Chains for use in conveying and elevating. These attachments lend themselves better to steel straight sidebar chain, but some can be adapted for offset construction. Many other styles of attachments are made to meet specific requirements.

The Advantages of Chains in Applications

Earlier in this chapter, usages of the steel chains covered in this book were resolved, in theory at least, to those transmitting power or force and those handling materials. Chains were classified as Roller, Silent, and Engineering Steel Chains.

There are excellent reasons why many kinds of chains exist to perform so many varied functions. Reviewing all the functions and all three types of the chains, several broad generalities relative to the advantage of the chains, over other equipment intended to do the same functions are apparent:

Chain has controlled flexibility in only one plane.

The action of chain is positive over sprockets; no slippage takes place.

Chains can carry very heavy loads with negligible stretch.

Because of large internal mechanical advantages of links in flexure, the action of a chain joint passing around a sprocket approaches 100% efficiency.

Flexure takes place between bearing surfaces with high hardness designed specifically for extended wear life.

Chains can be operated satisfactorily in adverse surroundings such as elevated temperatures or where subject to moisture or foreign materials.

Chains can be manufactured of special steels to resist specific environments.

Chains have unlimited shelf life; they do not deteriorate with age or with sun, oil, or grease.

Typical bucket elevator chain

Bar-link chain with special attachments for steel mill cooling

Hollow pin-type straight sidebar roller conveyor chain.

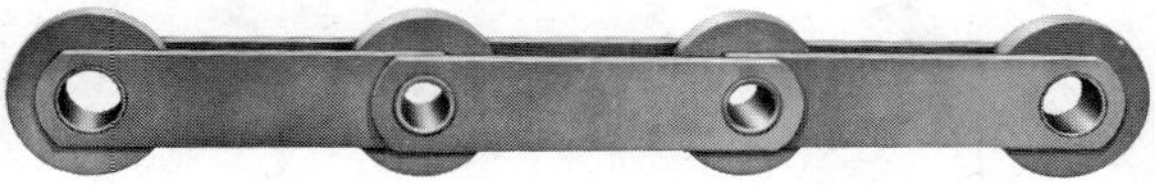

Straight sidebar conveyor chain with flanged rollers and central bushing.

Straight sidebar steel roller chain with holes for central bushings.

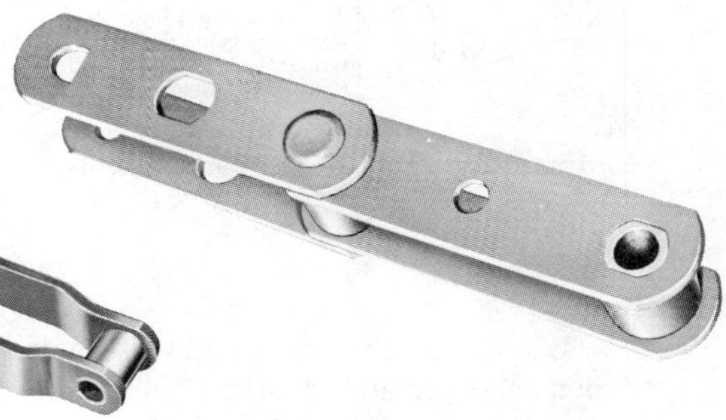

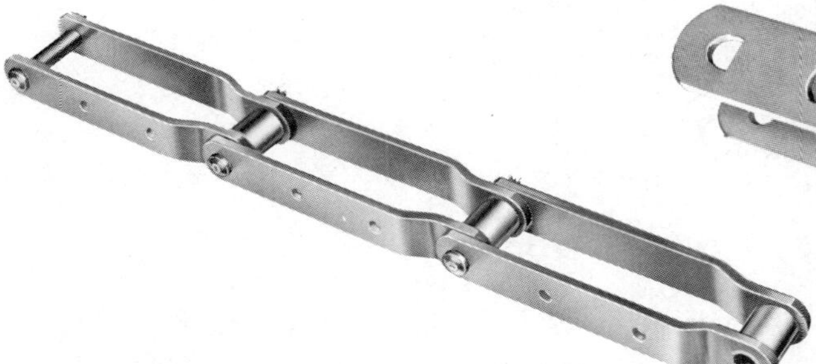

Offset link, long-pitch conveyor chain.

Figure 2-24 Types of Engineering Steel Chains.

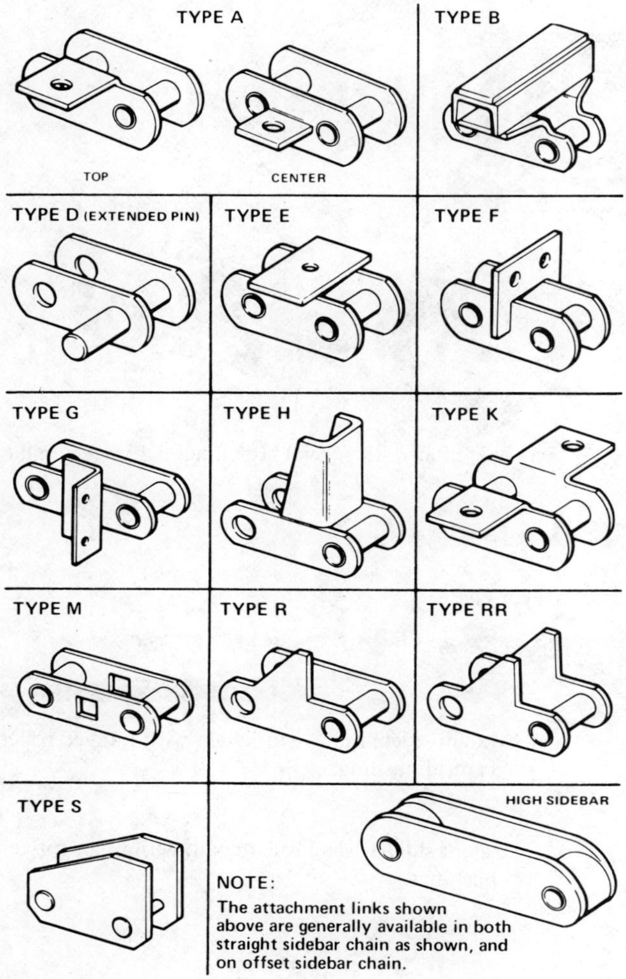

Figure 2-25 Typical chain attachments.

Examining the advantages of chains from the standpoint of functions, drive chains have certain advantages, in addition to those just named, over other types of power transmission equipment. The major advantage of all drive chains is the ability to bridge relatively large gaps between shafts.

Chain drives often provide a simpler, less costly, and more practical arrangement than gear drives because the center-to-center shaft distance is much less restrictive for chain drives than for gear drives.

Chain drives do not occupy the space that is required for a belt drive. For any given horsepower and speed, chain drives will be narrower than belt drives, and the sprockets will have a smaller diameter than the pulleys.

Chains do not depend on friction as belts do; therefore, their minimum arc of contact is smaller. This results in a shorter shaft center distance being required for a given drive ratio.

Chains are more easily installed than gears or belts. All drive media require careful alignment and installation, but the machinery built for a chain drive need not be held to as close a tolerance as that required for a gear drive. Chains are less costly to install and often a savings in time is realized.

Later changes in requirements are more easily accomplished with a chain drive than with gears, due to the flexibility of a chain installation.

Chain drives do not slip or creep as do belt drives. Therefore, chains maintain a positive speed ratio between the driving and the driven shafts, and are more efficient since no power is lost through slippage.

Bearing friction losses and maintenance are less with a chain than with a belt, since a chain requires much less slack-side tension, thereby reducing the load on the shaft and bearings.

Chain is much more practical for slow-speed power transmission.

The elongation of chain due to normal wear is a slow process when the chain is adequately lubricated and does not affect satisfactory operation of the drive, as does stretch in a belt. Therefore, adjustments are more frequent in a belt drive than in a chain drive.

Advantages of Silent Chain Drives

Silent Chain combines the flexibility and quietness of a belt and the positive action and durability of gearing with the convenience and high efficiency of a chain.

Silent Chains can be designed to handle very precise specific speed ratios, and the same joint design that reduces noise by minimizing impact also minimizes harmonic chordal "surges." In addition, Silent Chain has most of the advantages of Roller Chain, as noted previously.

Advantages of Engineering Steel Chains for Drives

In Engineering Steel drive chains, the ability to bridge large spaces between parallel shafts is especially significant. Such chains are usually quite heavily loaded, but travel at relatively slow speeds. The span covered by the chain, expressed in terms of gearing, bearings, and shafting (and protection for them from elements the chain can ignore), is not only economically significant but equally saving of space.

Chains of this type are very rugged, and have the toughness and elasticity to absorb heavy shock loads.

Advantages of Chains on Conveyors and Bucket Elevators

1. Chain conveyors often provide a simpler, less expensive, and more practical arrangement than belt conveyors on short to medium shaft center distances.
2. Chain conveyors will generally operate at much slower speeds than will belt conveyors.
3. Chain conveyors are most easily adapted for assembly and processing applications, permitting index movement of parts and packages. Chain conveyors can be designed for storage and holding stations, to permit movement of materials to processing stations as required.

4. Chain conveyors can be designed to handle more abrasive materials at higher elevated temperatures than can belt conveyors.
5. Chain conveyors can be designed to accommodate higher impact and shock loads at loading stations than can belt conveyors.
6. Chain conveyors offer more flexibility than do belt conveyors for alterations required for future requirements.
7. The elongation of chain conveyors due to normal wear is generally a slow process and does not affect satisfactory operation as does stretch in a belt conveyor. Thus a chain conveyor will generally require less adjustment than will a belt conveyor.

Advantages of Using Chain in Elevator Materials Handling

Chain elevators are especially advantageous because:

1. Many special types of material can be handled, such as material that is very hot or of large lump bulk.
2. Packing of material between elevator buckets and the belt can be a problem in a belt-type elevator; it is much less so in chain elevators.
3. Large chain elevators for handling heavy degradable materials can be operated at speeds that are too slow for comparable belt elevators.
4. Maintenance and cost factors are important in any installation. Initial replacement and maintenance costs of chain elevators are in many instances less than those for belt elevators.

3
Chain Components and Design Consideration

The first two chapters related how steel chains developed, and reviewed the rather bewildering variety of chains existing today. Some general explanation was made of how standard steel chains are classified and what use may be made of the various types.

This chapter first takes up the components from which the chains are assembled. The second portion of the chapter discusses the design considerations that determine why the parts were made and assembled into chains of the assorted shapes and types, expected to perform in as many different ways.

It is quite apparent that many different chain types differ only by the inclusion or omission of a specific component. Probably the most obvious example is roller versus rollerless chains.What is not as obvious is that the design of conjugate parts are usually somewhat different, to compensate in certain ways for the "missing" component.

For example, the rollers of a roller chain have the major function of making contact with the teeth of the sprockets. The bushing on which the roller turns has many other functions, but contact with the sprocket is not one of them. But in rollerless chain, such is not the case; a major function of the bushings of rollerless chain is to work on the teeth of sprockets. Consequently, whereas roller chain bushings, as will be seen later in this chapter, are usually thin-walled cylinders, the bushings of one style of rollerless chain are made in a shouldered design with the diameter of the shouldered portion, in essence, duplicating that of a roller, by occupying the approximate space a roller would have required. As a result, the part dispensed with in the rollerless design implies only that the function of roller rotation has been done away with. The other functions normally performed by the roller are retained, relative to dimensions and wear life; in fact, they are probably augmented since the whole thing was done for a purpose.

The shapes and styles of the component parts of a chain are tightly bound to the overall design and intended function of the final product. An analytical examination of an individual part usually reveals the functional usage relative to loads and type of operation and, perhaps, the environment in which it is to operate.

Both sections of this chapter take up the three basic ANSI types of chain considered in this book in the same order: Roller Chain, Silent Chain, and Engineering Steel Chains.

Roller Chain Components and Construction

Construction of Roller Chain

A Roller Chain is made up of two kinds of links, roller links and pin links, alternately spaced throughout the length of the chain (Figs. 2-7 and 3-1).

Roller Links. A roller link is an assembly of two roller link plates, two bushings, and two rollers (Fig. 3-2). The link plates are of steel, figure-eight in shape, with a hole at each end to receive the bushings. The center distance between holes is kept within a close tolerance to maintain uniform pitch in the assembled chain. The link plate holes are accurately sized for a press fit on the bushings, so that the bushing cannot turn in the link plates as the chain articulates.

Each bushing is a hollow cylinder, with its outside and inside surfaces hardened to resist wear. Bushing diameters are held to close limits to provide proper fit and smooth operation of the pins and rollers.

The rollers are also hollow cylinders, hardened and finished to precise diameters. Each is of the proper length to permit it to turn freely between the link plates.

Pin Links. A pin link is an assembly of two pin link plates and two pins (Fig. 3-3).

Pins are heat-treated to provide a tough center for resisting shock and a hardened surface for resisting wear. The riveted end prevents the link plate from being forced off the pin; a press fit between the pin and its link plates prevents rotation. The cotter-type pins have a hole in one end. They extend beyond the link plate on one side, where they are secured by cotter pins.

Chain Assembly. In assembling the chain, the pins of a partially assembled pin link are inserted in one bushing of each of two roller links. The remaining link plate of the pin link is then pressed onto the two exposed pin ends, and is secured in place.

Types of Connectors. At installation on the sprockets, the chain is made endless by means of a connecting link or an offset link or an offset section with a connecting link.

It is desirable that the chain, when installed, have an even number of pitches. If the design of the drive requires an uneven number of pitches, an offset link or an offset section must be used, preferably the latter.

Connecting Links. With an even number of pitches, to provide for easy installation or later removal, a special type of pin link, called a *connecting* or *coupling* link (Fig. 3-4), is used to connect the two ends of the chain. The partially assembled connecting link consists of two pins press-fitted and riveted in one link plate. The pin holes in the free link plate are sized for either a slip fit or a light press fit on the exposed pins, and the plate is secured in place by cotter pins or a spring clip.

Offset Links. If an odd number of pitches is required, an offset link (Fig. 3-5) may be used at one end of the open chain, and connected with a connecting link to that of the other end to make the chain endless.

An offset link is a combined pin-and-roller link consisting of two offset link plates, a single bushing with roller, and a single pin.

Figure 3-1 Pin link and roller link—parts.

The link plates are offset to accommodate the difference between the widths of the pin and roller links. The offset portion of the plate is designed to provide proper clearance with the adjacent link plates. The roller end of the link plate is similar to the end of a normal roller link plate.

The pin has a head on one end and a flat on the other, to prevent rotation, and corresponding holes in the plates are sized for a slip fit on the pin, which is secured in place by a cotter.

By using a removable pin, the offset link can also serve as a connecting link.

Horsepower ratings are based on the use of interference-fit connecting links and interference-fit offset sections. Chain using slip-fit connecting links or slip-fit offset links may not meet these ratings.

Figure 3-2 Roller link.

Offset Sections. An offset section is an assembly of a standard roller link and an offset link, which are connected by a riveted pin. An offset section is used in chains having an odd number of links (Fig. 3-6). Its use provides greater stability than does a single offset link. Since the pin of the offset section is press-fitted and riveted in place, a standard connecting link must be used with an offset section to connect with the free end of the chain.

Multiple Strand Roller Chain Construction. A multiple strand roller chain is essentially an assembly of two or more single strand chains, placed side by side (Fig. 3-7). The roller links are identical with those of single strand chains.

The pin link plates carry pins that extend through the entire width of the chain, and thus maintain uniform alignment of the different strands, as shown by Figure 3-8.

Figure 3-3 Pin links.

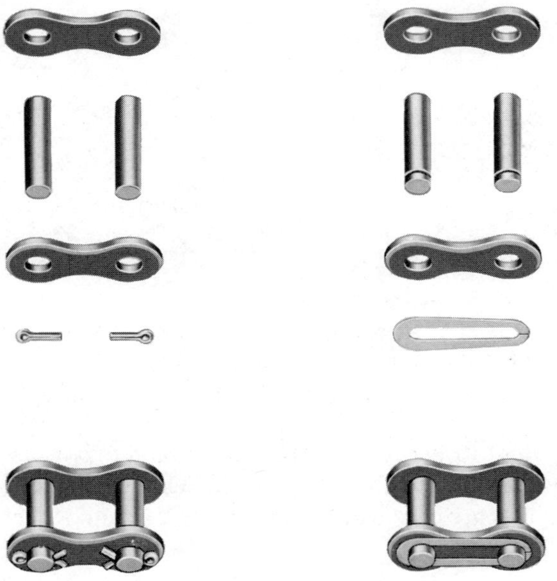

Figure 3-4 Connecting links.

Figure 3-5 Offset link.

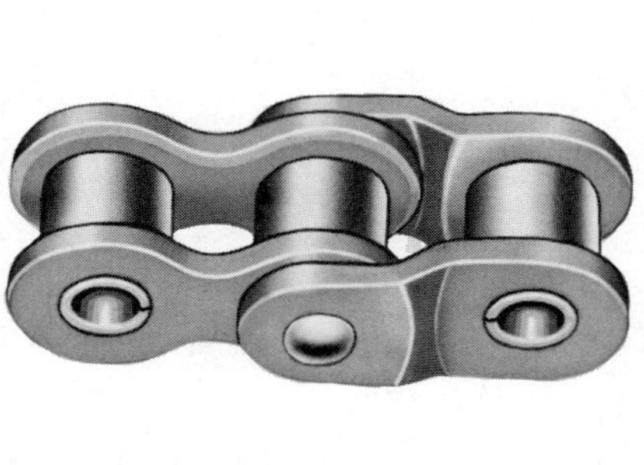

Figure 3-6 Offset section.

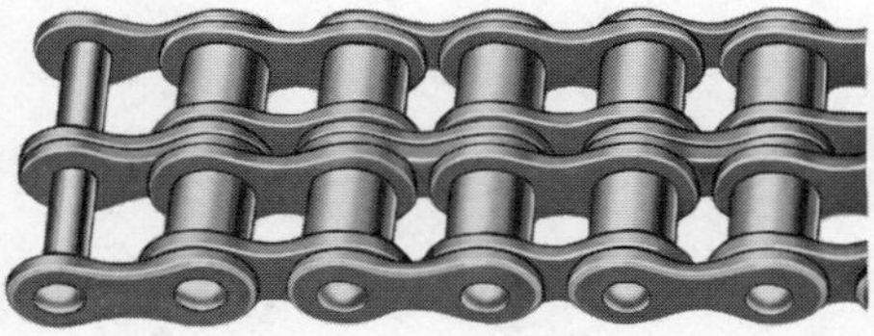

Figure 3-7 Typical double-strand Roller Chain.

Link plates of pin links for multiple strand chains are not necessarily identical to those used in single strand Roller Chain. Therefore, pin links of Roller Chains should always be supplied on a unit link basis rather than attempting to make up such links from miscellaneous parts.

Connecting links for multiple strand Roller Chain, offset links, and offset link sections, shown in Figures 3-8 and 3-9, are used with multiple strand chains in exactly the same way in which their counterparts are used with standard single strand Roller Chain.

Standard Double-Pitch Roller Chain Construction. Standard double-pitch Roller Chains use the same pins and bushings as the standard series of Roller Chains. However, the sidebars are double the pitch and may be supplied in either of two styles, and the rollers may be either standard or be larger-diameter carrier rollers. These components are shown in the chains illustrated in Figures 2-9 to 2-12.

The double-pitch link plates of the standard figure-eight shape are for the double-pitch power transmission series as specified in ANSI Standard B29-3. These are made in the standard shape for the usual reason, to provide clearance over hubs of small

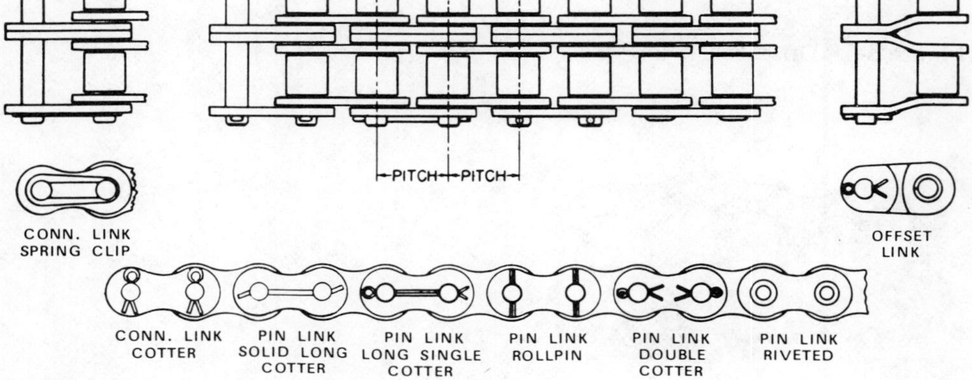

Figure 3-8 Components of multiple strand Roller Chains.

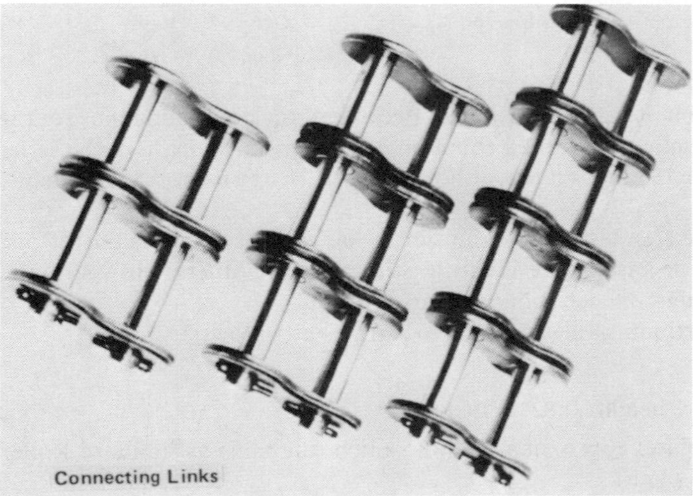

Figure 3-9 Multiple strand Roller Chain offset sections and links.

sprockets. The straight-sided link plates are for double-pitch conveyor series Roller Chains, specified in ANSI Standard B29-4.

The larger rollers are normally used for conveying service to provide clearance between link plates and tracks on which the conveyors roll. Standard rollers would be used in a conveyor expected to slide or which is supported by other means.

Application of Roller Chains of this type to conveyor is discussed in Chapter 10.

Leaf Chains

Figure 3-10 is a drawing of a typical leaf chain. Figure 3-11 shows its components and their relationships to one another. Many types of leaf chain exist, but only the heavy BL series is listed in ANSI Standard B29.8. Until 1975, a lighter AL series was also

Figure 3-10 Typical leaf chain with 4×4 lacing.

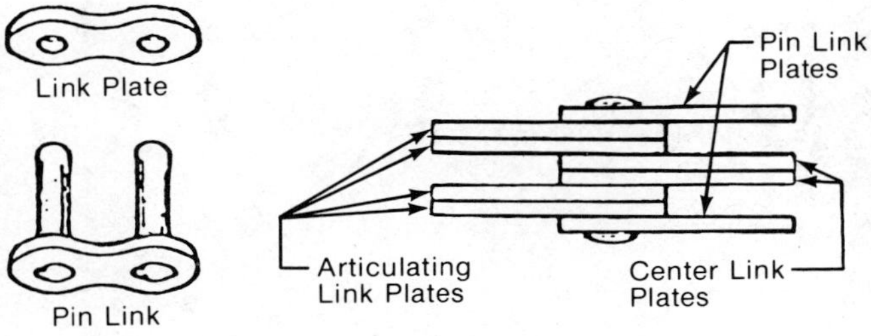

Figure 3-11 Leaf chain components.

listed in the standards, but it has been dropped. Because of the predominant use of the heavier series in new applications and a common desire for simplification, the series was eliminated from the 1977 standards, although it is still listed in most manufacturers' catalogs.

The purpose of the standard is to assure interchangeability, and chains of one manufacturer will not necessarily be identical with those of others. However, the standard does assure interconnectability and substitution.

The general proportioning rules of the two series is as follows:

AL (lighter) series:

1. Maximum link plate height: 0.82 × pitch
2. Thickness of link plate: approximately 1/8 × pitch (the same as standard Roller Chains of identical pitch)
3. Pin diameter: approximately 5/16 × pitch

PL (heavier) series:

1. Maximum link plate height: 0.95 × pitch
2. Thickness of link plate: approximately equal to standard Roller Chain of the next larger pitch, with the link plate thickness for 2-in. pitch being 9/32 in.
3. Pin diameter: 3/8 × pitch

Silent Chain Components and Construction

Silent Chain (Fig. 3-12) consists of toothed link plates and a variety of joint components. These and longitudinal guide plates are the basic components of Silent Chains. Attachments for Silent Chain are available on special order, but are infrequently used because Silent Chain is used primarily for power transmission and timing applications.

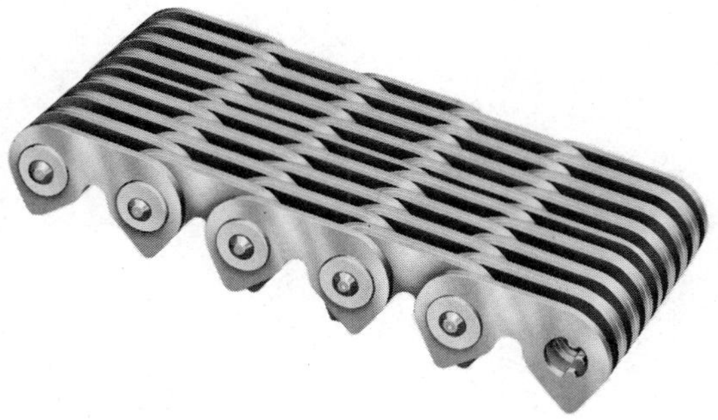

Figure 3-12 Typical Silent Chain.

Silent Chain Link Plates

Figure 3-13 shows typical variations in the configuration of the apertures of the link plates. The notches, flats, and projections in the apertures are designed to position the joint components in the link plates. Cylindrical bushings are assembled with interference fits to prevent rotation.

Guide Plates

Guide plates are provided in Silent Chains to prevent lateral movement. The guide plates differ from the tooth plates in that they have no crotch (see Fig. 3-36) to clear the sprocket teeth. A typical guide plate is shown in Figure 3-14.

Silent Chain Joint Components

Manufacturers of Silent Chains have developed different types of chain joints in an attempt to provide smooth and quiet operation at high speeds accompanied by long service life. Some of these types are described and illustrated next.

Round Pin with Bushing. In the design shown in Figure 3-15, the joint components consist of a plain cylindrical pin with a bushing added to increase the bearing area of the chain joint.

Round Pin and Segmental Bushing. Pictured in Figure 3-16, the components of this style of joint are a plain cylindrical pin with segmental bushings. Each bushing covers less than half the circumference; lugs between the bushing segments prevent rotation in the link plates.

Rocker Pin. In the rocker pin design, shown in Figure 3-17, two pins, differently shaped and keyed, are in contact in such a way that a rocking motion occurs between the pins when the chain flexes at the joint.

Rolling Pins. This design is shown in Figure 3-18. The joint components are two pins with cylinderical surfaces that roll on each other.

Chain Assemblies

Center Guide. A center guide chain is assembled with a row of guide plates in the center of the chain (Fig. 3-19). A circumferential groove in the sprocket teeth pattern provides a track for the guide plates. For wide chains, two rows of guide plates and two grooves are used. The spacing of the guides is specified in an ANSI standard.

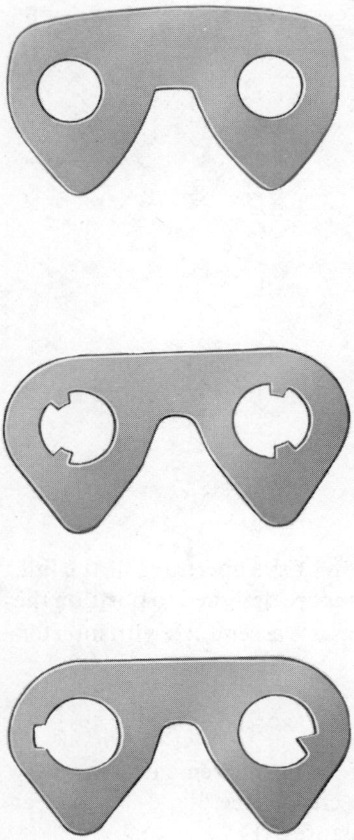

Figure 3-13 Typical Silent Chain link plates.

Figure 3-14 Typical Silent Chain guide plate.

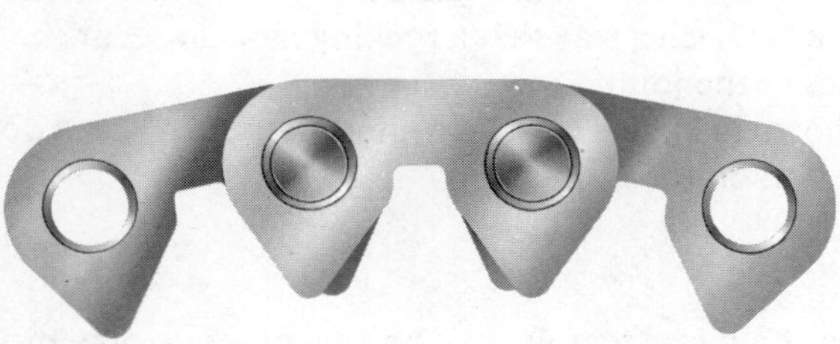

Figure 3-15 Round pin with bushing joint.

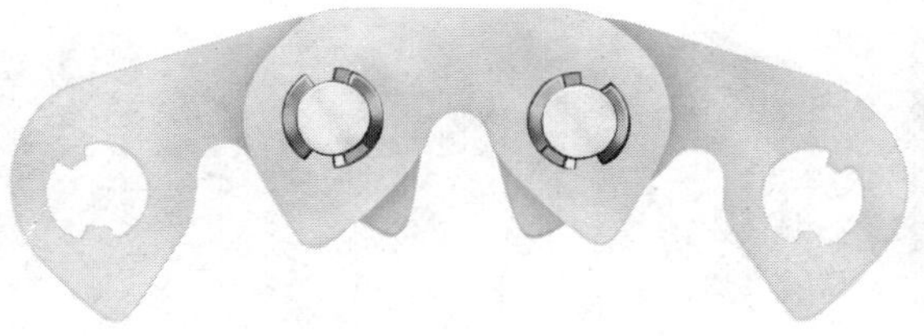

Figure 3-16 Silent Chain round pin and segmental bushing joint.

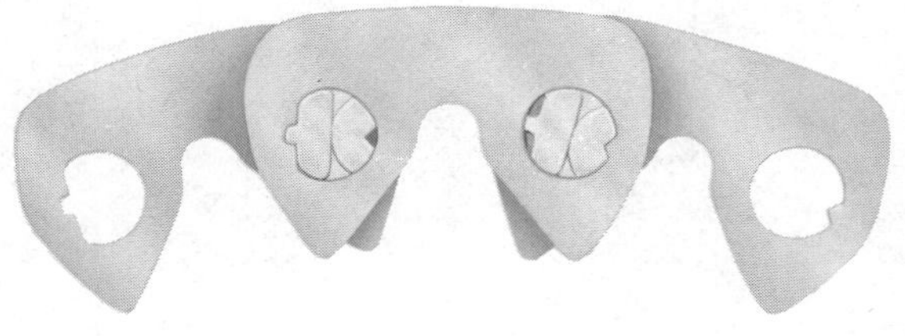

Figure 3-17 Silent Chain rocker pin joint.

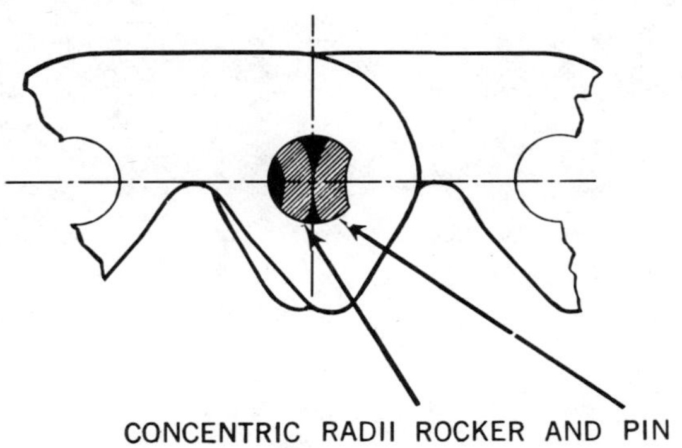

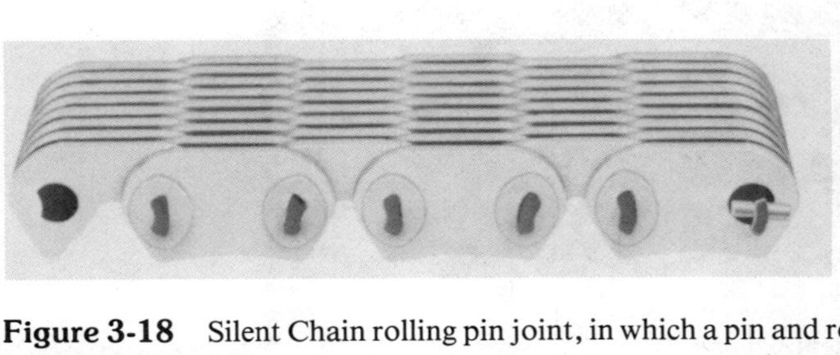

Figure 3-18 Silent Chain rolling pin joint, in which a pin and rocker with identical curvature roll on one another.

Figure 3-19 Silent Chain with center guide.

Figure 3-20 Silent Chain with side guides.

Figure 3-21 Duplex Silent Chain.

Figure 3-22 Reversible Silent Chain.

Figure 3-23 Silent Chain offset section.

Side Guide. A side guide chain has a row of guide plates on each side of the chain, as shown in Figure 3-20. These guides straddle the sprocket teeth and do not require a guide groove in the sprocket.

Duplex. Duplex Silent Chain is shown in Figure 3-21. This type of chain has teeth projecting from both sides in every pitch. It is designed for use where the sprockets must make contact with both sides of the chain, such as in serpentine drives or drives that require an idler sprocket.

Reversible. Reversible chain is a special application chain in that it is designed for use, like duplex chain, where sprockets must make contact with both sides of the chain. It differs from duplex chain in that its teeth project to both sides in every other pitch. An example of reversible Silent Chain is shown in Figure 3-22.

Offset Sections

An offset section, shown in Figure 3-23, is factory assembled and is made up in three or more pitches, one of which consists of offset plates. It is used in a chain having an odd number of pitches, so that the two ends can be coupled.

Engineering Steel Chain Components and Construction

In Chapter 2, assorted types of Engineering Steel Chains for which ANSI standards exist or are contemplated, plus a few for which there are no standards, were described and general areas of usage given. Describing the chains was done in terms of sidebars, pins, bushings, rollers and such, but details of these components and their relationships to one another in the various designs were reserved for this chapter.

From the discussions in Chapter 2, it may be seen that, in general, Engineering Steel Chains are made up of combinations of mostly similar parts, their functions also being similar but not necessarily identical. These similarities and differences will be apparent with an understanding of each of the components, as given in the following section.

Engineering Steel Chain Components

Engineering Steel Chains normally consist of pins, bushings, sidebars, and rollers, as shown in Figures 3-24 to 3-26. The pins and bushings provide the bearing surfaces for

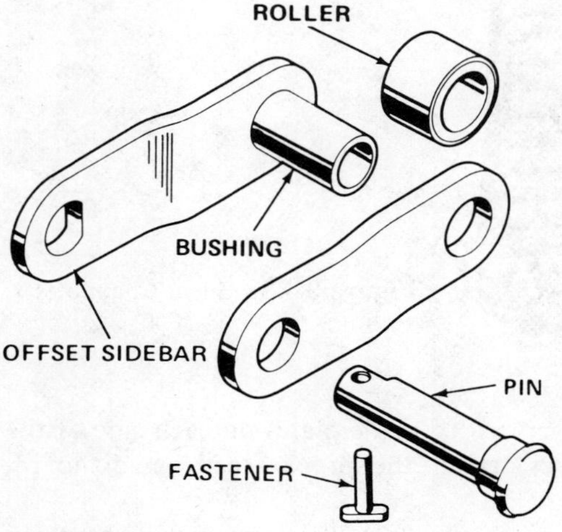

Figure 3-24 Link components, offset sidebar roller chain.

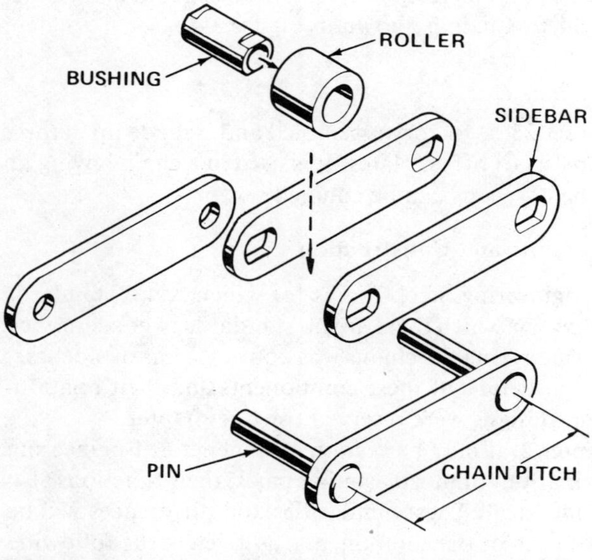

Figure 3-25 Link components, straight sidebar roller chain.

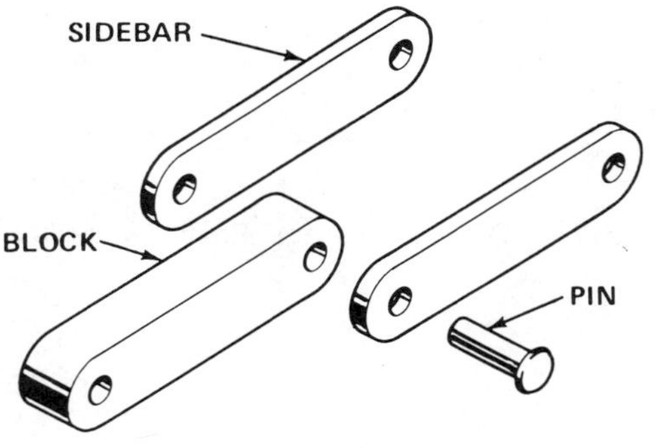

Figure 3-26 Link components, bar link chains.

the chain to articulate over sprockets. The pin-bushing bearing area is commonly referred to as the *chain joint.* The sidebars connect these bearing surfaces into chain. The sidebars also establish the chain pitch, which is the distance between chain joints. Rollers provide smooth rolling action during engagement with the sprocket teeth. Many conveyor chains have rollers larger in diameter than the sidebar height to provide the mechanical advantage of rolling friction in place of sliding friction.

Figures 3-24 to 3-26 show the components of three basic chains and their relationships to one another. Figure 3-27 shows similar detail for welded steel chain prior to welding. Straight sidebar and offset steel bushed rollerless chain have the same general relationship of parts as shown in Figures 3-24 and 3-25, except that there are, of course, no rollers, and a bushing of the solid shouldered type shown in Figure 3-29 is often used.

Pins. The chain pin forms one part of the chain joint. In addition, the pin connects one link to another and is the "shear" member between each set of inner and outer sidebars. The pin is locked laterally in the sidebars by a pin head or fastener on one end and by a fastener or riveting on the opposite end. To prevent rapid chain elongation from wear, the articulation must take place between the pin and bushing.

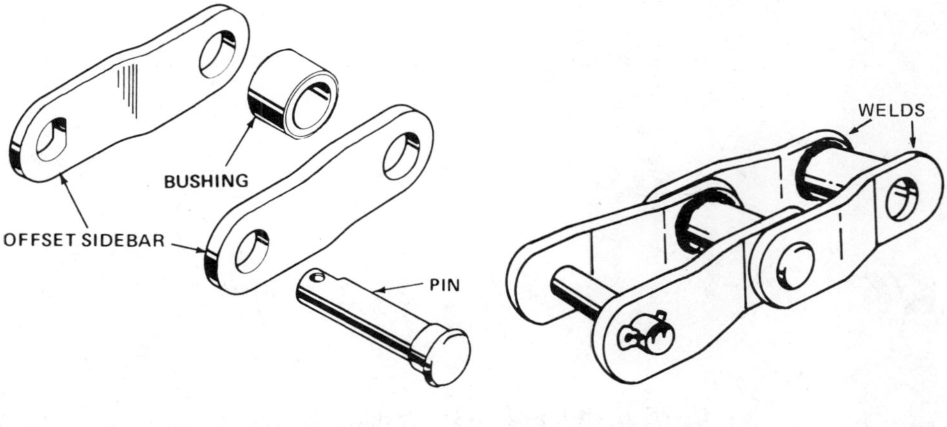

Figure 3-27 Exploded view, welded steel link and complete links.

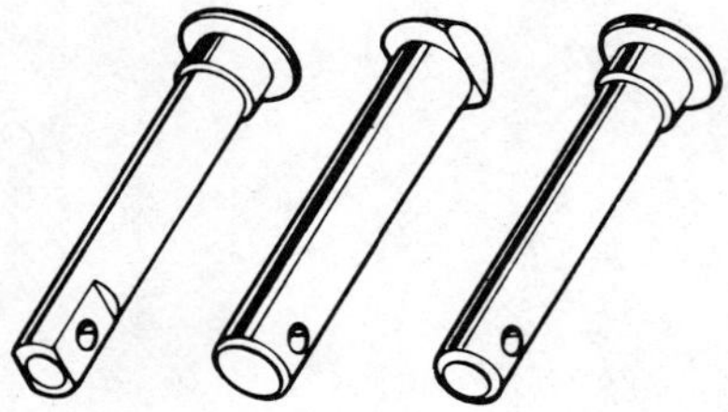

Figure 3-28 Typical pins for Engineering Steel Chains.

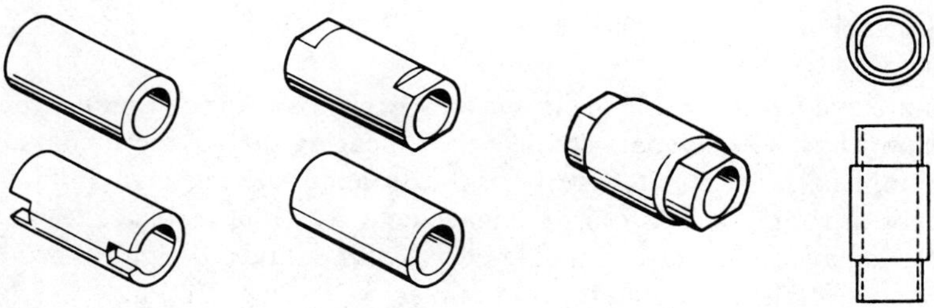

Figure 3-29 Bushings for Engineering Steel Chains.

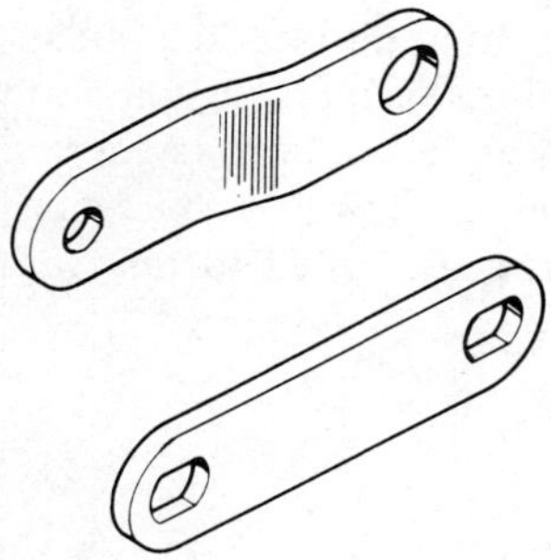

Figure 3-30 Typical offset and straight sidebars.

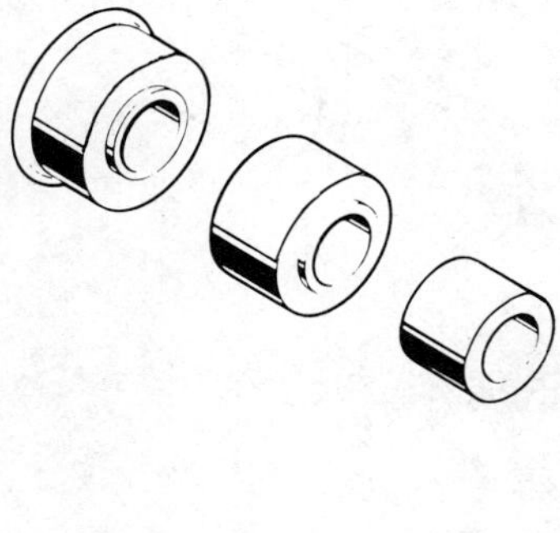

Figure 3-31 Typical rollers for Engineering Steel Chains.

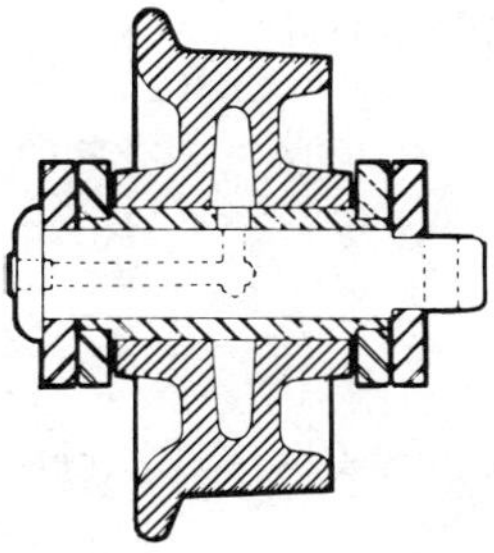

Figure 3-32 Cross section through joint of steel roller conveyor chain, showing special features to permit lubrication.

Therefore, the pin has to be prevented from rotating in the outer sidebars. This is accomplished by press fits between the pin and sidebars or by mechanical locking devices. Typical pin styles are shown in Figure 3-28.

Bushings. The chain bushing forms the other part of the chain joint. The bushings are locked laterally by the use of press fits and are prevented from rotating in the sidebars by press fits between the bushing and sidebars or by mechanical locks. The bushing outside diameter also provides a bearing area for the chain roller or sprocket contact on a rollerless chain. Typical bushing styles are illustrated in Figure 3-29.

Sidebars. The chain sidebars are the tensile members connecting the chain joints. As indicated elsewhere, the chain sidebars also establish the pitch of the chain. Figure 3-30 illustrates sidebar styles.

Rollers. The chain roller provides rolling action (reduces friction and smooths action) as the chain enters and leaves the sprocket. In conveyor-type applications, rollers that extend beyond the sidebars are used to provide rolling action and to reduce friction. Typical rollers are shown in Figure 3-31.

Figure 3-32 shows a cross section through the joint of a roller conveyor chain designed for periodic lubrication with grease. Several features of the parts are designed to provide a passage for the lubricants. The pin is equipped with a grease fitting at the head, and is drilled through and across to permit the passage of grease to the bushing, which is provided with holes to permit grease to reach the roller. The roller is a flanged, cored cast iron style, the recess in which is sometimes used as a reservoir for lubricants.

Attachments for Engineering Steel Chains were shown in Figure 2-25.

Chain Design—Considerations and Principles

This section reviews the considerations for the design of Roller, Silent, and Engineering Steel Chains relative to the principles used for the designs. The considerations differ as the functions of each type differ.

Conceptually, chain has been described as a series of connected bearings forming a flexible rack serving as a tensile member, usually in contact with toothed wheels.

Functionally, the intended uses of chain are:

1. To transmit power
2. To convey
3. To convert rotary motion to linear motion, or linear to rotary motion
4. To synchronize or to time motion

The subject under scrutiny here is handled in a substantially different way for each class of chain. The concept of what each chain is intended to accomplish should be examined from that aspect. In all cases, the philosophy of the designers has been to produce a product that will perform well in a specific kind of application. Each application has different needs to fulfill and different kinds of problems to solve—which is what design is all about.

Probably, no section of this book will bring out more clearly the differences in viewpoints among the producers of the various kinds of chains, as represented by their products, as will this section.

Roller Chain, in spite of a great deal of usage in other applications, has always been centered on power transmission service between shafts, at speeds probably best described as "moderate," but ranging from drives for high-speed motorcycles to relatively low speed machinery drives handled by large multiple strand chains. In the same way, Silent Chain is centered on power transmission, but at relatively high rotational speeds of the shafting. The concept, right from its beginnings in England, as described in Chapter 1, also recognized a need for a drive that would transmit a great deal of horsepower at low noise levels. Engineering Steel Chains have been the product intended for materials handling, but have also been the "catchall" of chain needs. Chain speeds are low, and loads (even at times in drives considered as chain tension rather than power) are comparatively high compared to those imposed on chains meant for long service at higher speeds.

Roller Chain—Design Considerations and Principles

Roller Chain is built to withstand a wide variety of speed and load conditions. Rotational speeds up to 10,000 rpm may be handled. The component parts of a chain drive are designed to proportions and tolerances based on (1) a detailed study of the types of loading to which each part is subjected and (2) empirical rules that long experience and research have proved to be successful.

A discussion of the empirical design proportioning rules establishing the series of standard Roller Chains was given in Chapter 2. This is explained further in terms of design in the ensuing pages of this chapter.

Considerations. In service, the components of a Roller Chain are subjected to wear, fatigue, impact, and galling.

Wear. Normal chain wear is caused by articulation of the pins in the bushings. In a properly lubricated chain drive, as described in Chapter 12, a separating film of fluid lubricant is formed in the operating chain joints.

Without adequate lubrication, wear in the chain joints is usually the limiting factor in the life of the chain. Such wear results in elongation of the chain; in other words, the chain pitch is increased. This increase in pitch allows the chain to ride out on the sprocket teeth, which are designed to permit moderate pitch elongation. When excessive elongation occurs, the chain must be replaced before it overrides the sprocket teeth. Obviously, proper lubrication of the bearing surfaces of the chain joints is of major importance in retarding wear.

In power transmission drives, there is only minor rotation of rollers under load, and roller wear is of little design importance. In conveyor applications, however, there is continuous rotation under load when the roller is the supporting member and, consequently, roller wear becomes of definite importance.

Improper alignment, whether due to faulty static alignment or dynamic misalignment, will cause abnormal wear in both chains and sprockets.

Installation of a new chain on badly worn sprockets may result in excessive chain wear and serious shortening of the life of the drive.

Fatigue. As each link traverses the tight side of the drive, it is subjected to maximum tension, and this tension decreases to a minimum value as each link reaches the slack side. This resulting cyclic variation in tension is taken into consideration in the capacity ratings. Where drives are operated beyond their rated capacity, however, link plate fatigue failure may seriously reduce drive life expectancy.

Impact. Engagement of the chain rollers with the teeth of the sprocket results in an impact. The magnitude of impact is reduced with an increase in the number of teeth on the sprocket or a reduction in drive tension or speed. A properly designed Roller Chain drive, based on rated chain capacities (tables in Chapter 5), takes these factors into account. If the drive is operated in excess of the rated chain capacity, impact can ultimately result in breakage of the rollers and/or bushings.

Galling. Operation of chain drives in excess of the speed and load limits shown in the horsepower rating tables (Chapter 5) can result in objectionable galling, regardless of the amount of lubricant applied to the chain. Galling is brought about by metal seizure between pin and bushing during joint articulation, and can result in greatly accelerated wear.

It is recommended that the Roller Chain manufacturer be given the opportunity to evaluate the conditions of operation of chain drives applied in the range of the shaded areas of the tables, since some degree of galling may occur under these operating conditions.

Chain Tension. Tension in a transmission chain is caused by:

1. The pull required for transmission of power
2. The centrifugal force generated as the chain travels around the sprocket, and also that caused by its travel over a curved path between sprockets
3. The catenary effect caused by the weight of that portion of the chain between sprockets
4. The dynamic effects of chordal action

Item 1, the pull required for transmission of power, is the basic force to be computed in designing the chain. It is equal to the useful tension in the tight side of the chain.

Item 2, tension from centrifugal force, results from the circular path followed by that portion of the chain that rotates with the sprocket. The resultant centrifugal force may be treated as a series of forces concentrated at each of the chain joints. Each such concentrated force is balanced by a pair of equal tension forces, one acting along the longitudinal centerline of each of the two adjacent links. These forces increase the total chain tension but do not affect the load on the sprocket teeth or on the shaft bearings. The tension due to centrifugal force is proportional to the weight of the chain per foot and the square of its linear velocity, and is independent of the sprocket radius. Standard chains, built to provide for the power pull, can safely carry the additional tension caused by centrifugal effects. Only at chain speeds exceeding 3000 ft/min does the centrifugal tension become a significant factor. Allowance has been made for this force in the horsepower capacity charts.

Item 3, the catenary effect, is produced by the approximate catenary form assumed by the slack side of the chain during operation—that is, the shape of a perfectly flexible cord suspended from two points. The weight of the chain sets up a tension that is approximately uniform throughout its length, and thus causes an additional load on the chain and sprockets of the drive. The significance of catenary tension is confined to the slack side of the chain; on the tight side, it is overridden by the working tension. Only in drives between extremely long centers is consideration of this tension required.

Regarding item 4, when a chain engages a sprocket, the centers of the chain joints lie on the pitch circle of the sprocket, and the center line of each link (the line between joint centers) forms a chord of this circle. As roller A (Fig. 3-33) approaches the sprocket, it tends to follow the chordal line to the position of roller B. However, engagement with the sprocket forces it to follow the arc of the pitch circle. When roller A has advanced to the position shown in the second part of Figure 3-33, it has been raised through the distance $PR - r$, as shown. This *chordal action* sets up pulsations in the following links, in terms of both movement perpendicular to the path of the chain and in chain velocity fluctuation.

The magnitude of chordal action depends on the number of teeth on the sprocket. The greater the number of teeth, the less will be the speed variation, the flatter will be the arc through which the link swings, and the smaller the pulsations.

Excessive chordal action will result in a noisy drive, and will tend to cause variations in the rotative speed of the sprockets. However, with proper selection of drive components, the speed irregularities will be absorbed satisfactorily by the elastic qualities of the chain.

Approximate Ratios of Part Dimensions to Chain Pitch. The approximate standard proportions for chain dimensions are:

1. Roller diameter D_r, approximately 5/8 of the pitch.
2. Chain width W, minimum distance between roller link plates, approximately 5/8 pitch.
3. Pin diameter D_p, approximately 5/16 pitch.
4. Thickness of link plates LPT, approximately 1/8 pitch.
5. Maximum height of roller link plates H_r, 0.95 pitch.
6. Maximum height of pin link plates H_p, 0.82 pitch.
7. Minimum ultimate tensile strength of standard chains, in pounds, $12{,}500 \times (\text{pitch})^2$.
8. Measuring load, in pounds, $125 \times (\text{pitch})^2$, with a minimum of 18 lb. The length of the assembled chain must be measured under this load to take up the clearance slack between the many pins and bushings.
9. The link plates of an offset link must be shaped to provide clearance for the link plates of the adjacent pin and roller links. Referring to the diagram of Figure 3-34, for standard offset links, the minimum radii of the flats on the link plate are:

$$x\ (\text{minimum}) = 0.41P + (0.001\ \text{in.} + 0.004P)$$

$$y\ (\text{minimum}) = 0.475P + (0.001\ \text{in.} + 0.004P)$$

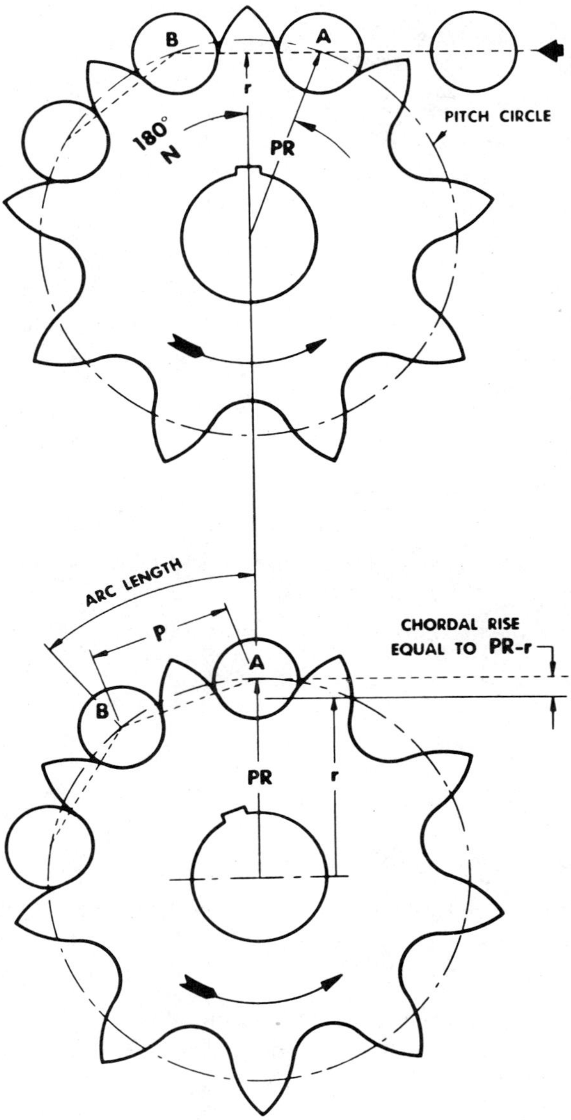

Figure 3-33 Chordal action of chain.

where P is the pitch.

Note: The parenthetical dimension is never less than 0.002 in. or more than 0.008 in.

Tolerance and Ultimate Dimensional Limits. The manufacturing tolerances specified in the standard are intended as limiting tolerances to ensure interchangeability; however, most manufacturers work to closer tolerances.

The specified tolerances are based on the following:

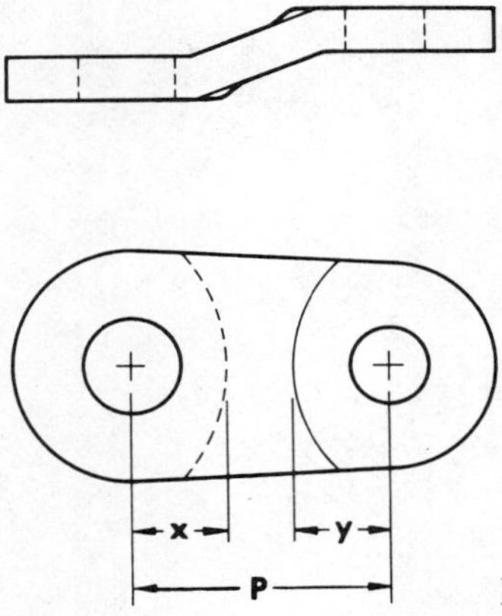

Figure 3-34 Offset link dimensions.

Maximum pin diameter = nominal pin diameter plus 0.0005 in.

Minimum diameter of hole in bushing = nominal pin diameter plus 0.0015 in.

Maximum width of roller link (width over outside of link plates) = width between roller link plates plus (2.12 × nominal link plate thickness)

Minimum distance between pin link = maximum width of the roller link plus 0.002 in.

New Roller Chains, under standard measuring load, must not be underlength. Overlength tolerance is as follows:

$$\text{Tolerance, in inches per foot} = \frac{0.001}{(\text{chain pitch, in inches})^2} + 0.015$$

Roller Chain length tolerances are given in Table 3-35.

The design of sprockets for Roller Chain is covered in Chapter 4.

Silent Chain—Design Considerations and Principles

Silent Chain is designed to transmit power between shafting rotating at high speeds. It is expected to do this quietly and without slippage, which it does. The key to the process is the comparable ease of joint action that occurs in the engagement of Silent Chain with its sprocket. When the chain is wrapped around the sprocket, the chain links may simultaneously engage both sides of the sprocket teeth.

Earlier in this chapter, several different types of joint construction were described and illustrated (Figs. 3-15 to 3-18). These designs are not a subject of ANSI standards, but most are intended to produce a maximum of rolling contact and a minimum of sliding contact in the joints and between the chain and the sprocket teeth. Rolling rather than sliding contact produces extended wear life, and in Silent Chain, a great deal of emphasis is placed on design relative to this factor.

An ideal chain design is one in which the components all wear out after the same predictable service life. This would be a balanced design in which the apertures in the link plates are small enough so that the link has high tensile strength but are large enough to accommodate joint components having high shear strength with sufficient bearing area to provide the desired wear life. A balanced design will also result in a chain that will not fail from metal fatigue before reaching its wear limit.

General Principles. Silent Chain power transmission capacity is expressed as horsepower per inch of chain width, and varies with the operating speeds and size of the smallest sprocket in the drive.

Basically, all capacity ratings depend on the allowable chain working load, which, in turn, depends on the following conditions.

The chain links must have sufficient strength to withstand the working tension. This strength is a function of the plate thickness, the smallest section of the plate, and the amount of metal surrounding the holes for the joint components.

Joint components are designed on the basis of allowable bearing pressure on the moving areas. This fact affects the link plate dimensions, since the size of the joint components determines the size of the holes in the plate.

Standard Specifications. ANSI Standard B29.2 for Silent Chains includes the following specifications for 3/8-in. pitch and larger chains.

Link plate contours may vary, but must engage standard sprocket teeth so that the joint centers lie on the pitch circle of the sprockets.

The minimum plate crotch height (Fig. 3-36) must be

$$\text{MCH} = 0.062\text{P} \tag{3-1}$$

where MCH is the minimum crotch height in inches and P the chain pitch in inches.

Table 3-35 Length Tolerances for Standard Roller Chains

Standard chain number	Pitch (in.)	Length tolerance (in./ft)
25	¼	0.031
35	3/8	0.022
40	½	0.019
41	½	0.019
50	5/8	0.018
60	3/4	0.017
80	1	0.016
100	1¼	0.016
120	1½	0.015
140	1¾	0.015
160	2	0.015
180	2¼	0.015
200	2½	0.015
240	3	0.015

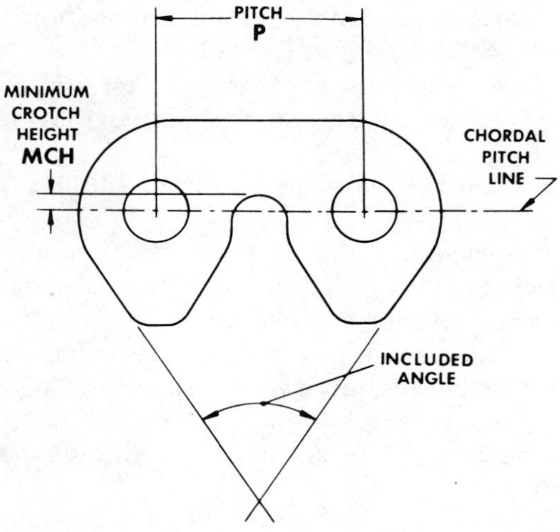

Figure 3-36 Silent Chain link plate—standard dimensions.

The maximum overall width of the chain W_M, (see Fig. 4-37, Chapter 4) is

$$W_M = W_N + 0.25P \qquad (3\text{-}2)$$

where W_N is the nominal chain width.

The measuring load M_L (load under which chain length is to be measured) is

$$M_L = 25PW_N + 20\,\text{lb} \qquad (3\text{-}3)$$

The tolerances for the length of new chain are: plus 1/32 in. per foot of nominal length, minus zero.

Sprocket design for Silent Chain is covered, including ANSI standards, in Chapter 4.

Engineering Steel Chains—Design Considerations and Principles

Engineering Steel Chains are intended for a wide variety of applications. Whether such applications are conveyors, elevators, drives, or tension linkages, the design considerations and principles of design used are about the same. This is because such depend primarily on the spectra of load, speed, and operating conditions and the anticipated number of cycles of operation considered reasonable under those conditions.

As will be seen, design considerations and principles reflect a different approach from that taken in designing and developing Roller and Silent Chains. But the approach taken is just as appropriate.

With the standard engineering class chains, the intent is to furnish a product that will carry the anticipated load and perform well under less than ideal conditions of operation. Inadequate or no lubrication; conditions where corrosive or abrasive materials, unfriendly to steel chain wear and life, are known to be present, with frequent or heavy shock loads—all these are known to the designer to be probabilities as often as possibilities.

Metal fatigue failure, from the standpoint of several million cycles under a known load spectrum with predictable results, is not much considered for this class of chain. The possibilities of corrosion fatigue together with abrasive wear are considered thoroughly, but the general design premise is that chains should wear out long before corrosion, and certainly metal fatigue, takes a toll.

Often, calculations of loadings, maximum and average, are no more than approximations based on meager data submitted. The approach in designing Engineering Steel Chains is well proven under such circumstances.

Fundamental Stress Considerations. In operation, chain components are subjected to tensile, shear, and compression stresses. An analysis of the major stresses in the various chain components is illustrated in Figure 3-37. (Actual chain ultimate tensile failures are usually a combination of the types of stresses shown.) Proper knowledge of these stresses enables the chain engineer to evaluate the performance of a chain as affected by ultimate, yield, and/or fatigue strength.

Sidebar. The critical stresses in the sidebar occur at the sidebar holes, as shown in Figure 3-37.

Pin. Chain tension results in a shear stress on the pin between each set of inner and outer sidebars. In addition, pin bending causes tensile stress on the pin surface opposite the bearing area.

Bushing. Both longitudinal and circumferential tensile and compressive stresses result from bending or deformation of bushings under sprocket contact loads.

Roller. The chain roller is subject to tensile stresses on the inside diameter during contact with the sprocket teeth. In conveyor and elevator applications, rollers are also subject to friction loads.

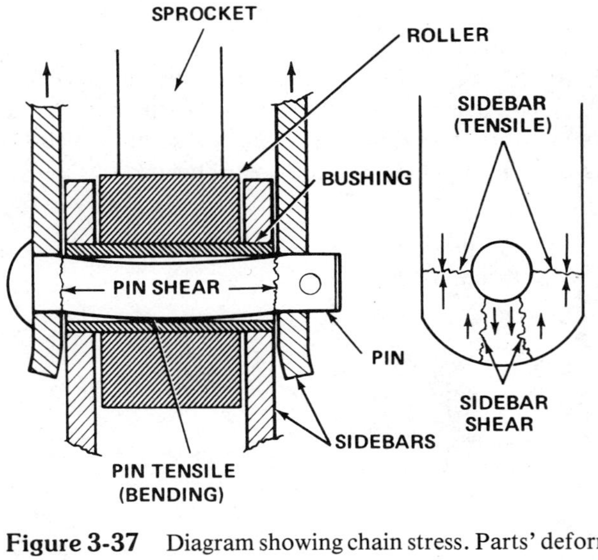

Figure 3-37 Diagram showing chain stress. Parts' deformations are exaggerated to illustrate the reactions to the load.

Chain Strength

Ultimate Strength. The ultimate strength of a chain is the single maximum load that breaks the chain. Chain failure in ultimate strength is normally due to one of the following:

1. Sidebar break (tensile) or shear
2. Pin break (tensile) or shear

In addition, bushings and rollers can fail in ultimate strength. Failure of the bushings and rollers does not result in chain parting but will drastically reduce the life of the chain or may make it inoperative.

Yield Strength. The elastic limit or yield strength of a chain is the load that causes permanent elongation of the chain. Above this load, the chain continues to elongate permanently and becomes too long for proper operation on the sprockets.

The yield strength of a chain is normally 40 to 60% of its ultimate strength. When a chain yields, it is normally the sidebars and pins that actually yield. As pins yield and permanently bend, the bearing area is drastically reduced and stiff joints can result because the bent pins take up the clearance between pins and bushings.

Fatigue Strength. Fatigue is the phenomenon leading to fracture under repeated or fluctuating stresses having a maximum value less than the tensile strength of the material. Fatigue strength is the maximum stress that can be sustained for a specified number of cycles without failure.

Since chain is a tensile member that is repeatedly stressed in service, fatigue is an important consideration. It is especially important in high-speed short center applications such as drives, where many stress cycles occur in a short period of time. All major chain components are subject to fatigue.

Fundamental Wear Considerations. Chain is a flexible assembly of moving parts that must articulate over sprockets or wheels. During articulation, wear occurs between the pins and bushings. This normal wear will result in chain pitch elongation. The allowable chain pitch elongation is limited by the amount of elongation the sprockets can tolerate. Analysis of this wear must be based on how the sprocket "sees" the chain. Figure 3-38 illustrates a sprocket with new and worn chain. Note that if the chain pitch elongates any more than that shown on the sketch, it will no longer fit the sprockets. The rate of chain pitch elongation wear depends on:

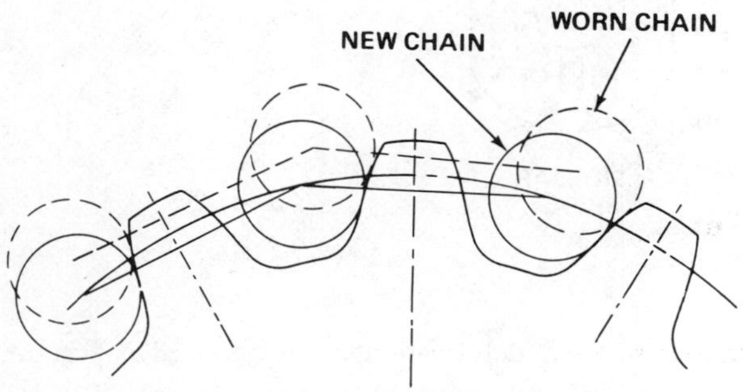

Figure 3-38 Diagram showing wear on new and worn chain.

1. Chain tension.
2. Projected bearing area. This is the contact surface between pin and bushing and is equal to the pin diameter times the bushing length.
3. Smoothness and hardness of contacting surfaces.
4. Lubrication.
5. Frequency and degree of articulation in the chain joints.
6. Environment.
7. Chain vibration.

Characteristics of Wear. Wear is thought of as a loss of material due to the chain components working either against each other or against some foreign material. The actual amount of wear on a given chain depends on several variables, including materials and heat treatment of the chain components, load and speed, lubrication, the presence of abrasive and corrosive substances, and the configuration of the sprockets. Most sliding or dragging chains have wear on the edges of the sidebars and often on the outer surfaces of the bushings on rollerless chains. In drive and elevator applications, wear in the chain joints is usually the limiting factor in the life of the chain. In bushed rollerless chains, wear is caused by the movement of the pins in the bushings and by the bushings rubbing on the sprocket teeth. In Engineering Steel Chains with rollers, wear occurs on the outside diameter of the rollers, between the rollers and bushings, and between the bushings and pins.

Chain Elongation Wear. Since a chain must articulate over a sprocket, wear occurs between the chain bushings and the pins and causes elongation of the chain or lengthening of the pitch (see Fig. 3-39). Sprockets are designed to accept a reasonable amount of elongation (3 to 6% for many styles) from wear, but when the chain elongates beyond this point it no longer fits the sprockets and the system will not operate properly. Chain wear elongation can be reduced by:

1. Selecting a chain with reduced bearing pressure between pins and bushings (increasing the bearing area or reducing the chain tension)
2. Selecting a chain with increased hardness of the chain joint members (pins and bushings)
3. Improved lubrication
4. Increasing sprocket size (increasing the number of teeth in the sprocket), which will result in a smaller angle of flexure

As chain-joint wear occurs, the resultant elongation is slightly different for straight sidebar chain compared to offset sidebar chain as the sprocket "sees" it. Figure 3-39 illustrates this difference. For the offset link, elongation is uniform from link to link ($A_3 = B_3$). However, on straight sidebar worn chain, the pitch across the bushing link (A_2) is the same as the unworn link (A_1). The pitch across the pin link (B_2) shows the resultant wear from two joints.

On conveyors, the chain is sliding or rolling between sprockets. This results in wear between the bushing and roller on rolling-type chain or wear of the sidebar on sliding-type chains. In each case, wear life is improved with increased wearing area, increased hardness of the wearing parts, or improved lubrication.

Wear of Bushing or Roller Due to Sprocket Contact. As a chain flexes onto a

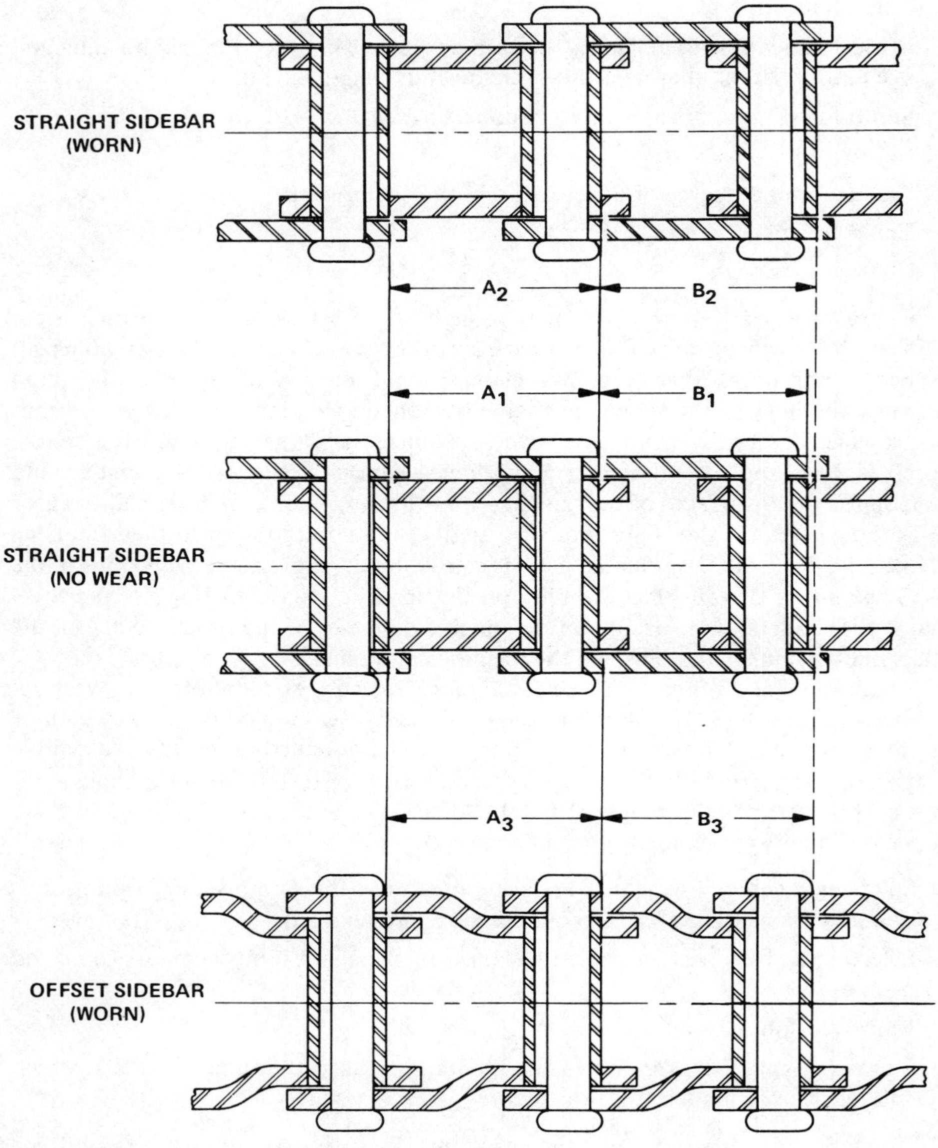

Figure 3-39 Diagram showing chain elongation due to wear.

sprocket and as the chain links pass around the sprocket, movement takes place that can result in the wear of the bushing outside diameter on a rollerless-type chain, or in the wear of the roller outside diameter and roller-bushing bearing area on the roller-type chain.

Rolling Wear. On roller chains used in conveyor service, the roller-bushing bearing area has a high load and a high bearing pressure of short duration as the chain flexes on the sprocket, and a lower bearing pressure of longer duration as the chain is carrying or rolling its conveyed load between sprockets. Hence the bearing areas are subjected to wear through the length of the conveying run rather than just when the links are engaging the sprocket teeth (see Fig. 3-40).

Sliding Wear. In applications where the chain slides (see Fig. 3-41), loss of material on the sidebar edge can develop to the point where the strength of the sidebars is substantially reduced. In sliding applications, the resistance to sliding wear can be improved by:

1. Selecting a chain with reduced bearing pressure by increasing the sliding bearing area or reducing the down load
2. Selecting a chain with increased hardness of the chain sliding bearing area
3. Improving the lubrication at contact surfaces

Working Load of Chains. Engineering Steel Chains are applied on the basis of a working load which considers both the strength and the wear life. A direct result of the design process, working loads are a basic criterion of chain selection, as will be seen in subsequent chapters.

Working loads are such that those listed are always considerably less than the yield strength of the chain.

Environment. Environment of operation is an important factor in the application and selection of chain. As a potential for consideration in the design of standard "off-the-shelf" chains, it is also important. Chains expected to work in an essentially hostile environment must be designed to produce reasonable life in the intended application.

The major environmental problems to be considered in the design area are corrosion and abrasive dusts. Temperature, either abnormally high or low, is also important.

Corrosion. Chains made from conventional materials are expected to operate in mildly corrosive conditions. Most industrial applications fall into this category, but some industries, such as fertilizer and chemicals, provide highly corrosive conditions.

The most important aspect of corrosion other than accelerating wear is that of corrosion fatigue in pins. Pins of chains expected to encounter such conditions are often provided with special plating or coating or are made of special materials to combat the problem.

Abrasion. The relative hardness of material to be handled by a chain system is a

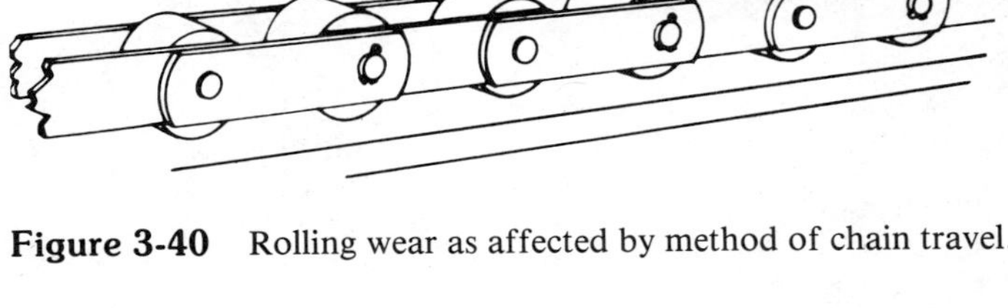

Figure 3-40 Rolling wear as affected by method of chain travel.

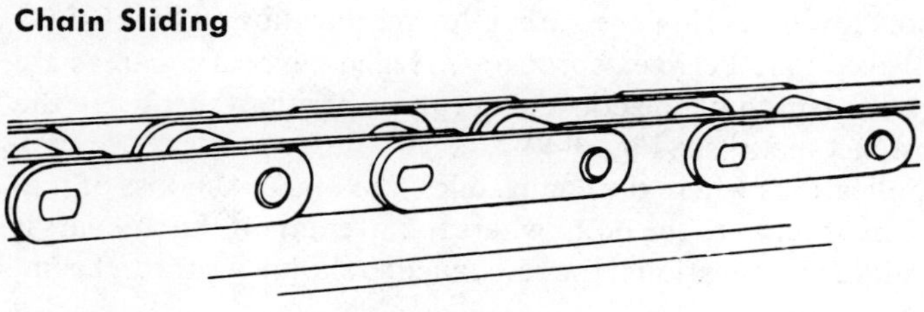

Figure 3-41 Sliding wear as affected by method of chain travel.

primary factor in selection of the proper chain. Highly abrasive materials such as sand, gravel, stone, or slag impose different conditions for chain selection than do flour or grain, for example. Abrasive conditions seriously affect chain wear unless the chain components are heat treated to resist the abrasive wear. For best results, the hardness of the chain parts should be harder than the Moh hardness of the conveyed material, if possible.

Chains of conveyors or bucket elevators exposed to dust or abrasion may be equipped with pressure-lubricating fittings on the rollers or pins (see Fig. 3-42). Lubricant applied in this manner also serves to flush dirt and debris out of the chain rather than wash such foreign material into it.

Lubrication, as a maintenance process, is covered in Chapter 12. However, proper planning for adequate lubrication of chain in a hostile or difficult application is an essential design consideration.

Temperature. If the application environment involves extremely high or low temperatures, special consideration must be given to the selection of chain that will withstand these temperatures. Component parts can be made of special material selected for its ability to withstand high- and low-temperature conditions. If such conditions exist, the user should check with the chain manufacturer.

Sprockets. Design considerations for sprockets for Engineering Steel Chains are covered in Chapter 4.

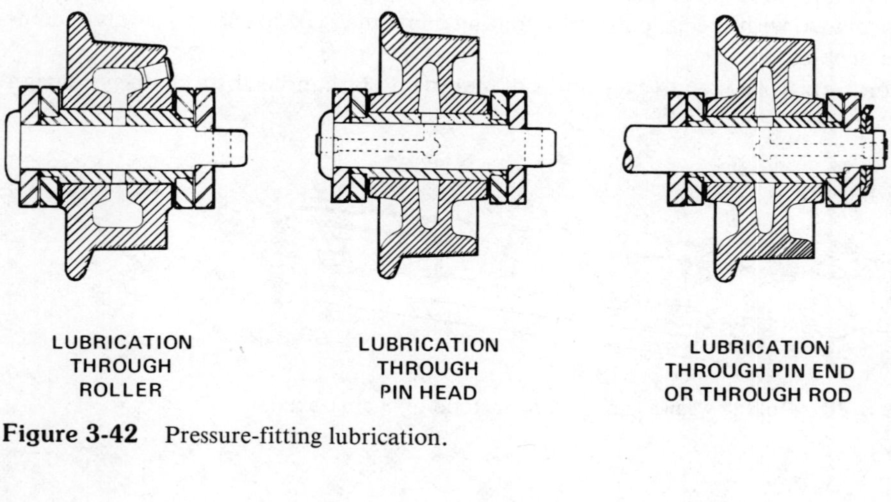

Figure 3-42 Pressure-fitting lubrication.

Research-Laboratory and Field Data

In the normal operation of a chain, each part is called on to play its part efficiently. This applies to all the types of chain considered here, whether used for drives, conveyors, or tension linkages. In addition to the normal loading, each part must also absorb and withstand impact loads, cyclic fatigue loads, corrosive and abrasive influences, and other problems.

Continued research by the manufacturers in their own laboratories, supplemented by consultants and others, has resulted in improvements in drive chain design and manufacturing to meet ever-increasing requirements.

Experimentation and field testing, as well as long experience, has been equally useful in the conveyor chain field, where better materials and manufacturing techniques have enabled chain manufacturers to meet applications that were previously troublesome.

Research, field data, and improved materials and manufacturing techniques have all played important parts in producing the improved chains of today.

4
Sprockets

Because the performance of a whole conveyor or drive installation depends largely on sprocket-chain interaction, choosing the right sprocket is as important as choosing the right chain. It is necessary, first, that the designs of chain and sprockets be compatible.

Standard sprockets are classified as either semiprecision (cast tooth or flame-cut tooth) or precision (machine-cut tooth). Although today's technology permits production of cast or flame-cut semiprecision sprockets with an accuracy approaching that of machine-cut teeth, the differences between precision and semiprecision really represent much more than just accuracy of manufacturing.

The types of sprockets used in any installation are really a function of the type of chain used. Obviously, both are a function of the purpose of the installation. Generally speaking, only precision sprockets are used with Roller and Silent Chains. Accurately cut teeth are necessary for the Roller Chain to accomplish the purpose for which it was installed. The tooth geometry of machine-cut sprockets differs in a great many ways from that used with, for example, a slow-moving long-pitch conveyor chain. The geometric path and velocity of the Roller Chain links around the drive sprocket will also differ in many ways from those functions of the long-pitch conveyor chain with its sprocket. The finish of the sprocket teeth will be different, as will the functions of the respective installations. In each case, the design and degree of precision and finish of the portions of the sprocket that interact with the chain must be matched to that chain.

For this reason, unless chains and sprockets are manufactured to a tight and accepted standard such as ANSI B29.1, it is recommended that sprockets normally be obtained from the manufacturer of the chain involved. As the effect of wear is alteration of tooth configuration and the elongation of chain, worn sprockets should always be replaced when new chain is installed (or vice versa).

Types of Sprockets

Sprockets are manufactured in various ways and from many materials. Manufacturing methods and materials used are reflected in the precision of the formation of the teeth and chain contact surfaces of the sprockets, and in the surface finishes. The materials used and subsequent processing, such as heat treatment, are reflected in performance in terms of load-carrying capacity, the rotative speeds at which the sprockets can be operated, and the service life of the sprockets.

The two basic types of sprockets, precision and semiprecision, are so identified in terms of contact and operation with the chains. Such identification may be further subdivided in terms of materials and manufacturing methods, which include economic factors together with the degree of accuracy required by the chains with which the sprockets are paired.

Sprockets made from highly finished steel or cast iron with machine-cut teeth are normally used with Roller Chain and are always used with Silent Chain. Engineering Steel Chains are usually operated over flame-cut or cast sprockets. These statements are, however, generalities; exceptions are not hard to find, particularly in the Engineering Steel chain field. However, regardless of how the sprocket teeth may have been formed, the most important criterion of sprocket selection is that the chain and sprockets have been designed to operate with one another.

Sprockets with machine-cut teeth are smooth and quiet in operation and are commonly used in high-speed drives. Sprockets with cast or flame-cut teeth are economical and give adequate life when used with Engineering Steel Chains. The various kinds of sprockets of precision and semiprecision types, as furnished by manufacturers, are reviewed in the following sections.

Cut-Tooth Sprockets for Roller Chain

ANSI Standard B29.1 defines the dimensions, tolerances, and tooth forms used in cut tooth sprockets. The general equations by which the standard Roller Chain tooth is proportioned is described in the design section later in this chapter.

ANSI Sprocket Types

Four sprocket types are included in ANSI Standard B29.1: type A is a plain plate sprocket without hub, type B has the hub on one side only, type C has the hub on both sides, and type D has a detachable hub. Types A, B, and C are shown in Figure 4-1; type D is shown in Figure 4-2.

Another design of a demountable hub-type sprocket is shown in Figure 4-3. It is manufactured in both the B and C sprocket styles and is a combination of a tapered split bushing with a tapered bore sprocket. The design provides for a positive grip fit to the shaft and a choice of finished bore sizes within a single sprocket selection.

Between the hub and the rim, a sprocket may be one of several forms, according to the size. The smaller sprockets are usually made solid, except for the shaft bore, since their weight is not excessive. For medium sizes, which are usually cast, webbed forms are used, having cored holes between the hub and the rim to reduce the weight (Fig. 4-4). The larger sprockets are usually cast and made in the form of spoked wheels to obtain the maximum possible reduction in weight (Fig. 4-5).

Figure 4-1 Roller Chain sprockets, types A, B, and C.

Figure 4-2 Roller Chain sprockets, type D.

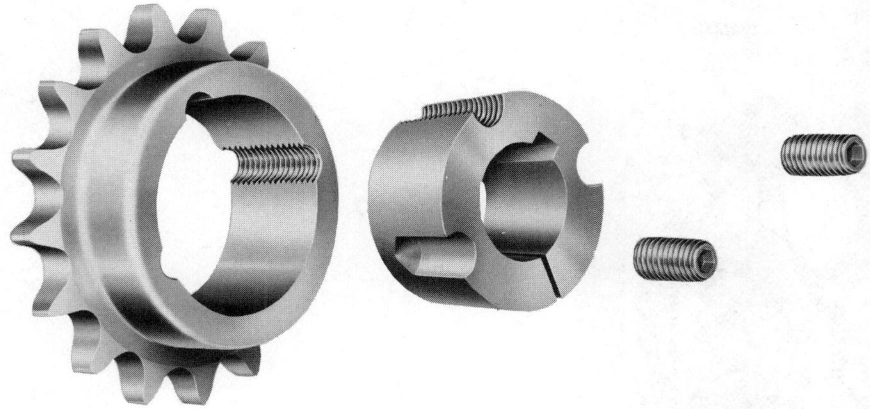

Figure 4-3 Taper lock bushing and sprocket.

Where the sprocket must be mounted between bearings, split forms (Figs. 4-6 and 4-7) are used to permit installation or removal without disturbing the shafting.

Some type D sprockets consist of a solid hub and a split plate (Fig. 4-2). Other type D sprockets are made with both the hub and the plate split (Fig. 4-7). (In the manufacture of any split sprocket, care must be taken to prevent any variation of the pitch at the split line. Figure 4-7 shows special dowels on the split for this purpose.)

Shear pin and slip clutch sprockets are designed to prevent damage to the drive or to other equipment, caused by overloads or stalling.

A shear pin sprocket consists of a hub, keyed to the shaft, and a sprocket that is

Figure 4-4 Typical webbed sprocket, type C.

Figure 4-5 Typical spoked sprocket, type C.

Figure 4-6 Typical split sprocket, type C.

Figure 4-7 Typical Roller Chain split sprocket with split demountable hub.

free to rotate either on the shaft or on the hub when the shear pin breaks (Fig. 4-8). The pin is designed to shear under a specified load. The shear pin is designed to transmit up to a predetermined amount of torque and to shear when overloaded. When the pin shears, the sprocket is free to rotate either on the shaft or on the hub which is keyed to the shaft. The shear pin is made of a material with known shearing strength, and its failure point can be established with reasonable accuracy by machining it with a groove of a calculated diameter at the shear plane. The assembly can be returned to working condition by inserting a new shear pin.

Sprockets for Double-Pitch Roller Chain. There are two forms of sprockets, designated as *single-cut* (Figure 4-9) and *double-cut* (Figure 4-10). Note the difference in tooth form. The term "effective teeth" indicates the number of teeth that engage with the chain during one revolution of the sprocket. Obviously, the number of effective teeth on a single-cut sprocket is equal to the number of actual teeth.

For a double-cut Roller Chain sprocket, the number of effective teeth is just half that of the actual teeth. For a sprocket with an odd number of teeth, the number of effective teeth is fractional. For example, with 25 actual teeth there are 12½ effective teeth. The advantage of an odd number of actual teeth is that each tooth will engage the chain only on every other revolution, thus practically doubling the life of the sprocket. With an even number of actual teeth, the same effect can be obtained by periodically advancing the chain by one tooth on the sprocket.

Note that a double-cut sprocket cannot be made for conveyor series chain with carrier-type rollers, as insufficient space exists between the rollers for the "extra" teeth.

Sprockets for double-pitch chain have different pitch diameters and bottom diameters than sprockets of the corresponding base chain of the ANSI series. When more than 35 teeth are involved, the differences become negligible.

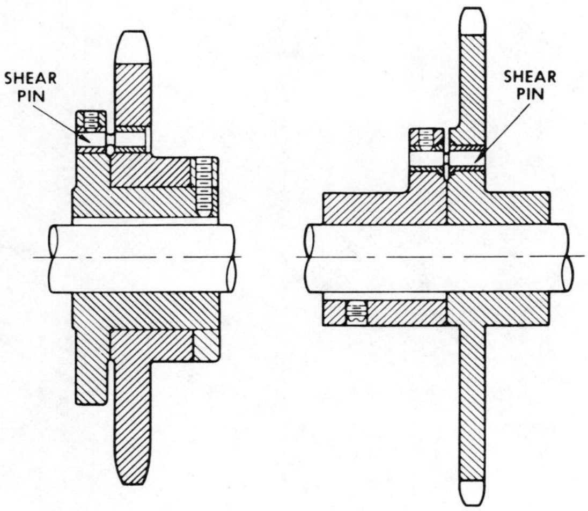

Figure 4-8 Shear pin hub sprocket.

Figure 4-9 Single-cut sprocket for double-pitch Roller Chain.

Figure 4-10 Double-cut sprocket for double-pitch Roller Chain.

Although the pitch diameters of sprockets for double-pitch Roller Chains with carrier rollers will be the same as for other Roller Chains of the same size, the bottom diameters will be smaller, to accommodate the larger rollers.

Sprockets for Multiple Strand Roller Chains. Sprockets for multiple strand Roller Chains are furnished as integral wheels with correct spacing for the strands and alignment of teeth. Typical multiple strand Roller Chain sprockets are shown in Figure 4-11.

Manufacturing of Cut-Tooth Sprockets. Cast Roller Chain sprockets have cut teeth, and the rim, hub, face, and bore are machined. The smaller sprockets are generally cut from steel bar stock and are finished all over. Sprockets cut from bar stock are limited in size by the maximum diameter of the commercially available stock.

Sprockets are often made from forgings or forged bars. The extent of finishing depends on the particular specifications that are applicable.

Many sprockets are made by welding a steel hub to a steel plate. This process produces a one-piece sprocket of desired proportions and one that can be heat-treated.

Hardness of tooth surfaces and toughness of body are factors important to sprocket life. These requirements can be met by the use of proper materials and heat treatment.

Materials for Cut-Tooth Sprockets. The most commonly used sprocket materials are iron and steel. Where corrosion exposure is great, stainless steel or bronze may be used. Formica, nylon, and various plastic materials may also be used.

For large Roller Chain sprockets, cast iron is commonly used. This material is especially applicable to drives with large speed ratios, since the teeth of the larger

Figure 4-11 Typical sprockets for multiple strand Roller Chain.

sprocket are subjected to fewer chain engagements in a given period. For severe service, cast steel or steel plate is preferred.

The smaller sprockets of a drive are usually made of steel. With this material, the body of the sprocket can be heat-treated to produce toughness for shock resistance, and the tooth surfaces can be hardened to resist wear.

Cut-Tooth Sprockets for Silent Chains

Except for the tooth forms, Silent Chain sprocket types parallel the general run of types of Roller Chain sprockets.

As may be noted from the typical Silent Chain sprockets pictured in Figure 4-12, the sprockets have the same general appearance as a gear, as the teeth are much more like gear teeth than like teeth for Roller Chain or Engineering Steel chains. The tooth configurations are covered by ANSI Standard B29.2 and are discussed further in the design section of this chapter.

Not all Silent Chain sprockets have teeth exactly like gearing, as will be seen by the sprockets described in the following sections.

Sprockets for Duplex Silent Chain

For a duplex chain (Fig. 3-21), the sprocket teeth must be grooved to allow passage of the reversed link plates (Fig. 4-13).

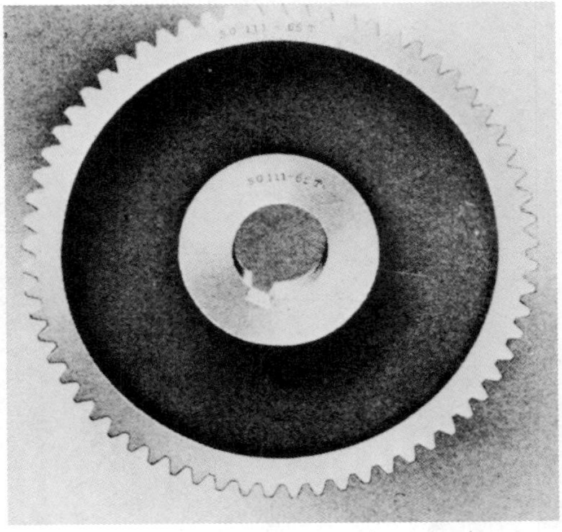

Figure 4-12 Sprockets for standard Silent Chain. Note the gearlike appearance.

The transverse spacing and widths of the sprocket grooves are determined by the assembly of the chain link plates for each individual drive. The groove depth must be sufficient to provide clearance for the back edges of the reversed plates.

Sprockets for duplex chains do not require grooves for guide plates, since the reversed link plates serve as guides.

Sprockets for Reversible Silent Chain

For a reversible chain (Fig. 3-22), every other row of sprocket teeth is omitted to allow clearance for the row of reversed link plates (see Fig. 4-14). Sprockets for reversible chains do not require grooves for guide plates since reversible chains are furnished only as side guide chain.

Figure 4-13 Sprocket for duplex Silent Chain.

Figure 4-14 Sprocket for reversible Silent Chain.

Sprockets for Engineering Steel Chains

Sprockets for Engineering Steel Chains are made of various metals and in various forms and shapes according to the features required for the particular application. Standard sprockets are available for most operating conditions.

Just as with Engineering Steel Chains, which are offered in many versions and styles not included in ANSI standards, so are there many such sprockets. Most of these are basically similar to standard wheels, and many are to work with standard chains where a standard sprocket will not answer the needs of the application.

This section concerns itself more with styles of sprockets available for the standard chains or chains with slight variations from standards, but listings of specific sprockets must be taken from manufacturers' catalogs.

Materials Used in Sprocket Manufacture

Sprockets for Engineering Steel Chains are usually made of the following materials:

1. Cast iron: often the most economical and selected for moderate conditions.
2. Chill iron: these sprockets have the teeth cast in a *chill.* The chill results in rapid cooling of the tooth surface upon casting, which results in a smooth hard tooth that can resist ordinary abrasion with a depth of hardness between 1/16 and 3/8 in., depending on the sprocket size.
3. Cast or fabricated steel: the cast steel sprockets have a cast tooth, whereas the fabricated utilize a flame-cut tooth. Both are designed to sustain heavy torque and shock loads. For the most demanding work in a highly abrasive condition, the teeth are flame or induction hardened.
4. Stainless steel: used where sprockets are exposed to corrosion.
5. Special materials, such as iron and steel alloys, bronze, and aluminum, can be used to satisfy special requirements.

Cast Iron Sprocket Types

Arm centers (Fig. 4-15) are used on cast sprockets whenever possible to reduce their weight, lower their cost, and make them easier to handle. The same effect can be accomplished with fabricated steel sprockets by flame cutting lightening holes in the plate, as shown in Figure 4-16.

Solid centers (Fig. 4-17) are for sprockets that are too small to accommodate arms; they are also used when allowable chain pull exceeds the strength of arm sprockets or when hub diameters are very large.

If the sprocket must be mounted between bearings, split sprockets (Fig. 4-18) are sometimes used to allow installation and removal without disturbing the shafting and other terminal equipment.

Segmented removable rims (Fig. 4-19) are bolted to the sprocket center. These sprockets are designed to be installed and replaced without necessitating removal of the chain, hubs, bearings, or shafts.

Traction Wheels

Traction wheels (see Fig. 4-20) are generally used to advantage on the head shaft of single-strand bucket elevators subjected to frequent obstructions or dangerous overloads. Traction wheels, however, must be used with care, as the frictional grip between the chain and head wheel must be sufficient to transmit the power required to drive the unit under normal load.*

If slipping occurs, heat and sparks may be generated. Accordingly, it is important that traction wheels not be used in circumstances where sparks could result in an explosion.

*Consult the chain manufacturer for allowable conditions for traction wheels.

Figure 4-15 Typical cast iron arm center sprocket.

Figure 4-16 Fabricated steel sprocket with lightening holes.

Figure 4-17 Typical cast iron solid or plate center sprocket.

Figure 4-18 Typical split cast iron spoked sprocket.

Figure 4-19 Sprocket with chilled iron segmental removable rims.

Figure 4-20 Typical traction wheel with chilled iron segmental rims.

Special-Purpose Sprockets for Engineering Steel Chain

Figure 4-21 shows a group of sprockets with special features that provide a variety of specific services. These are usually made from cast iron, often with chilled wearing surfaces.

A cast iron shear pin hub sprocket is shown at the left of Figure 4-21. Operation is exactly like that of the similar Roller Chain component shown in Figure 4-8. These are often used in conveyor drive service.

The two center wheels depicted in Figure 4-21 are used on drag conveyors with welded steel drag chain. Both wheels are drive units, the extensions on the wide flange or drum sprocket being used to prevent spillage when discharging the conveyor over the head shaft (see Chapter 9).

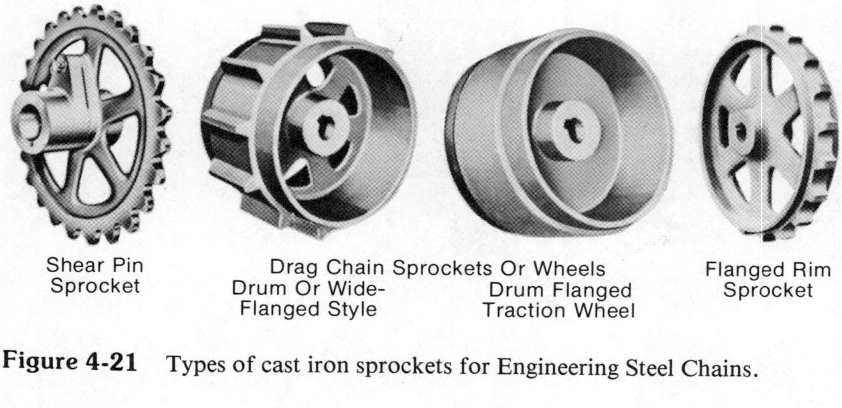

Figure 4-21 Types of cast iron sprockets for Engineering Steel Chains.

The sprocket at the right, with flanged rims to support the chain by contact with the sidebars (which extends the service life of both), is a type usually used on sanitary conveyors. These chilled iron wheels are often furnished in a "hunting tooth" design, similar in function to the double-cut Roller Chain sprockets shown in Figure 4-10. In this type of design, the sprocket tooth pitch is slightly greater than half the pitch of the chain, the sprocket being cast with an odd number of teeth. Since the links only engage every other tooth at a given instant, this means that each tooth on the sprocket will engage the chain only on every other revolution, thus nearly doubling the sprocket life. (This concept was first used in wooden gearing in the early nineteenth century.)

Another type of special-purpose sprocket is shown in Figure 9-38. This is the head shaft compensating sprocket used on drives of conveyors having long-pitch Engineering Steel Chains.

Sprockets for Engineering Steel Offset Drive Chains

Sprockets for Engineering Steel Offset Drive Chains are usually made of cast steel or flame cut from steel plate, the hubs and other parts being fabricated. The recommended tooth forms are given in ANSI Standard B29.10. Figure 4-22 shows the sprockets for a double-strand special version of 6.00-in.-pitch ANSI B29.10 No. 4824 Offset Engineering Steel Drive Chain. These sprockets show the ANSI tooth form and were machine cut to assure alignment of the double strand chain under heavy loads, the basic construction being by fabrication from alloy steel and heat treatment.

Sprockets for these chains range over a wide variety of designs.

Figure 4-22 Steel fabricated, cut-tooth sprocket used as drive for heavy double-strand Engineering Steel offset drive chain.

Legend

- ● Normally used.
- ○ Often used.
- □ Infrequently used.
- ☒ Not appropriate.

TYPES OF SPROCKETS

TYPES OF CHAINS	With Cut teeth; and finished, Machined Contact Surfaces.				With Flame Cut Teeth and/or Contact Surfaces.			With Cast Teeth and/or Chain Contact Surfaces. (Mostly available in splt designs with bltd hubs.)					
	Bar Stock, steel.	Steel fabricated, with or w/o Lightener holes.	Cast Iron spoked or web center, with or w/o lightener holes.	Cast Iron, split design.	Fabricated from steel plate, with or w/o lightener holes.	Fabricated with Segmental Rims; Cast or Fabricated center.	Traction Wheels, H.T., with Solid or Segmented Rim.	Cast Iron, with solid web, or with spokes or light'r holes.	Same, with chilled, hard contact surfaces.	Cast chilled iron Segmental Rims, bolted to cast or fab'd ctrs.	Cast Steel; Carb. or Alloy, H.T.	Cast Iron Traction Wheels, with or w/o Chilled Rims.	Same; Seq't'l Rims: Chilled C.I. or Cast, H.T. Alloy St'l.
Standard Roller Chains													
Roller Chains, B29.1	●	●	●	□	□	□	☒	□	□	☒	□	☒	☒
Same, Heavy or Offset versions, B29.1	●	●	●	□	□	□	☒	□	□	☒	□	☒	☒
Multiple Strand Roller Chains, B29.1	●	○	●	□	□	□	☒	☒	☒	☒	☒	☒	☒
Same, Double Pitch, B29.1	●	●	●	□	□	□	☒	□	□	☒	□	☒	☒
Same; Double Pitch, Conveyor B29.1	○	●	●	○	○	□	☒	□	□	☒	□	☒	☒
Same, Double Pitch, Conveyor Carrier Rollers B29.1	□	●	●	○	○	□	☒	□	□	☒	□	☒	☒
Leaf Chains. (if spkts used)	●	□	○	□	□	□	☒	□	□	☒	□	☒	☒
Silent Chains													
All Types, B29.2.	●	□	●	□	☒	☒	☒	☒	☒	☒	☒	☒	☒
Engineering Steel Chains													
Straight Sidebar Chains, Rollerless	☒	☒	☒	☒	●	○	○	○	●	○	○	○	○
Same, Roller Chains	□	□	□	□	●	○	☒	●	●	○	○	☒	☒
Offset Sidebar Chains, Rollerless.	□	□	□	□	●	○	○	○	●	○	○	○	○
Offset Sidebar Chains, Roller Conveyor.	□	□	□	□	●	○	☒	●	●	○	○	☒	☒
Welded Steel Chains, Mill.	□	□	□	□	●	□	□	○	○	□	□	□	☒
Welded Steel Chains, Drag.	☒	☒	☒	☒	○	□	□	●	○	□	□	□	☒
Open Barrel Steel Pintle Chains	□	□	□	□	○	□	□	●	○	□	□	□	☒
Bar Link Chains.	□	○	○	□	●	○	☒	○	○	□	□	☒	☒
Offset Steel Drive Chains, B29.10.	☒	□	□	□	●	□	☒	○	●	□	○	☒	☒

Table 4-23 Styles of Sprockets Used with Standard Chains

Sprockets for Types of Engineering Steel Chains

From the discussions in this section it should be apparent that a wide range of styles of sprockets are available for most Engineering Steel types of chain. Aside from the obvious requirement of a proper match-up of tooth configuration and chain type and dimensions, it might seem that the chains of this class can be operated over any type of wheel. Such is not the case, however; certain wheel styles will not work with certain chains in some applications. For example, a roller chain will not work on a traction wheel.

Conversely, some styles of sprockets are normally used for some chains, with all others used only in special cases.

Table 4-23 tabulates the various styles of sprockets supplied by manufacturers versus the types of chains of the Engineering Steel category described in this book. The tabulation is not intended as a recommendation of what should be used. Rather, it is intended as a guide to what is *normally* used and what is not, and then, only in general terms. As with most generalities, exceptions may be found to the tabulations.

In the key to the usages, round and square symbols are used. Basically, the round symbols indicate that the sprocket type is fairly routinely used with the chain noted and that standard wheels will often be found so listed in manufacturers' catalogs. Here, "normally" and "often used" can usually be separated by differences in cost and, sometimes, in quality. The square symbols indicate that usage of the particular wheel with the chain indicated is either infrequent or inappropriate. The major reason for "infrequent" usage is generally economics, but "inappropriate" usually is because of purely technical factors. Needless to say, any proposed installation suggesting chain-sprocket match-ups as indicated by the square symbols should only be done in consultation with a manufacturer of the chain involved.

Sprocket Tooth Forms—Design Considerations

The major function of a sprocket is to transmit torque. Therefore, the major design considerations in sprocket tooth form design are:

1. The sprocket teeth must transfer or absorb the tension load of the chain to or from the chain. This should be accomplished smoothly and consistently whether the sprocket be a driver or a driven wheel or operated as an idler wheel.
2. The sprocket must accommodate a reasonable amount of wear, in the form of pitch elongation of the chain or of tooth and bottom (or root) diameter wear of the sprocket.
3. The design must be compatible with the expected conditions of operation of the applications, including the possibility of degrees of foreign materials being interposed between chains and sprocket teeth.

The teeth of sprockets for Roller Chain, Silent Chain, and the various Engineering Steel Chains have great similarities and yet there are important design and conceptual differences. There are also basic differences in the terminologies of design factors and even in the way the teeth are expected to interface with their chains. Probably the major differences arise from the intended basic function of each type. An example is the means used to absorb chain pitch elongation due to wear for Roller Chain compared to Engineering Steel conveyor roller chain. In Roller Chain sprockets, elongation is absorbed by radial movement of the rollers on the sprocket tooth flanks as the links pass around the sprockets. A trace of the path would be vaguely parabolic; the link would move in and seat briefly in the sprocket bottom diameter and then move out again. The design of teeth for Engineering Steel Chain sprockets usually includes a factor not used on Roller Chain sprockets: pitch line clearance. This factor provides a short section of sprocket bottom diameter in the gaps between the teeth. With this type of wheel, the roller of a slowly moving Engineering Steel roller conveyor chain can move circumferentially along the bottom diameter of the sprocket to compensate for differences or changes of chain pitch and the corresponding placement of sprocket teeth. Other chains and their sprockets may be designed in other ways to absorb these differences.

Chain-sprocket interrelationships, very complex in theory, can be far more so when the effects of both wear and elastic deformation of chain and sprocket components under load are considered. A discussion of these is beyond the scope of this book, which must necessarily be limited to a discussion of how the teeth are laid out and dimensional considerations, as specified by ANSI B29 standards and manufacturers' practices.

The following sections take up the basic tooth designs in turn for Roller Chain, Silent Chain, and Engineering Steel Chains.

Sprocket Teeth for Roller Chain

As a logical approach to the subject of sprocket design, this section lists and defines the governing dimensions, and follows with an outline of the general principles of their application. Following this, the methods for the detailed design are presented.

Definitions Relative to Roller Chain Teeth. Sprocket diameter designations are indicated on Figure 4-24. These and certain other sprocket features are defined in the following paragraphs, and are discussed in more detail later.

Pitch Diameter. Pitch diameter is the diameter of the pitch circle that passes through the centers of the link pins as the chain engages the sprocket. Since the chain pitch is measured on a straight line between the centers of adjacent pins, the chain pitch lines form a series of chords of the pitch circle.

Bottom Diameter. The bottom diameter is the diameter of a circle tangent to the curve (called the *seating curve*) at the bottom of the tooth gap.

Caliper Diameter. The caliper diameter is the same as the bottom diameter for a sprocket with an even number of teeth. For a sprocket with an odd number of teeth, it is defined as the distance from the bottom of one tooth gap to that of the nearest opposite tooth gap. Some manufacturers term this the caliper "dimension."

Outside Diameter. The outside diameter is the diameter over the tips of the teeth.

Tooth Form. The tooth form is the outline of the working surface of the tooth on which the roller engages.

Tooth Profile. The tooth profile is the outline of the tooth as projected on a plane through the sprocket axis and the center of the tooth (Fig. 4-29).

Hub Dimensions. The hub and groove diameter designates either the outside diameter of the hub or the diameter of a groove cut in its outside circumferential surface for chain clearance.

The hub length is the distance between the end surfaces of the hub (Fig. 4-24).

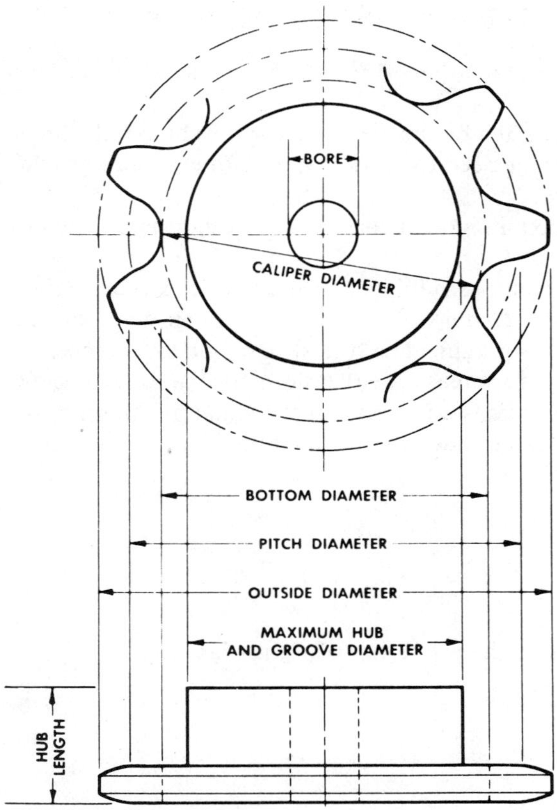

Figure 4-24 Roller Chain sprocket diameters.

Sprocket Forms. The types of sprockets are defined and the general forms are described in earlier portions of this chapter.

General Principles of Roller Chain Sprocket Construction. To provide smooth operation, all chain sprockets must be manufactured with reasonable static balance.

The tooth form must provide for:

1. Smooth engagement with the chain
2. Distribution of the load so that each tooth engaged with the chain will carry its share of the load
3. Provision for normal elongation of the chain within reasonable limits

The standard specifies detailed proportions for the tooth form to meet these requirements.

The tooth profile must be proportioned to provide proper clearance between the roller link plates.

A standard chamfer radius at the sides of the teeth is specified to permit easy engagement.

For multiple strand chains, the transverse pitch, dimension K (Fig. 4-29), is specified to accommodate the inner link plates.

Hub Dimensions. The important hub dimensions are the outside diameter, the bore, and the length.

The maximum hub diameter is limited by the need to clear the chain link plates.

The maximum bore is limited by the need for sufficient hub thickness to provide for a keyway and set screws.

Except for the limitations on maximum diameter, the hub design is governed by the maximum torque requirements.

Hub length must be sufficient to ensure stability of the sprocket on the shaft.

Dimensional Formulas for Roller Chain Sprockets. Certain sprocket dimensions are determined by geometrical requirements. Others, based on experience, are specified in the standard. The required formulas are either developed or quoted below.

Pitch Diameter. The pitch diameter is a function of the chain pitch and of the number of teeth in the sprocket; its value is determined as follows: Let

PD = pitch diameter, in.
PR = pitch circle radius, in.
P = chain pitch, in.
N = number of teeth in the sprocket

From Figure 4-25 it is seen that, due to the properties of right triangles,

$$PR = \frac{P/2}{\sin(180°/N)}$$

Therefore,

$$PD = \frac{P}{\sin(180°/N)} \tag{4-1}$$

Bottom Diameter. To permit the chain roller centers to follow the sprocket's pitch circle, the bottom diameter must be

$$BD = PD - D_r \tag{4-2}$$

where

BD = bottom diameter
PD = pitch diameter
D_r = roller diameter

all in inches.

Caliper Diameter. This important measurement is used in the quality control methods of manufacture to check the precision to which the bottom diameter has been cut. For a sprocket with an even number of teeth, the caliper diameter is the distance measured by calipers across the bottoms of the seating curves of two opposite tooth gaps; it is obviously equal to the bottom diameter.

With an odd number of teeth, no two tooth gaps are diametrically opposite, and

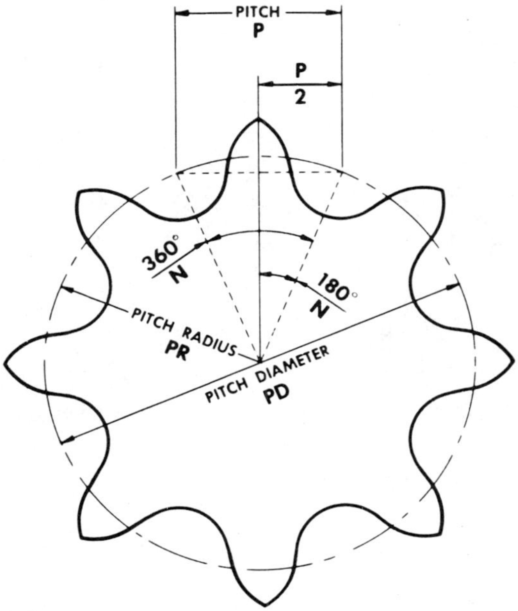

Figure 4-25 Computation of Roller Chain sprocket pitch diameter.

the "caliper diameter" is not a true diameter. Therefore, it is called the *caliper dimension* by many manufacturers. However, its value for a given odd-tooth sprocket can be computed as follows: Let

CD = caliper diameter for an odd number of teeth

D_r = chain roller diameter

In Figure 4-26, of the right triangle abc, angle

$$acb = \frac{90°}{N} \quad \text{and hence} \quad bc = ac \cos \frac{90°}{N}$$

$$= PD \cos \frac{90°}{N}$$

Therefore,

$$CD = PD \cos \frac{90°}{N} - D_r \tag{4-3}$$

Since accurate caliper measurement over concave surfaces is difficult, measurement over gauge pins is often used to check the bottom diameter. A gauge pin of proper diameter is placed in each of the two tooth gaps (Fig. 4-27) and the distance between their outer circumferences is calipered. The values for this measurement are

$$CD_g = PD + D_g \text{ for an even number of teeth} \tag{4-4}$$

and

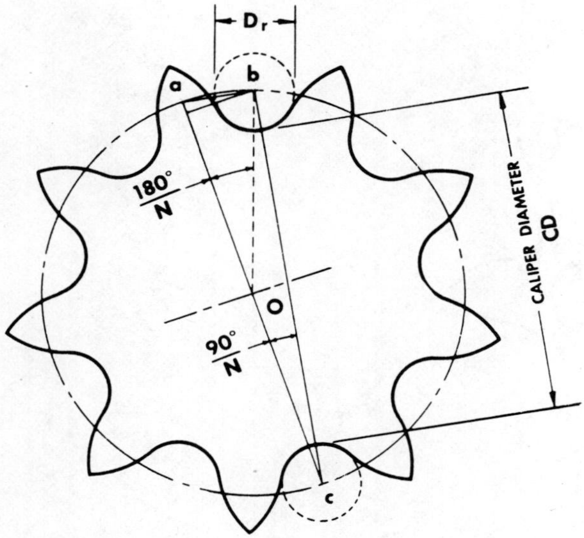

Figure 4-26 Computation of caliper diameter for sprocket with odd number of teeth.

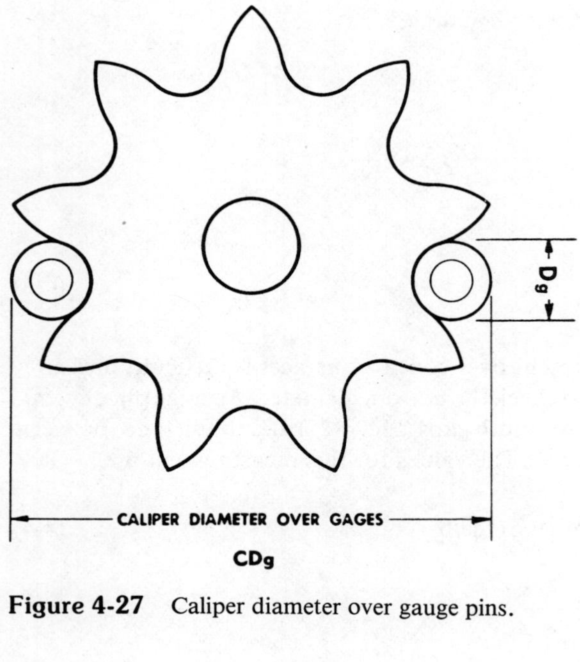

Figure 4-27 Caliper diameter over gauge pins.

$$CD_g = PD \cos \frac{90^\circ}{N} + D_g \text{ for an odd number of teeth} \tag{4-5}$$

where D_g, the diameter of the gauge pin $= D_r$.

Tolerance for all caliper diameter measurements is plus 0.000, minus $(0.001P\sqrt{N} + 0.003)$.

Outside Diameter. For the method of computing the outside diameter, with either pointed or truncated teeth, refer to Figure 4-28 for layout of the standard tooth form.

The maximum hub and groove diameter MHD (see Fig. 4-29) to ensure clearance for the standard chain link plates is specified as

$$MHD = P\left(\cot \frac{180^\circ}{N} - 1\right) - 0.030 \tag{4-6}$$

Design Procedure for Roller Chain Sprockets

Basic Data. The basic data for design of a roller chain sprocket are:

1. Number of teeth
2. Chain pitch
3. Roller diameter
4. Roller width
5. Height of roller link plates
6. Thickness of link plates of multiple strand chains

For an ANSI standard chain, items 3 to 6 are specified for each chain pitch and are given in Table 2-6.

Tooth Form. The theoretical tooth form is generated from the path of the chain roller as it moves from the chain pitch line of the tight strand into the pitch circle of the sprocket. The layout of the tooth form in terms of the chain pitch P, number of sprocket teeth N, and roller diameter D_r is shown in Figure 4-28. Note that radius R of the seating curve and radius F of the topping curve include the necessary clearance allowances between the sprocket and the engaging rollers. The formula for height of teeth H is given, as well as other dimensions required to draw the tooth form.

Tooth Profile. The standard tooth profiles for single and multiple width Roller Chains are shown on Figure 4-29. The dimensions in this figure are computed by the following formulas where

P = chain pitch

W = nominal chain width

LPT = nominal link plate thickness

MHD = maximum "hub and groove diameter" (measured at the bottom of the groove for a grooved hub)

The maximum tooth flange thickness (when machined)

$$t = 0.93W - 0.006 \text{ for single-width chains} \tag{4-7}$$

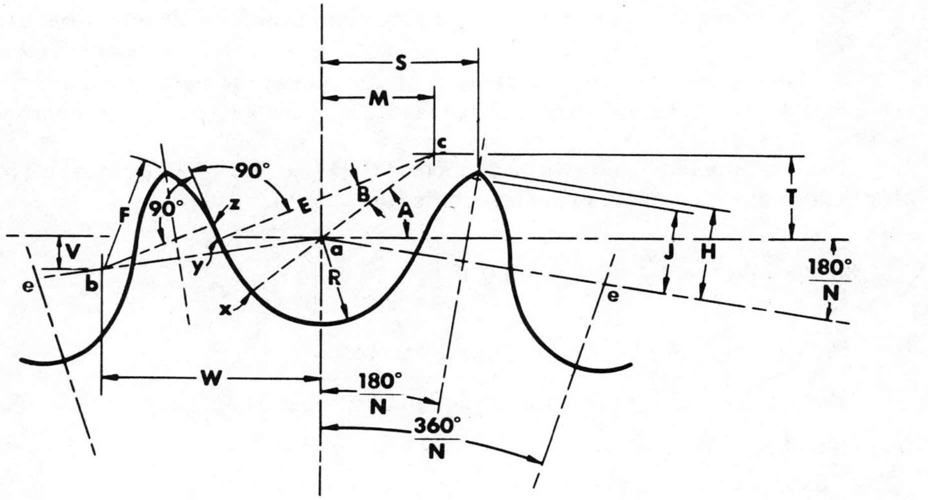

P = CHAIN PITCH (ae)

N = NUMBER OF TEETH

D_r = ROLLER DIAMETER

D_s (SEATING CURVE DIAMETER) $= 1.005\ D_r + 0.003$

$$R = \frac{D_s}{2} = 0.5025\ D_r + 0.0015$$

$$A = 35^\circ + \frac{60^\circ}{N}$$

$$B = 18^\circ - \frac{56^\circ}{N}$$

$$ac = 0.8\ D_r$$

$$M = 0.8 D_r \cos\left(35^\circ + \frac{60^\circ}{N}\right)$$

$$T = 0.8\ D_r \sin\left(35^\circ + \frac{60^\circ}{N}\right)$$

$$E = 1.3025\ D_r + 0.0015$$

CHORDAL LENGTH OF ARC $xy = (2.605\ D_r + 0.003) \sin\left(9^\circ - \frac{28^\circ}{N}\right)$

$$yz = D_r\left[1.4 \sin\left(17^\circ - \frac{64^\circ}{N}\right) - 0.8 \sin\left(18^\circ - \frac{56^\circ}{N}\right)\right]$$

$$ab = 1.4\ D_r$$

$$W = 1.4\ D_r \cos\frac{180^\circ}{N}$$

$$V = 1.4\ D_r \sin\frac{180^\circ}{N}$$

$$F = D_r\left[0.8 \cos\left(18^\circ - \frac{56^\circ}{N}\right) + 1.4 \cos\left(17^\circ - \frac{64^\circ}{N}\right) - 1.3025\right] - .0015$$

$$H = \sqrt{F^2 - \left(1.4\ D_r - \frac{P}{2}\right)^2}$$

$$S = \frac{P}{2} \cos\frac{180^\circ}{N} + H \sin\frac{180^\circ}{N}$$

APPROXIMATE OUTSIDE DIAMETER OF SPROCKET WHEN J IS 0.3P

$$= P\left(0.6 + \cot\frac{180^\circ}{N}\right)$$

OUTSIDE DIAMETER OF SPROCKET WHEN TOOTH IS POINTED

$$= P \cot\frac{180^\circ}{N} + \cos\frac{180^\circ}{N}(D_s - D_r) + 2H$$

THE PRESSURE ANGLE FOR A NEW CHAIN IS $xab = 35^\circ - \frac{120^\circ}{N}$

THE MINIMUM PRESSURE ANGLE IS $xab - B = 17^\circ - \frac{64^\circ}{N}$

THE AVERAGE PRESSURE ANGLE $= 26^\circ - \frac{92^\circ}{N}$

Figure 4-28 Design of standard Roller Chain sprocket teeth.

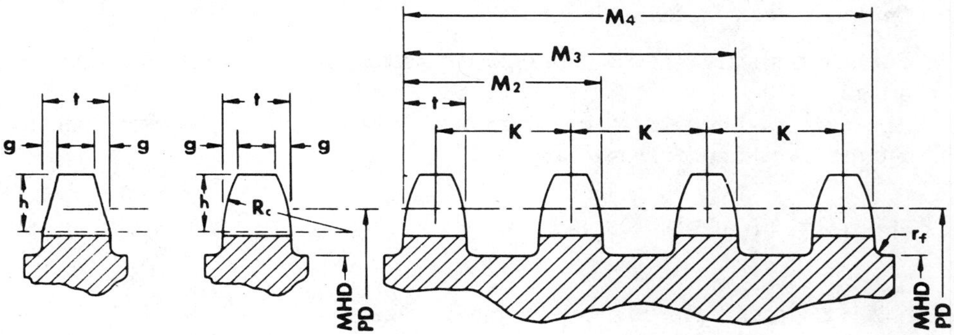

Figure 4-29 Roller Chain sprocket tooth profiles.

$$= 0.90W - 0.006 \text{ for double- or triple-width chains} \tag{4-8}$$

$$= 0.86W - 0.012 \text{ for quadruple- and wider-width chains} \tag{4-9}$$

Flange tolerance for sprockets when machined is plus 0.000, minus (0.01W + 0.006).

The minimum chamfer radius

$$R_c = 1.063P \tag{4-10}$$

The depth of chamfer

$$h = 0.5P \tag{4-11}$$

and

$$g = \text{approximately } 1/8P\text{, but not more than } \frac{W}{3} \tag{4-12}$$

The sprocket chamfer dimensions are noncritical and are given only as a guide for general design proportions.

The transverse pitch for multiple width chain is

$$K = W + 4.22LPT \tag{4-13}$$

The overall tooth flange width for multiple strand chain is

$$M_2, M_3, M_4, \text{etc.} = K(\text{width multiple} - 1) + t \tag{4-14}$$

with a tolerance of plus 0.000, minus (0.01W + 0.006).

The maximum fillet radius with maximum hub diameter is

$$r_f = 0.04P \tag{4-15}$$

Sprocket Wheel Design

Bar Steel Sprocket Wheels. For typical bar steel sprocket wheel proportions, refer to Figure 4-30.

The hub wall thickness H for such sprockets is a function of the pitch diameter PD and the bore diameter D as follows:

$$H = Z + \frac{D}{6} + 0.01\ PD \tag{4-16}$$

where Z is a constant depending on the pitch diameter.

For PD up to 2 in., Z = 0.125 in.

For PD 2 to 4 in., Z = 0.187 in.

For PD 4 to 6 in., Z = 0.25 in.

For PD above 6 in., Z = 0.375 in.

normal hub length (L) = 3.3H, with a minimum of 2.6H (4-17)

hub diameter HD = D + 2H, with a maximum allowable value as given by equation (4-6) (4-18)

Single-Strand Cast Sprocket Wheels. Typical cast sprocket wheel sections are shown in Figure 4-31, where P is the chain pitch and W the nominal chain width, with other dimensions as shown.

Acceptable dimensions are approximately as follows:

$$H = 0.375 + \frac{D}{6} + 0.01PD \tag{4-19}$$

$$L = 4H \text{ for semisteel castings} \tag{4-20}$$

$$C = 0.5P \tag{4-21}$$

$$C' = 0.9P \tag{4-22}$$

$$E = 0.625P + 0.93W \tag{4-23}$$

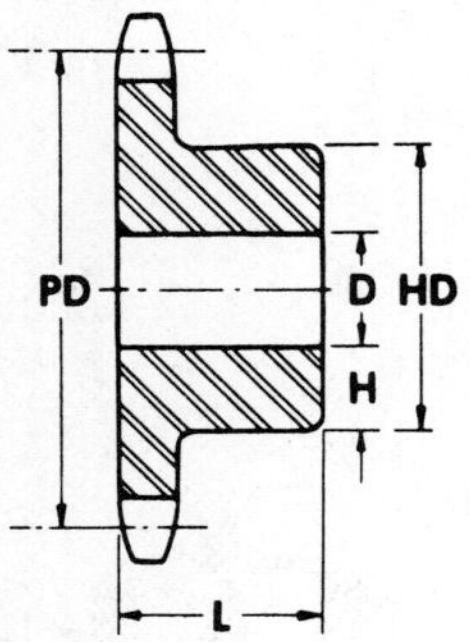

Figure 4-30 Roller Chain bar steel sprocket wheel dimensions.

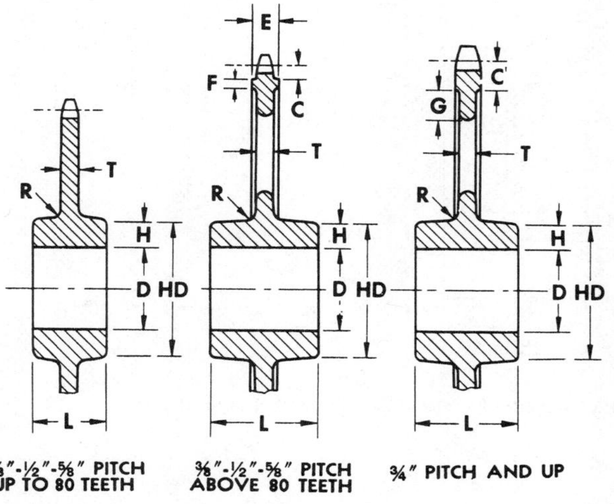

Figure 4-31 Single strand Roller Chain cast sprocket wheel dimensions.

$$F = 0.156 + 0.25P \tag{4-24}$$

$$G = 2T \tag{4-25}$$

$$R = 0.04P \tag{4-26}$$

The web thickness T is a function of the chain pitch as given in Table 4-32.

Multiple Strand Cast Sprocket Wheels. For typical proportions of multiple strand cast sprockets, refer to Figure 4-33.

The various proportions are the same as those given for single-strand cast sprocket wheels except as follows:

$$R = 0.5T \tag{4-27}$$

Table 4-32 Web Thickness T of Single-Strand Cast Roller Chain Sprocket Wheels as a Function of Chain Pitch

P	T	P	T
⅜	0.312	1½	0.625
½	0.375	1¾	0.750
⅝	0.406	2	0.875
¾	0.437	2¼	1.000
1	0.500	2½	1.125
1¼	0.562	3	1.250

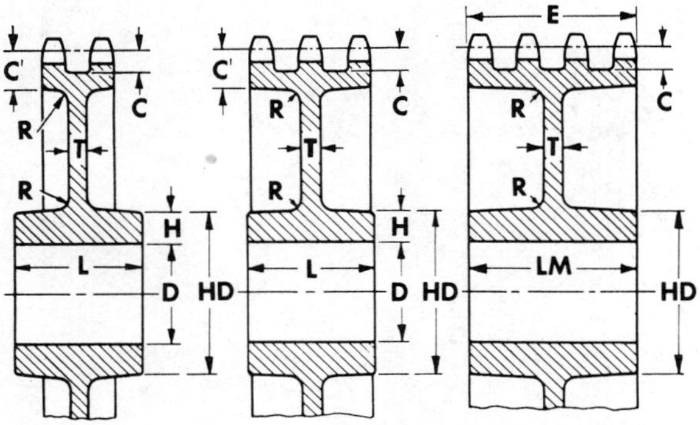

Figure 4-33 Multiple strand Roller Chain cast sprocket wheel dimensions.

$$\text{LM for quadruple strands} = \text{E} \tag{4-28}$$

Values for T are given as a function of chain pitch P in Table 4-34.

Spoked Sprocket Wheels. The design of spoked (see Fig. 4-5) sprocket wheels is usually based on the following assumptions:

1. That the maximum torque load acting on a sprocket is the chain tensile strength times the sprocket pitch radius
2. That the torque load is equally divided between the arms by the rim
3. That each arm acts as a cantilever beam

The arms are generally elliptical in cross section, with the major axis twice the minor axis.

Sprocket Teeth for Silent Chains

Except for the tooth forms, Silent Chain sprocket design parallels the general design practice of roller chain sprockets as described in the preceding section. The general appearance and styles of sprockets for Silent Chain were also covered in an early section.

Table 4-34 Web Thickness T of Multiple Strand Roller Chain Sprocket Wheels as a Function of Chain Pitch

P	T	P	T
3/8	0.375	1 1/2	0.750
1/2	0.406	1 3/4	0.875
5/8	0.437	2	1.000
3/4	0.500	2 1/4	1.125
1	0.562	2 1/2	1.250
1 1/4	0.625	3	1.500

Sprockets manufactured to the standard have teeth with straight-line working faces. The tops of the teeth may be either rounded or square. Bottom clearance below the working face is not specified, but must be sufficient to clear the chain teeth. The standard tooth form is designed to mesh with link plate contours having an included angle of 60° (see Fig. 4-36).

Sprocket Diameters. Figure 4-35 shows the essential sprocket diameters and their symbols.

If N is the number of teeth in the sprockets, the various diameters are as follows (all values in inches):

$$PD = \frac{P}{\sin(180°/N)} \tag{4-29}$$

$$D_P = 0.625P \tag{4-30}$$

$$\text{OD (for rounded teeth)} = P\left(\cot\frac{180°}{N} + 0.08\right) \tag{4-31}$$

$$G = P\left(\cot\frac{180°}{N} - 1.16\right) \tag{4-32}$$

$$\text{MHD (for hobbed teeth)} = P\left(\cot\frac{180°}{N} - 1.33\right) \tag{4-33}$$

$$\text{MHD (for straddle-cut teeth)} = P\left(\cot\frac{180°}{N} - 1.25\right) \tag{4-34}$$

$$\text{OGD (for even number of teeth)} = PD - 0.125P \csc\left(30° - \frac{180°}{N}\right) + 0.625P \tag{4-35}$$

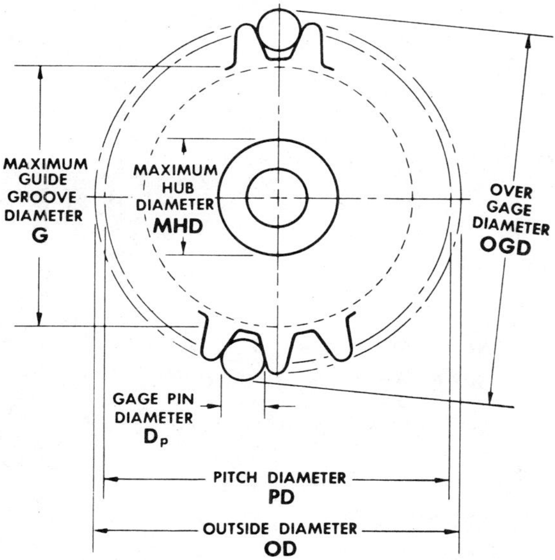

Figure 4-35 Silent Chain sprocket diameters—standard dimensions.

and

OGD (for odd number of teeth)

$$= \cos \frac{90°}{N} \left[PD - 0.125P \csc \left(30° - \frac{180°}{N} \right) \right] + 0.625P \tag{4-36}$$

Note that the OGD for Silent Chain finished sprockets is measured over gauge pins of specified diameters.

Tolerances for Sprockets. Standard tolerances in inches for sprocket diameters are as follows: for G, plus 0.000, minus 1/32 in.; for OGD, plus 0.000, minus (0.004 + 0.001 P$\sqrt{N}$); and for maximum eccentricity (total indicator reading), 0.001PD; but not less than 0.006 for pitch diameters less than 6, and not more than 0.032 for pitch diameters over 32 in.

Sprocket Tooth Forms for Silent Chain

Standard Tooth Form. The layout of the tooth form for standard sprockets in terms of the chain pitch and number of teeth is shown in Figure 4-36.

Minimum Number of Teeth. From Figure 4-36 it will be seen that the angle between the faces of a given tooth (60° – 720°/N) is an inverse function of the number of

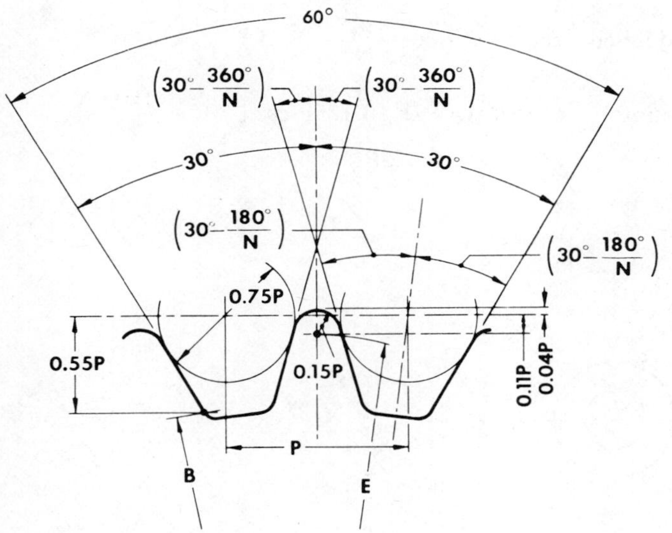

P = CHAIN PITCH

N = NUMBER OF TEETH

E = DIAMETER TO CENTER OF TOPPING CURVE

B = DIAMETER TO BASE OF WORKING FACE

$E = P \left(\cot \frac{180°}{N} - 0.22 \right)$

$B = P\sqrt{1.515 + \left(\cot \frac{180°}{N} - 1.1 \right)^2}$

NOTE 1. TEETH MAY BE EITHER ROUNDED (TOPPED) OR SQUARE (TURNED).

2. SHAPE OF ROOT LINE BELOW WORKING FACE MAY VARY WITH TYPE OF CUTTER.

Figure 4-36 Layout of standard sprocket tooth form for Silent Chain.

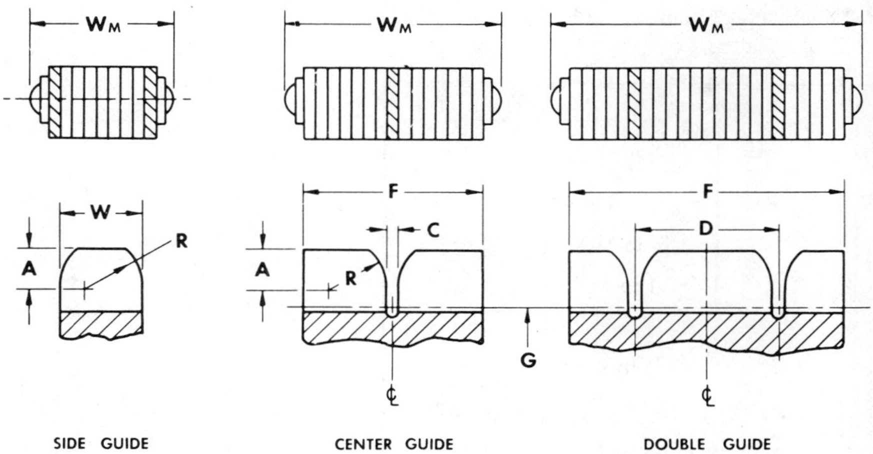

Figure 4-37 Standard sprocket tooth profiles for Silent Chain.

teeth. For a 12-tooth sprocket, this angle will be zero, that is, the tooth faces will be parallel. For smaller sprockets, the tooth would be undercut.

For best results, 21 or more teeth are recommended; fewer than 17 teeth should not be used.

Tooth Profiles. Tooth profile and dimensions for standard Silent Chain sprockets are shown in Figure 4-37. For center and double guide chains, the sprocket tooth width is F = nominal chain width; plus 0.125, minus 0.000. Other dimensions shown on Figure 4-37 are given in Table 4-38; the standard chain widths for the different types of chain guides are shown in Table 4-39.

Table 4-38 Standard Dimensions for Silent Chain Sprocket Profiles (see Fig. 4-37)

Chain Pitch (in.)	DIMENSIONS (in.)				
	A	R ± 0.003	C ± 0.005	*D ± 0.010	W + 0.010 − 0.000
3/8	.133	.200	.100	1.000	.410
1/2	.133	.200	.100	1.000	.410
5/8	.177	.250	.125	2.000	. . .
3/4	.274	.360	.180	4.000	. . .
1	.274	.360	.180	4.000	. . .
1 1/4	.274	.360	.180	4.000	. . .
1 1/2	.274	.360	.180	4.000	. . .
2	.274	.360	.218	4.000	. . .

*For double guides only.

Table 4-39 Standard Chain Guide Types for Various Silent Chain Widths

Chain Pitch (in.)	Range of Chain Widths (in.)		
	Side Guides	Center Guides	Double Guides
3/8	1/2	3/4 to 2 1/2	3 to 6
1/2	1/2	3/4 to 3 1/2	4 to 8
5/8	. .	1 to 4	5 to 10
3/4	. .	1 to 6	7 to 12
1	. .	2 to 6	7 to 16
1 1/4	. .	2 1/2 to 7	8 to 20
1 1/2	. .	3 to 7	8 to 24
2	. .	4 to 7	8 to 30

Referring to Table 4-38, the sprocket tooth width W for the side-guide chain is specified as 0.410 in. for 1/2-in.-wide chains with 3/8- and 1/2-in. pitches. No value of W for other chain sizes is specified in the standard.

Sprocket Teeth for Engineering Steel Drive Chains

The sprocket tooth forms for Engineering Steel drive chains, as specified in ANSI Standard B29.10, differ from those for Roller Chain in having pitch line clearance and an undercut bottom diameter slightly smaller than the theoretical root diameter (see Fig. 4-40). These differences permit use of a less precisely made tooth form for drive service than the machine-cut Roller Chain sprocket tooth.

The surfaces of the teeth for sprockets for Engineering Steel drive chains are characteristic of the sand cast steel and flame-cut finishes of typical sprockets of this class. Drives using these chains are often operated in locations where mud, dirt, ore, and rock dusts and such get into the chain and the drive is exposed to the weather. Lubrication in such circumstances is minimal, and cast steel and fabricated steel sprockets are appropriate to such applications. Pitch line clearance and the undercut bottom diameter both help provide proper chain-sprocket action under such adverse conditions. The action of these chains and their sprockets is essentially the same as that of a Roller Chain drive.

When the chain and sprocket are interacting properly, the working face is the only functioning surface on the sprocket tooth. Therefore, the angle (pressure angle) and length of this working face are the primary bases for tooth form design. In general, the pressure angle must be small enough to absorb the required amount of working chain tension to assure transmission of torque, while the working face should be long enough to accommodate chain wear without interfering with the roller flex path.

Symbols used in the text are defined below; see Figure 4-40 for other pertinent terminology.

C_b = undersize compensation, in.
C_p = pitch line clearance, in.
D = pitch diameter, in.
D_b = bottom diameter, in.
D_{rt} = root diameter, in.
L_f = length of working face of tooth, in.

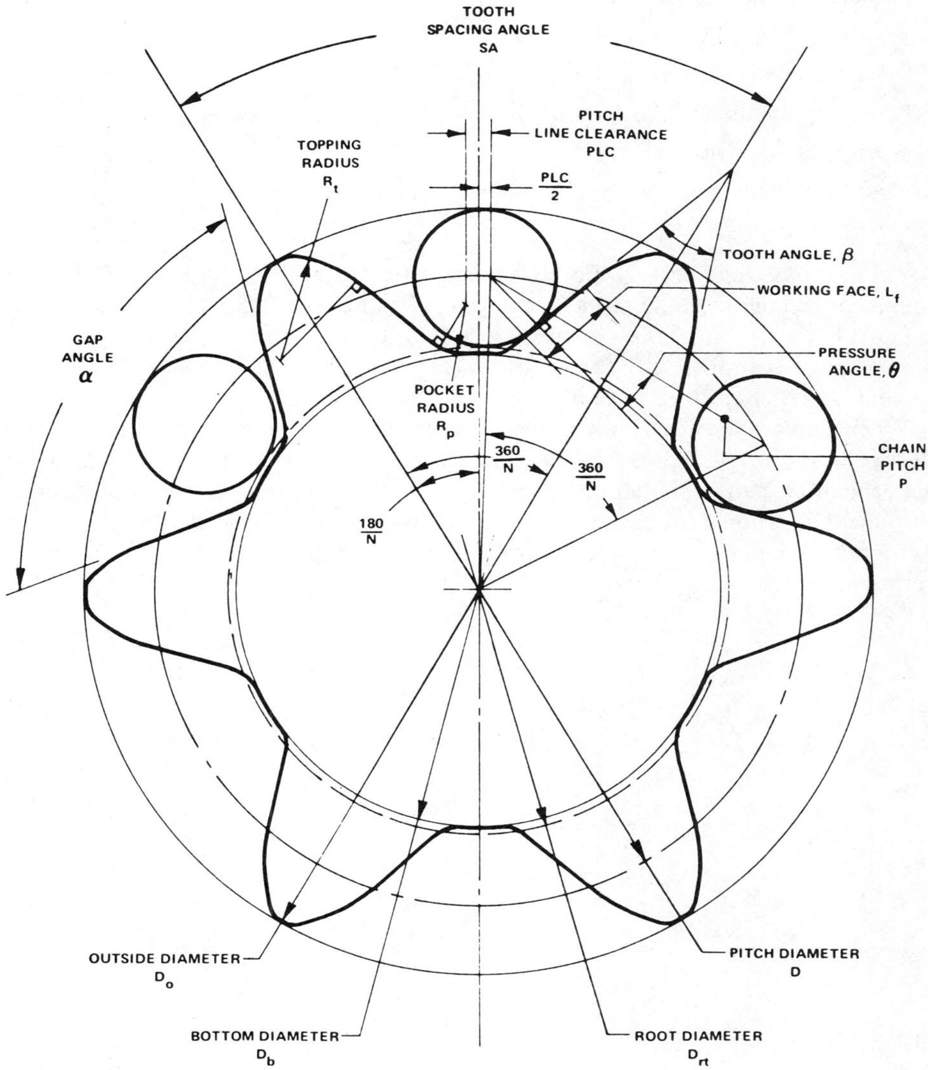

Figure 4-40 Nomenclature for Engineering Steel drive chain sprocket tooth forms.

N = number of teeth in sprocket
n = number of teeth in contact with chain
p = chain pitch, in.
R_p = pocket radius, in.
R_t = topping radius, in.
T = chain tension, lb
T_L = link tension, lb
T_r = roller ejection force, lb
T_s = slack strand tension, lb
T_t = tooth force, lb
T_u = unabsorbed chain tension, percent
T_1, T_2 = components of T, lb
α = gap angle, deg
β = tooth angle, deg
θ = pressure angle, deg

Tooth and Link Forces. Force is transmitted in a chain and sprocket drive through contact between the chain barrel, or roller, and the working face of the tooth (see Fig. 4-41). The tooth face is at an angle to the pitch line of the chain. Therefore, as chain tension T is applied, two forces result: (1) tooth force T_t, normal to the working face, and (2) roller ejection force T_r, parallel to the tooth face.

When a link to the next tooth pocket is added, its link tension must counteract the roller ejection force T_r. Since the link tension force T_L is not exactly opposite to T_r, it must be greater than T_r so that component $T_2 = T_r$ to keep the system in equilibrium. Component T_1 reduces tooth force T_t. Friction, which tends to reduce the roller ejection force and the link tension, is not considered here. Forces T_t and T_L can be easily determined from

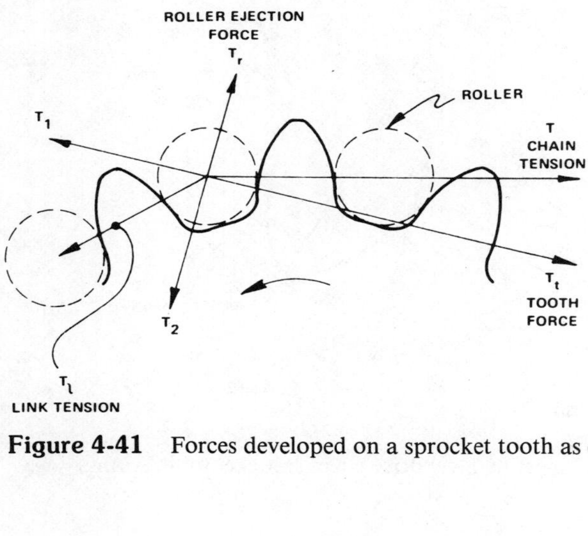

Figure 4-41 Forces developed on a sprocket tooth as chain tension is applied.

$$T_t = T \frac{\sin(360/N)}{\sin(360/N) + \theta} \tag{4-37}$$

$$T_L = T \frac{\sin \theta}{\sin(360/N) + \theta} \tag{4-38}$$

The force diagram shown in Figure 4-41 is repeated at each tooth in contact with the chain. Each tooth absorbs the same percentage of the remaining tension. Link tension at any tooth is the unabsorbed chain tension at that tooth. Although link tension T_L decreases as the link progresses around the sprocket, this tension theoretically never reaches zero. Neglecting friction, the percentage of chain tension not absorbed by the teeth can be found from

$$T_u = 100 \left[\frac{\sin \theta}{\sin(360/N) + \theta} \right]^n \tag{4-39}$$

Tooth Surfaces and Angles. Since variations in sprocket tooth shape will affect the performance of the sprocket-chain system, it is important to understand the function of the various surfaces and angles.

Working Face. Under load, roller ejection force T_r pushes the roller center out to the pitch diameter. As the chain wears and the pitch increases, the roller moves outward on the sprocket teeth until it finds the greater pitch diameter corresponding to the elongated pitch (see Fig. 4-42). But whether the chain is new or worn, almost all the working chain tension is absorbed by the sprocket teeth between B and A. Slack strand tension T_s is absorbed between C and A. At A the unabsorbed tight strand tension and unabsorbed slack strand tension are equal. Thus A is called the *balance point.* When the system is functioning properly, the working face is the only sprocket or tooth surface with which the roller makes contact, except at the balance point.

The working face may be of any shape, as long as the correct pressure angle is approximately maintained. The face must be of sufficient length to accept the desired amount of chain elongation, but must not be so long that it interferes with the roller as the chain flexes into or out of the tooth pocket. As a roller swings out of the tooth pocket, it moves away from the working face until the link centerline is normal to this working face (see Fig. 4-43). The roller then converges toward the working face plane.

While the working face does not normally extend beyond the point of maximum divergence of the roller flex path, this length may be exceeded if the roller flex does not interfere with the tooth.

Pressure Angle. The ability of the sprocket to transmit torque depends on:

1. Ratio of working chain tension to slack side tension
2. Number of teeth in the sprocket
3. Number of teeth in engagement
4. Pressure angle

All these items, except pressure angle, are dictated by the application and by space limitations. If these are fixed, the only controllable variable is the pressure angle.

Generally, each tooth in contact on a sprocket should share the load. Therefore, sprockets with few teeth should have small pressure angles, whereas those with many

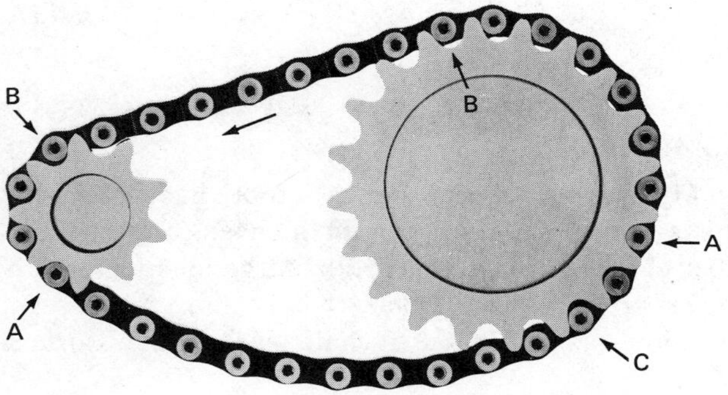

Figure 4-42 Sprocket with worn chain. Chain has crept upward on the working face of the sprocket tooth, giving a larger effective pitch diameter. This worn chain shows clearly where the balance point A is located.

teeth may have much larger pressure angles. As the pressure angle increases, the tooth force decreases, causing an increase in link tension.

Bottom Diameter. The roller does not make contact with the bottom diameter except at the balance point (see Fig. 4-42). The bottom diameter should not exceed the root diameter, and should preferably be smaller, to accept any manufacturing variations or possible buildup of material in the tooth pockets. Bottom diameter can be found from

$$D_b = D_{rt} - C_b \tag{4-40}$$

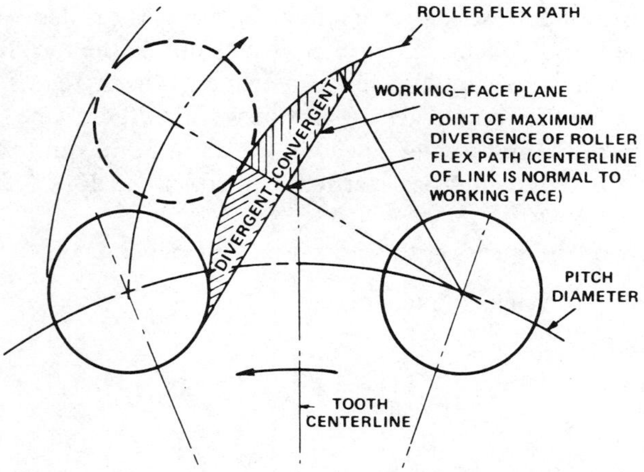

Figure 4-43 Path of a roller during operation. Maximum divergence of the roller from the tooth takes place when the centerline of the flexing link is normal to the working face.

where C_b is a compensating factor to allow for buildup of foreign material in the pocket.

Pocket Curve (or Radius). The primary purpose of the pocket curve or radius (see Fig. 4-40) is to reduce stress concentration between the working face and the bottom diameter. Intersection of the radius of the pocket curve and working face should not be above the point where the loaded roller makes contact with the tooth. Although, theoretically, pocket radius can equal roller radius, the best approach in designing sprockets used for Engineering Steel Chains is to have a pocket radius smaller than the roller radius.

Topping Curve. This is the curve described by the topping radius R_t. The primary purpose of the topping curve is to provide a guide path for the roller should malfunctioning take place. Generally, the topping curve should diverge from the roller flex path.

Pitch Line Clearance. This clearance serves the following purposes:

1. Reduces the possibility of material being trapped in a tooth pocket
2. Accommodates minor manufacturing inaccuracies
3. Reduces the possibility of certain malfunctions

When a sprocket has pitch line clearance, the pressure angle can be calculated from

$$\theta = \frac{\beta}{2} - \sin^{-1}\frac{C_p}{D} \tag{4-41}$$

With only a small C_p,

$$\theta \cong \frac{\beta}{2} \tag{4-42}$$

Pitch line clearance must be considerably more than 10% of the chain pitch before this relationship varies by more than 1 or 2°.

The effect of backlash on sprockets with pitch line clearance can be eliminated by making the bottom diameter sufficiently undersized that the roller entering the sprocket from the tight strand is on one side of the tooth pocket while that entering from the slack strand is on the opposite side of the pocket (see Fig. 4-44).

It should be remembered that excessive pitch line clearance will reduce the tooth section, and may cause tooth strength problems.

Gap Angle. Gap angle depends on the pressure angle and the number of sprocket teeth. With a constant gap angle, the pressure angle increases with an increase in the number of sprocket teeth. For a given gap angle and degrees of wrap, approximately the same percentage of chain tension will be absorbed regardless of the number of teeth in the sprocket. It is therefore practical to design chills, cutters, and templates with a constant gap angle, and to use them over a wide range of numbers of teeth. With no pitch line clearance, gap angle will be

$$\alpha = \frac{360}{N} + 2\theta \tag{4-43}$$

Figure 4-44 Position of chain during backlash. Roller entering the sprocket from the tight strand A is on one side of the tooth; the roller entering the slack strand B is on the opposite side.

Pitch Diameter. Diameter of the pitch circle which exactly circumscribes a polygon whose sides equal the chain pitch. (See Table 4-45 for a cosecant table used to calculate sprocket pitch diameters for numbers of teeth from 4 through 100.)

Caliper Diameter. The bottom diameter cannot be directly measured on sprockets with an odd number of teeth. The obvious reason is that tooth pockets are not 180° from each other. In these cases, the caliper diameter is used to check the bottom diameter accuracy. The caliper diameter measures the bottom diameter in pockets as close to 180° as possible.

$$\text{Caliper diameter} = \text{bottom diameter} \times \left(\cos \frac{90}{N}\right) \tag{4-44}$$

where N is the number of teeth in the sprocket.

For more detail on sprocket selection and application, refer to one of the manufacturers' catalogs.

Pitch dia. = Pitch of chain × cosecant for number of teeth in sprocket

No. Teeth	Cosecant▲	No. Teeth	Cosecant▲	No. Teeth	Cosecant▲	No. Teeth	Cosecant▲	No. Teeth	Cosecant▲	No. Teeth	Cosecant▲	No. Teeth	Cosecant▲
4	1.4142	18	5.7588	32	10.2023	46	14.6536	60	19.1073	74	23.5620	88	28.0170
5	1.7013	19	6.0755	33	10.5201	47	14.9717	61	19.4255	75	23.8802	89	28.3355
6	2.0000	20	6.3925	34	10.8380	48	15.2898	62	19.7437	76	24.1984	90	28.6537
7	2.3048	21	6.7095	35	11.1558	49	15.6079	63	20.0618	77	24.5166	91	28.9723
8	2.6131	22	7.0266	36	11.4737	50	15.9260	64	20.3800	78	24.8349	92	29.2901
9	2.9238	23	7.3439	37	11.7917	51	16.2441	65	20.6982	79	25.1531	93	29.6082
10	3.2361	24	7.6613	38	12.1096	52	16.5622	66	21.0164	80	25.4713	94	29.9268
11	3.5495	25	7.9787	39	12.4275	53	16.8803	67	21.3346	81	25.7896	95	30.2447
12	3.8637	26	8.2962	40	12.7455	54	17.1984	68	21.6528	82	26.1079	96	30.5632
13	4.1785	27	8.6138	41	13.0635	55	17.5166	69	21.9710	83	26.4261	97	30.8815
14	4.4940	28	8.9315	42	13.3815	56	17.8347	70	22.2892	84	26.7442	98	31.1997
15	4.8097	29	9.2491	43	13.6995	57	18.1529	71	22.6074	85	27.0626	99	31.5177
16	5.1259	30	9.5668	44	14.0175	58	18.4710	72	22.9256	86	27.3807	100	31.8362
17	5.4423	31	9.8845	45	14.3356	59	18.7892	73	23.2438	87	27.6989		

▲The Cosecant is for an angle equal to $\frac{180°}{\text{number of teeth}}$

Table 4-45 Pitch Diameter of Sprockets as a Function of Chain Pitch and Number of Teeth

Typical Tooth Forms for Types of Engineering Steel Chains. Figure 4-46 illustrates typical tooth forms used for:

1. Engineering Steel drive chain sprockets
2. Engineering Steel conveyor chain sprockets
3. Bar-link chain sprockets

Actually, although the tooth profiles shown in Figure 4-46 are termed typical, sprocket tooth forms used for Engineering Steel Chains can have quite a large variety of shapes and forms. For example, some of the sprockets for long-pitch conveyor chains have a superficial resemblance to the bar-link chain sprockets of Figure 4-46, but have pockets for the chain rollers at the corners, the flats being to clear the chain through-rod bushings in the center of the links.

Sprockets for Engineering Steel Chains should be purchased with the chains, relying on the experience of the manufacturers to furnish the sprocket tooth design that is right for each application.

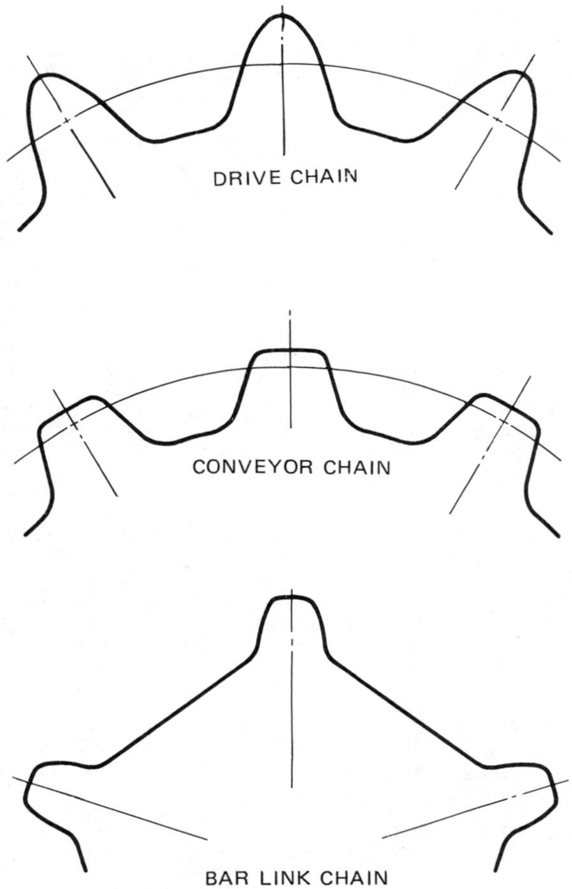

Figure 4-46 Typical tooth forms for Engineering Steel Chains.

Sprocket Hubs, Keys and Keyways, Set Screws, and Shafting Selection

Selection of the higher-speed sprocket of a chain drive frequently requires the smallest chain and sprocket compatible with shafting and sprocket hub sizes. (These considerations are discussed in Chapters 5 to 7.) Hub sizes must be large enough to include a standard keyway with set screws and yet be small enough to clear chain link plates or sidebars. Table 4-47 shows the maximum bore and hub sizes of Roller Chain sprockets with from 11 through 25 teeth. Often, for applications where single-strand Roller Chain is not adequate, a multiple strand chain can be used satisfactorily.

Keys, Keyways, and Set Screws

To secure sprockets to the shafts, both keys and set screws should be used. Selection of sizes is usually on the basis of industry standards, where keys are matched to shafting sizes for strength. Tables 4-48 and 4-49 list recommended sizes of set screws and keys of various types for standard shafting sizes.

The key is used to prevent rotation of the sprocket on the shaft. Keys should be carefully fitted in the shaft and sprocket keyways to eliminate all backlash, especially

Table 4-47 Maximum Bore and Hub Diameters of Roller Chain Sprockets (with Standard Keyways)

No. of Teeth	3/8" Pitch		1/2"		5/8"		3/4"		1"	
	Max. Bore	Maximum Hub Dia.	Max. Bore	Maximum Hub Dia.	Max. Bore	Maximum Hub Dia.	Max. Bore	Maximum Hub Dia.	Max. Bore	Maximum Hub Dia.
11	19/32	55/64	25/32	1 11/64	31/32	1 15/32	1 1/4	1 49/64	1 5/8	2 3/8
12	5/8	63/64	7/8	1 21/64	1 5/32	1 43/64	1 9/32	2 1/64	1 25/32	2 45/64
13	3/4	1 7/64	1	1 1/2	1 9/32	1 7/8	1 1/2	2 1/4	2	3 1/64
14	27/32	1 15/64	1 5/32	1 21/32	1 5/16	2 5/64	1 3/4	2 1/2	2 9/32	3 11/32
15	7/8	1 23/64	1 1/4	1 13/16	1 17/32	2 9/32	1 25/32	2 3/4	2 13/32	3 43/64
16	31/32	1 15/32	1 9/32	1 63/64	1 11/16	2 31/64	1 31/32	2 63/64	2 23/32	3 63/64
17	1 3/32	1 19/32	1 3/8	2 9/64	1 25/32	2 11/16	2 7/32	3 7/32	2 13/16	4 5/16
18	1 7/32	1 23/32	1 17/32	2 19/64	1 7/8	2 57/64	2 9/32	3 15/32	3 1/8	4 41/64
19	1 1/4	1 27/32	1 11/16	2 29/64	2 1/16	3 5/64	2 7/16	3 45/64	3 5/16	4 61/64
20	1 9/32	1 61/64	1 25/32	2 5/8	2 1/4	3 9/32	2 11/16	3 61/64	3 1/2	5 9/32
21	1 5/16	2 5/64	1 25/32	2 25/32	2 9/32	3 31/64	2 13/16	4 3/16	3 3/4	5 19/32
22	1 7/16	2 13/64	1 15/16	2 15/16	2 7/16	3 11/16	2 15/16	4 7/16	3 7/8	5 59/64
23	1 9/16	2 5/16	2 3/32	3 3/32	2 5/8	3 57/64	3 1/8	4 43/64	4 3/16	6 15/64
24	1 11/16	2 7/16	2 1/4	3 17/64	2 13/16	4 5/64	3 1/4	4 29/32	4 9/16	6 9/16
25	1 3/4	2 9/16	2 9/32	3 27/64	2 27/32	4 9/32	3 3/8	5 5/32	4 11/16	6 7/8

No. of Teeth	1 1/4"		1 1/2"		1 3/4"		2"		2 1/2"	
	Max. Bore	Maximum Hub Dia.	Max. Bore	Maximum Hub Dia.	Max. Bore	Maximum Hub Dia.	Max. Bore	Maximum Hub Dia.	Max. Bore	Maximum Hub Dia.
11	1 31/32	2 31/32	2 5/16	3 37/64	2 13/16	4 11/64	3 9/32	4 25/32	3 15/16	5 63/64
12	2 9/32	3 3/8	2 3/4	4 1/16	3 1/4	4 3/4	3 5/8	5 27/64	4 23/32	6 51/64
13	2 17/32	3 25/32	3 1/16	4 35/64	3 9/16	5 5/16	4 1/16	6 5/64	5 3/32	7 39/64
14	2 11/16	4 3/16	3 5/16	5 1/32	3 7/8	5 7/8	4 11/16	6 23/32	5 23/32	8 27/64
15	3 3/32	4 19/32	3 3/4	5 33/64	4 7/16	6 29/64	4 7/8	7 3/8	6 1/4	9 7/32
16	3 9/32	5	4	6	4 11/16	7 1/64	5 1/2	8 1/64	7	10 1/32
17	3 21/32	5 13/32	4 15/32	6 31/64	5 1/16	7 37/64	5 11/16	8 21/32	7 7/16	10 27/32
18	3 25/32	5 51/64	4 21/32	6 31/32	5 5/8	8 9/64	6 1/4	9 5/16	8 1/8	11 41/64
19	4 3/16	6 13/64	4 15/16	7 29/64	5 11/16	8 45/64	6 7/8	9 61/64	9	12 7/16
20	4 19/32	6 39/64	5 7/16	7 15/16	6 1/4	9 17/64	7	10 19/32	9 3/4	13 1/4
21	4 11/16	7	5 11/16	8 27/64	6 13/16	9 53/64	7 3/4	11 15/64	10	14 3/64
22	4 7/8	7 13/32	5 7/8	8 57/64	7 1/4	10 25/64	8 3/8	11 7/8	10 7/8	14 27/32
23	5 5/16	7 13/16	6 3/8	9 3/8	7 7/16	10 15/16	9	12 33/64	11 5/8	15 21/32
24	5 11/16	8 13/64	6 13/16	9 55/64	8	11 1/2	9 5/8	13 5/32	13	16 29/64
25	5 23/32	8 39/64	7 1/4	10 11/32	8 9/16	12 1/16	10 1/4	13 51/64	13 1/2	17 1/4

Table 4-48 Standard Set Screw Dimensions

Sprocket Bore & Shaft Diameter D	Size of Setscrew Used With Parallel Key
Inches	Inches
1/2 — 9/16	1/4
5/8 — 7/8	1/4
15/16 — 1 1/4	3/8
1 5/16 — 1 3/8	3/8
1 7/16 — 1 3/4	3/8
1 13/16 — 2 1/4	1/2
2 5/16 — 2 3/4	5/8
2 7/8 — 3 1/4	5/8
3 3/8 — 3 3/4	3/4
3 7/8 — 4 1/2	3/4
4 3/4 — 5 1/2	7/8
5 3/4 — 7 3/8	1
7 1/2 — 9 7/8	1 1/4
10 — 12 1/2	1 1/4

Table 4-49 Standard Keys and Keyways

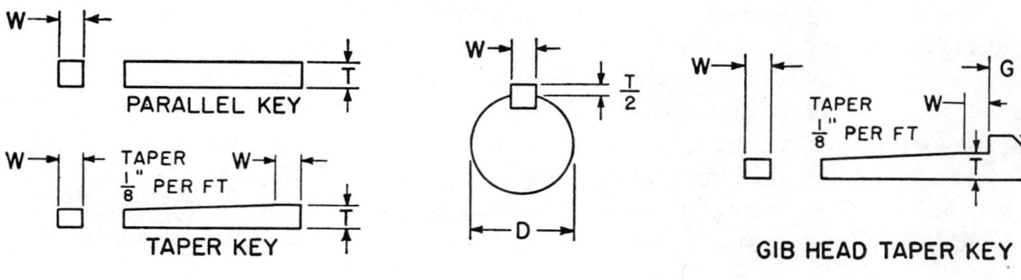

Sprocket Bore (= Shaft Diam.)	Keyway Dimensions — Inches				Key Dimensions — Inches					Gib Head Dimensions — Inches				Key Tolerances Taper and Gib Head	
	For Square Key		For Flat Key		Square		Flat		Tolerance on W and T (−)	Square Key		Flat Key			
Inches D	Width W	Depth T/2	Width W	Depth T/2	Width W	Height T	Width W	Height T		H	G	H	G	W (−)	T (+)
1/2 — 9/16	1/8 x	1/16	1/8 x	3/64	1/8 x	1/8	1/8 x	3/32	0.002	1/4	7/32	3/16	1/8	0.002	0.002
5/8 — 7/8	3/16 x	3/32	3/16 x	1/16	3/16 x	3/16	3/16 x	1/8	0.002	5/16	9/32	1/4	3/16	0.002	0.002
15/16 — 1 1/4	1/4 x	1/8	1/4 x	3/32	1/4 x	1/4	1/4 x	3/16	0.002	7/16	11/32	5/16	1/4	0.002	0.002
1 5/16 — 1 3/8	5/16 x	5/32	5/16 x	1/8	5/16 x	5/16	5/16 x	1/4	0.002	9/16	13/32	3/8	5/16	0.002	0.002
1 7/16 — 1 3/4	3/8 x	3/16	3/8 x	1/8	3/8 x	3/8	3/8 x	1/4	0.002	11/16	15/32	7/16	3/8	0.002	0.002
1 13/16 — 2 1/4	1/2 x	1/4	1/2 x	3/16	1/2 x	1/2	1/2 x	3/8	0.0025	7/8	19/32	5/8	1/2	0.0025	0.0025
2 5/16 — 2 3/4	5/8 x	5/16	5/8 x	7/32	5/8 x	5/8	5/8 x	7/16	0.0025	1 1/16	23/32	3/4	5/8	0.0025	0.0025
2 7/8 — 3 1/4	3/4 x	3/8	3/4 x	1/4	3/4 x	3/4	3/4 x	1/2	0.0025	1 1/4	7/8	7/8	3/4	0.0025	0.0025
3 3/8 — 3 3/4	7/8 x	7/16	7/8 x	5/16	7/8 x	7/8	7/8 x	5/8	0.003	1 1/2	1	1 1/16	7/8	0.003	0.003
3 7/8 — 4 1/2	1 x	1/2	1 x	3/8	1 x	1	1 x	3/4	0.003	1 3/4	1 3/16	1 1/4	1	0.003	0.003
4 3/4 — 5 1/2	1 1/4 x	5/8	1 1/4 x	7/16	1 1/4 x	1 1/4	1 1/4 x	7/8	0.003	2	1 7/16	1 1/2	1 1/4	0.003	0.003
5 3/4 — 7 3/8	1 1/2 x	3/4	1 1/2 x	1/2	1 1/2 x	1 1/2	1 1/2 x	1	0.003	2 1/2	1 3/4	1 3/4	1 1/2	0.003	0.003
7 1/2 — 9 7/8	1 3/4 x	7/8	. .	. .	1 3/4 x	1 3/4	. .	. .	0.004	3	2	. .	. .	0.004	0.004
10 — 12 1/2	2 x	1	. .	. .	2 x	2	. .	. .	0.004	3 1/2	2 3/8	. .	. .	0.004	0.004

Standard Keyway Tolerances: Straight Keyway — Width (W) + .005 / − .000 Depth (T/2) + .010 / − .000

Taper Keyway — Width (W) + .005 / − .000 Depth (T/2) + .000 / − .010

under fluctuating loads. A set screw should be located over a flat key to secure it against longitudinal displacement.

In addition, set screws may be used to prevent axial displacement of the sprocket on the shaft. For such use, set screws are usually located 90 or 180° from the keyway. The hub thickness must be sufficient to provide a set screw thread engagement at least equal to the screw diameter.

Selection of Shafting

Selection of shaft sizes for the larger sprockets, whether drive components or for a conveyor, usually precedes the selection of the proper hub size. The data and examples given in this section, although based on Engineering Steel Chain sprockets, can apply equally well to shafting selections for Roller Chain or Silent Chain drives.

Shafting is subject to two forces, a bending moment and a torsional moment. The bending moment is the force that tends to bend the shaft and the torsional moment is the force that tends to twist the shaft. The first step in shaft selection is to calculate these moments. In order to calculate them, we must first know the chain tension, the bearing locations, plus any other forces acting on the shaft. The chain tension must take into account not only the forces involved to do work but, as in an elevator, must consider the weight of the chain, attachments, buckets, and any takeup weights.

The bending moment can be calculated using the standard bending moment equation for the particular condition. The torsional moment equals the net chain tension times one-half the pitch diameter of the sprocket.

The charts and engineering data in the following discussion on the strength of shafting are based on the American Society of Mechanical Engineers' code.

For ordinary conditions of service, the selection of a shaft may be made using the service factors in Table 4-50 and the shaft selection charts, Figures 4-51 to 4-53. These shafting selection charts are based upon 6000 psi maximum permissible shear stress for commercial steel shafting with allowance for standard keyways. When a shaft is not weakened by keyways, the size obtained from the chart should be multiplied by 0.91 to obtain the proper diameter. When a material justifying a shear stress other

Table 4-50 Service Factors for Shafting Selection

	NATURE OF LOADING	SERVICE FACTOR For Bending	SERVICE FACTOR For Torsion
Stationary Shaft	Gradually Applied Loads	1.0	1.0
	Suddenly Applied Loads	1.5 to 2.0	1.5 to 2.0
Rotating Shaft	Gradually Applied or Steady Loads	1.5	1.0
	Suddenly Applied Loads Minor Shocks only	1.5 to 2.0	1.0 to 1.5
	Suddenly Applied Loads Heavy Shocks	2.0 to 2.5	1.5 to 2.5

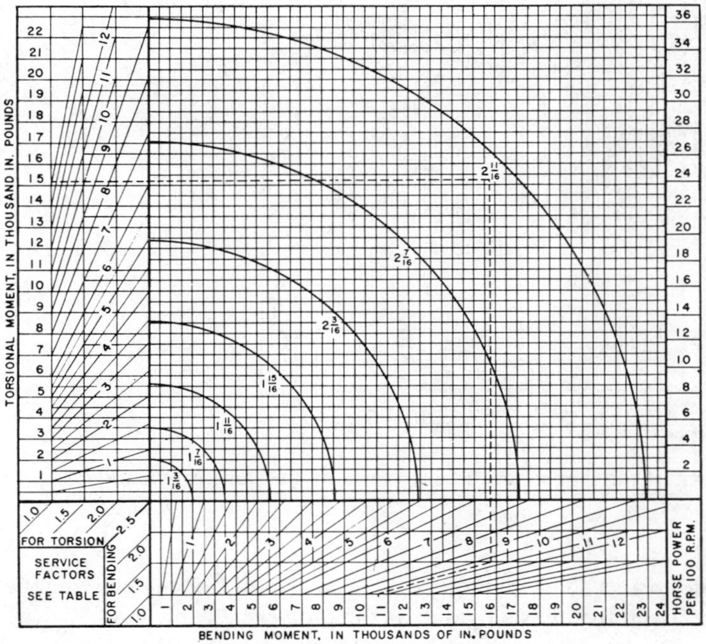

Figure 4-51 Selection chart for shafts from 1-3/16 to 2-11/16 in. in diameter, inclusive.

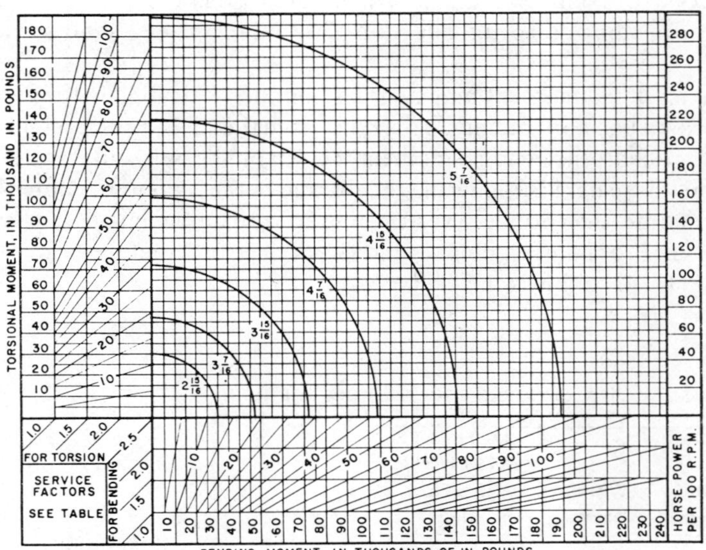

Figure 4-52 Selection chart for shafts from 2-15/16 to 5-7/16 in. in diameter, inclusive.

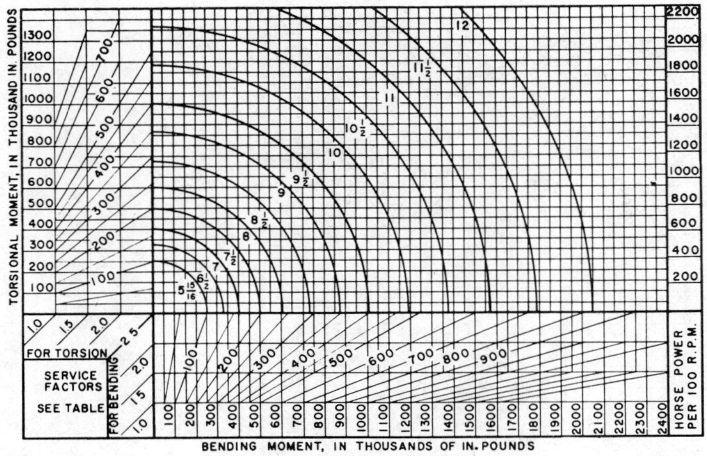

Figure 4-53 Selection chart for shafts from 5-15/16 to 12 in. in diameter, inclusive.

than 6000 psi is used, the correct shaft diameter can be determined by multiplying the shaft diameter obtained from the charts by one of the factors in Table 4-54.

The following method of shaft selection is general and applies to all cases of shaft design. The only limitation is that a selection made in this way does not consider the rigidity or the deflection of the shaft, which may be the controlling factor.

Example

A conveyor has a head shaft that is subjected to 15,200 in.-lb torsion and 10,200 in.-lb bending. The shaft is to be of commercial steel. The nature of the loading requires service factors of 1.5 for bending and 1.0 for torsion. To find the proper shaft size, proceed as follows.

On Figure 4-51 the torsional moment may be found on the vertical scale for the service factor of 1.0. Draw a line horizontally to the right from the 15,200 point on the scale. At the bottom of the chart the bending moment is given on the horizontal scale. Find the 10,200 point on this scale. Run a line diagonally and to the right until it intersects the 1.5 service factor line, then project a vertical line upward until it intersects the horizontal line first drawn. From the last point of intersection we find that the nearest standard shaft size is 2-11/16, so this is the proper size.

Table 4-54 Shear Stress and Strength Factors for Shafting Selection

SHEARING STRESS LBS PER SQ IN.	FACTOR With Keyway	FACTOR Without Keyway	ULTIMATE STRENGTH OF STEEL USED
6,000	1.00	.91	55,000
8,000	.91	.83	75,000
10,000	.84	.77	95,000

At the right on this same chart the horsepower ratings are given per 100 rpm based on the formula

$$HP = \frac{TS}{63{,}000} \qquad (4\text{-}45)$$

where T is the torque in inch-pounds and S the speed in rpm.

The horsepower is directly proportional to the speed of the shaft in rpm.

Hub and Key Size

From the working load of the chain and the pitch diameter of the sprocket, the class of hub required can be taken from the hub class chart, Figure 4-55. Assume, for example, a working load of 1500 lb and a pitch diameter of 20 in. The horizontal and vertical lines intersect at H, which is the required hub class.

Now turn to the upper section of Table 4-56 and draw one line down from H and another line out from bore size 2 11/16 in. (assuming the shaft size determined in the preceding example). The lines intersect at 5¼ in., which is the minimum diameter of the hub. At the same time, the table shows (top row) a maximum allowable torque of 17,000 in.-lb (15,200 being the actual torque in our example), a hub length of 3½ in. (bottom row), a key size of 5/8 in., and a set screw size of 5/8 in. The right-hand column shows that for a sprocket rotating on the shaft (such as a foot sprocket on a conveyor), a smaller hub is permissible—one 4¼ in. in diameter. The same holds for a hub set screwed to the shaft.

The bands of figures in Table 4-56 represent a balance between chain and shaft sizes. When the hub class and bore intersect in the blank space below and to the left of the band, it means that the chain and sprocket will not transmit the full torque value of the shaft. When the intersection is above and to the right of the band, the chain and sprocket are stronger than the shaft; the shaft may fail, or the designer has selected a heftier (and more expensive) chain than is needed. The charts, therefore, can also be used as a design check on the size of the shaft.

When the hub class and bore size intersect outside the band, however, the reason may be that the assembly, although correctly designed, has unusual features. For instance, a sprocket installed at one end of a long shaft might sustain a high bending moment in proportion to the torsional moment, and the shaft selected for it would be unusually large compared with the working load. This means that the bore size would be large and the hub class (which is computed by using the working load) would be small, and the lines would intersect below and to the left of the band. In such a case, calculate the hub dimensions this way, as shown in Table 4-56 (bottom). When the lines intersect above the band, use the first hub diameter left of the point of intersection. When they intersect below the band, use the first hub diameter to the right of the point of intersection. In both cases, the correct hub length is in the same column as the hub diameter.

Because the performance of the whole conveyor or drive installation depends largely on sprocket-chain interaction, choosing the right sprocket is as important as choosing the right chain.

It is necessary that the sprocket-chain design be compatible, and for this reason, as stated at the beginning of this chapter, it is recommended that sprockets normally be obtained from the manufacturer of the chain involved.

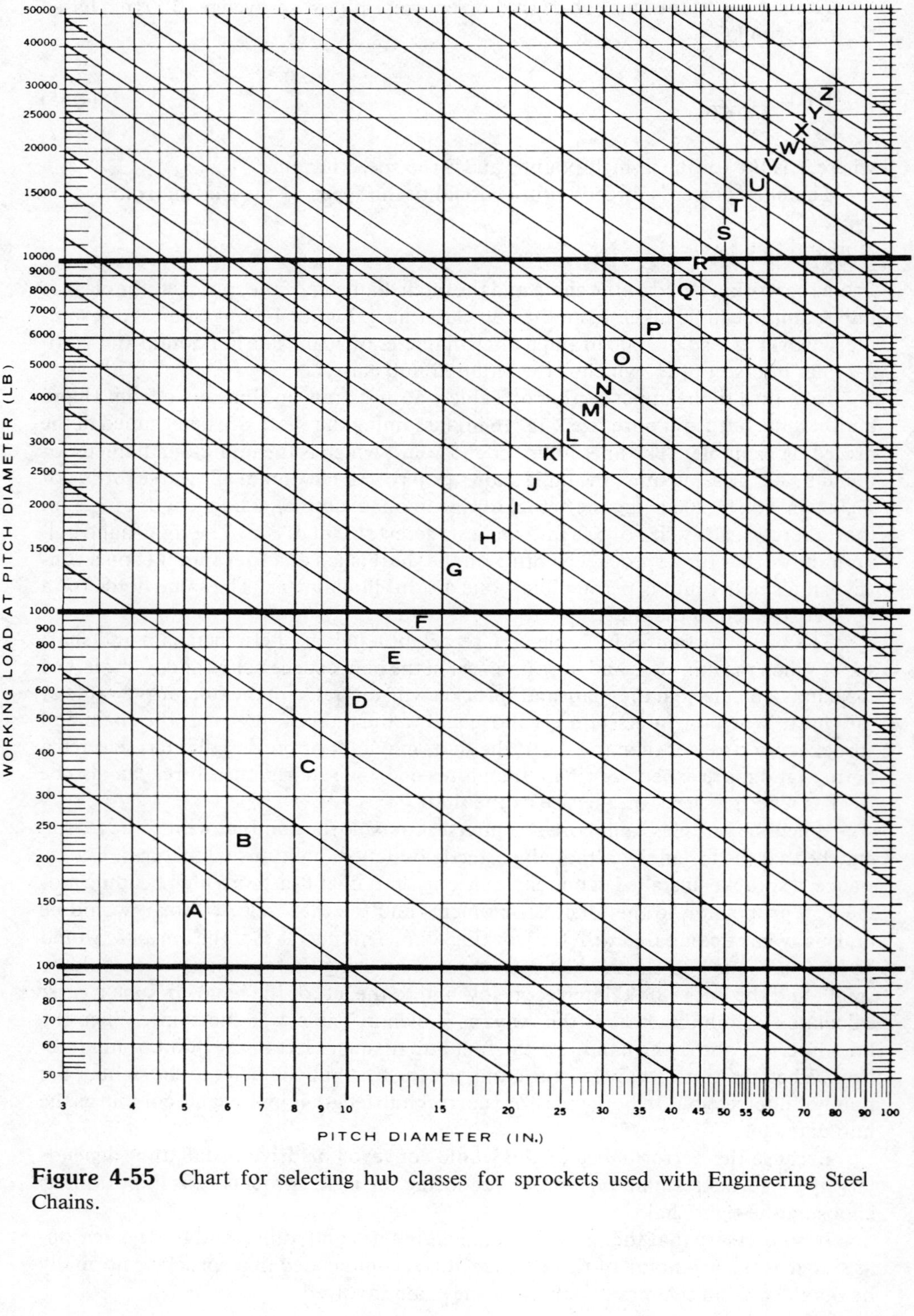

Figure 4-55 Chart for selecting hub classes for sprockets used with Engineering Steel Chains.

Standard Hubs for CAST IRON and CHILLED RIM Sprockets

Bore of Wheel	Sq. Key Size	Set Screw Size	500	1000	2000	3500	5600	8500	12500	17000	23000	30000	38000	47000	60000	70000	100000	140000	190000	245000	325000	400000	500000	600000	720000	850000	1000000	1250000	Minimum Hubs for Loose or Set-Screwed Sprockets: Diam.	Lgth.
			A	B	C	D	E	F	G	H	I	J	K	L	M	N	O	P	Q	R	S	T	U	V	W	X	Y	Z		
			Diameter of Keyseated Hubs																											
15/16	¼	⅜	1¾	2																									1¾	Length is determined same as for Key-seated Sprockets
1 3/16	¼	⅜	2	2¼	2¼																								2	
1 7/16	⅜	⅜	2½	2½	2¾	3																							2½	
1 11/16	⅜	⅜	2¾	2¾	3	3¼	3½																						2¾	
1 15/16	½	½	3¼	3¼	3¼	3½	3¾	4																					3¼	
2 3/16	½	½	3½	3½	3½	3¾	4	4¼	4½																				3½	
2 7/16	⅝	⅝	4	4	4	4	4¼	4½	4¾	5																			4	
2 11/16	⅝	⅝	4¼	4¼	4¼	4¼	4½	4¾	5	5¼	5½																		4¼	
2 15/16	¾	⅝	4½	4½	4½	4½	4¾	5	5¼	5½	5¾	6																	4½	
3 3/16	¾	⅝	4¾	4¾	4¾	4¾	5	5¼	5½	5¾	6	6¼	6½																4¾	
3 7/16	⅞	¾		5¼	5¼	5¼	5¼	5½	5¾	6	6¼	6½	6¾	6¾															5¼	
3 11/16	⅞	¾			5½	5½	5½	5¾	6	6¼	6½	6¾	7	7	7														5½	
3 15/16	1	¾				6¼	6¼	6¼	6¼	6½	6¾	7	7¼	7¼	7½	7½													6¼	
4 7/16	1	¾					6¾	6¾	6¾	6¾	7	7¼	7½	7½	8	8	8												6¾	
4 15/16	1¼	⅞						7½	7½	7½	7½	7½	8	8	8½	8½	8½	9											7½	
5 7/16	1¼	⅞							8½	8½	8½	8½	8½	8½	9	9	9½	9½	10										8½	
5 15/16	1½	1								9½	9½	9½	9½	9½	9½	9½	10	10	10½	11									9½	
6½	1½	1									10	10	10	10	10	10	10½	10½	11	11½	12								10	
7	1½	1										10½	10½	10½	10½	10½	11	11	11½	12	12½	13							10½	
7½	1¾	1¼											11½	11½	11½	11½	11½	11½	12	12½	13	13	14						11½	
8	1¾	1¼												12	12	12	12	12	12½	13	13½	13½	14	15					12	
8½	1¾	1¼													13	13	13	13	13½	14	14½	14½	15	15	16				13	
9	1¾	1¼														14	14	14	14	14½	15	15	15	16	16	17			14	
9½	1¾	1¼															15	15	15	15	15½	15½	15½	16	17	17	18		15	
10	2	1¼																16	16	16	16	16	16½	16½	17	18	18	19	16	
Length thru bore →			① 1¼	① 1½	1¾	2	2½	2¾	3¼	3½	4	4¼	4½	5	5¼	5½	6	6½	7	7¾	8½	9½	10	10½	11	11½	12	12½		

Allowable torque in inch pounds and hub class.

CHAIN AND SPROCKETS ARE STRONGER THAN SHAFTS

CHAIN AND SPROCKETS WILL NOT TRANSMIT FULL TORQUE VALUE OF SHAFT

Standard Hub Diameters for CAST STEEL and FABRICATED STEEL SPROCKETS

Bore of Wheel	Sq. Key Size	Set Screw Size	A	B	C	D	E	F	G	H	I	J	K	L	M	N	O	P	Q	R	S	T	U	V	W	X	Y	Z	Min. Hub Diam.	Lgth.
15/16	¼	⅜	1¾	1¾																									1¾	Length is determined same as for Key-seated Sprockets
1 3/16	¼	⅜	2	2	2																								2	
1 7/16	⅜	⅜	2¼	2¼	2½	2¾																							2¼	
1 11/16	⅜	⅜	2¾	2¾	2¾	3	3																						2¾	
1 15/16	½	½	3	3	3	3¼	3¼	3¼																					3	
2 3/16	½	½	3¼	3¼	3¼	3½	3½	3½	3¾																				3¼	
2 7/16	⅝	⅝	3¾	3¾	3¾	3¾	4	4	4¼	4¼																			3¾	
2 11/16	⅝	⅝	4¼	4¼	4¼	4¼	4¼	4¼	4½	4½	4½																		4¼	
2 15/16	¾	⅝	4½	4½	4½	4½	4½	4¾	4¾	[illegible]	5	5																	4½	
3 3/16	¾	⅝	4¾	4¾	4¾	4¾	4¾	5	5	5	5¼	5¼	5¼																4¾	
3 7/16	⅞	¾		5¼	5¼	5¼	5¼	5¼	5¼	5¼	5½	5½	5¾	5¾															5¼	
3 11/16	⅞	¾			5¼	5½	5½	5½	5½	5½	5¾	5¾	6	6	6														5½	
3 15/16	1	¾				6	6	6	6	6	6¼	6¼	6½	6½	6½	6½													6	
4 7/16	1	¾					6½	6½	6½	6½	6¾	6¾	7	7	7	7	7												6½	
4 15/16	1¼	⅞						7¼	7¼	7¼	7¼	7¼	7½	7½	7½	8	8	8											7¼	
5 7/16	1¼	⅞							8	8	8	8	8	8	8	8½	8½	8½	8½										8	
5 15/16	1½	1								9	9	9	9	9	9	9½	9½	9½	9½	9½									9	
6½	1½	1									9½	9½	9½	9½	9½	10	10	10	10	10	10								9½	
7	1½	1										10	10	10	10	10½	10½	10½	10½	10½	10½	11							10	
7½	1¾	1¼											11	11	11	11½	11½	11½	11½	11½	11½	12	12						11	
8	1¾	1¼												11½	11½	12	12	12	12	12	12	12½	12½	12½					11½	
8½	1¾	1¼													12	12½	12½	12½	12½	12½	12½	13	13	13	13½				12	
9	1¾	1¼														13	13	13	13	13	13	13½	13½	13½	14	14			13	
9½	1¾	1¼															13½	13½	13½	13½	13½	14	14	14	14½	14½	15		13½	
10	2	1¼																14½	14½	14½	14½	15	15	15	15½	15½	16	16	14½	
Length thru bore →			① 1¼	① 1½	1¾	2	2½	2¾	3¼	3½	4	4¼	4½	5	5¼	5½	6	6½	7	7¾	8½	9½	10	10½	11	11½	12	12½		

CHAIN AND SPROCKETS ARE STRONGER THAN SHAFTS

CHAIN AND SPROCKETS WILL NOT TRANSMIT FULL TORQUE VALUE OF SHAFT

① Hubs on one side only.

All Dimensions are in Inches

Table 4-56 Hub Selection Chart for Engineering Steel Chain Sprockets

5

Roller Chains for Drives

The design of a Roller Chain drive consists primarily of the selection of the chain and sprocket sizes. It also includes the determination of chain length, center distance, method of lubrication, and in some cases, the arrangement of chain casings and idlers.

This chapter outlines the theory of chain and sprocket selection, and gives horsepower rating data applicable to standard (ANSI) Roller Chains.

Design Factors

Chain and sprocket selection is based on:

1. The horsepower and type of drive
2. The speeds and sizes of the shafting
3. The surrounding conditions

Horsepower and Type of Drive

Although the horsepower is a basic factor in chain drive design, the conditions under which the drive operates must first be considered. Procedures for the use of service factors to adapt the design to satisfy the operating conditions are described in the following pages.

Speeds and Sizes of Shafting

The speed of the faster-running shaft limits the maximum chain pitch that can be used. In addition, since the shaft size determines the sprocket bore diameter, the sprocket size must be sufficient to provide adequate hub thickness, including the keyway over the required bore.

Surrounding Conditions

Adverse surrounding conditions, such as a dusty atmosphere, high ambient temperature, or corrosive fumes, tend to shorten the life of a drive. The effects of such conditions can be mitigated by careful attention to lubrication, the use of chain castings, or, in some applications, by the use of stainless steel or bronze chains. The latter are special and outside the scope of this book, but information on their use may be obtained from Roller Chain manufacturers.

Desired Service Life

The selection procedures and the chain ratings described later in this chapter provide for Roller Chain drives operating under proper conditions with a life expectancy of approximately 15,000 hr. A drive is said to be *overchained* if it is equipped with appreciably heavier chain than is called for by the rating. Similarly, an *underchained* drive has chain of lesser capacity than would normally be required.

Overchaining, which involves extra first cost, is employed only if extra-long life or other requirements make it economically desirable. However, it should be observed that oversize chains do not always provide improved life.

Underchaining is justified only if first cost is more important than chain life. For example, the life expectancy of some types of equipment is short. Obviously, the chain life need not greatly exceed that of the associated equipment. However, underchaining should never be used without careful analysis of all the factors involved.

Horsepower Formula Derivation

The horsepower capacity of well-lubricated Roller Chain drives is defined by both fatigue and joint galling. The specific limits are:

1. Link plate fatigue strength
2. Roller and bushing impact life
3. Galling between pins and bushings

These limits have been defined mathematically for Roller Chains of all pitch sizes (ANSI B29.1 chains) and numbers of teeth in the sprockets on the basis of 100-pitch chain length (for roller and bushing impact life), a service factor of 1 and a drive arrangement where two aligned sprockets are mounted on parallel horizontal shafts. Under these conditions approximately 15,000 hr of service life at full-load operation may be expected.

This derivation was the result of years of laboratory study and field observation in which the effects of speed, pitch, chain lengths, and number of teeth on the sprocket were correlated with chain life.

The following equations defining these limits, and the Roller Chain ratings, are based on whichever of these equations indicates the least amount of chain capacity for a given set of conditions.

1. Link plate fatigue strength

$$HP_s = K_s \times n^{1.08} \times R^{0.9} \times p^{(3.0-0.07p)} \tag{5-1}$$

where

HP_s = horsepower capacity due to link plate fatigure strength

K_s = 0.004 for all model chains except No. 41; for No. 41, K_s = 0.0022

n = number of teeth on the smaller sprocket

R = rpm of the smaller sprocket
p = chain pitch, in.

2. Roller and bushing impact life

$$HP_r = \frac{K_r \times n^{1.5} \times p^{0.8}}{(R/100)^{1.5}} \tag{5-2}$$

where

HP_r = horsepower capacity for roller bushing impact life of 15,000 hr
K_r = 29 for Nos. 25 and 35 chains
= 3.4 for No. 41 chain
= 17 for Nos. 40 to 240 chains

3. Galling limit defining limiting rpm

$$(R/1000)^{1.59 \log p + 1.873} = \frac{82.5}{7.95^p \times 1.0278^n \times 1.323^{L/1000}} \tag{5-3}$$

where L is the working tension per strand in pounds.

Solutions for the capacity ratings for link plates and rollers are presented in tabular form as maximum values in the horsepower rating tables. The galling limit equation defines speed and load limit where galling will start to appear. The horsepower tables allow speed somewhat beyond these calculated limits (as indicated by the shaded areas) because it has been found that chain drives can operate successfully in these speed ranges even though some galling does occur.

Use of Service Factors

The horsepower rating tables are for use under optimum drive conditions. For less favorable conditions, the specified horsepower must be multiplied by a "service factor" to obtain a "design" horsepower value, and this value used to enter the rating table. Accepted service factors are shown in Table 5-1.

Load Classification

Load classifications for some typical chain drives are shown in Table 5-2. An unlisted drive may be classified by its similarity to one that is listed.

Table 5-1 Service Factors for Roller Chain Drives

	TYPE OF INPUT POWER		
Type of Driven Load	Internal Combustion Engine with Hydraulic Drive	Electric Motor or Turbine	Internal Combustion Engine with Mechanical Drive
Smooth	1.0	1.0	1.2
Moderate Shock	1.2	1.3	1.4
Heavy Shock	1.4	1.5	1.7

Table 5-2 Load Classifications

Smooth Load	Moderate Shock Load	Heavy Shock Load
Agitators — Pure liquid	Clay working machinery — Pug mills	Clay working machinery — Brick press, briquetting machinery
Conveyors — Uniformly loaded or fed (apron, assembly, belt, flight, oven, screw)	Conveyors — Heavy duty and NOT uniformly loaded (apron, assembly, belt, bucket, flight, oven, screw)	Conveyors — Reciprocating and shaker
Fans — Centrifugal and light, small diameter	Cranes and hoists — Medium duty, skip hoists (travel motion and trolley motion)	Cranes and hoists — Heavy duty, including logging, lumbering, and rotary drilling rigs
Line shafts — Light	Dredges — Cable, reel and conveyor	Dredges — Cutter head drives, jig drives, screen drives
Machines — All types with uniform non-reversing loads	Food industry — Beet slicers, dough mixers, meat grinders	Hammer mills
Screens—Rotary (uniformly fed), traveling water intake	Grinders	Machine tools — Punch press, shears, plate planers
Sewage disposal equipment — Inside service (uniformly fed)	Laundry industry — Washers, tumblers	Machines — All types with severe impact shock loads or speed variations, and reversing service
	Line shafts — Heavy service	Metal mills — Draw bench, forming machines, slitters, small rolling mill drives, wire drawing or flattening
	Machine tools — Main and auxiliary drives	Mills — Rotary type (ball, cement kilns, rod mills, tumbling mills)
	Machine—All types with moderate shock and non-reversing loads	Paper industry — Mixers, calendars, rubber mill sheeters
	Screens — Rotary (stone or gravel)	Textile industry — Carding machinery
	Textile industry — Calenders, dyeing machinery, mangles, nappers, soapers, spinners, tenter frames	

Selection of Chain and Sprockets

In the design of a Roller Chain drive, the horsepower to be transmitted and the size and speed of one or both of the shafts are known specified quantities. The designer must select chain and sprocket sizes that will satisfy all the requirements. The designer must also investigate whether single or multiple strand chain best fits the job. Although an infinite number of solutions are mathematically possible, only a limited number are practical. Also, the maximum speed of the chain, for long service life, must be kept within the limits given in the horsepower chart and rating tables.

Selection of Chain and Small Sprocket

Chain Pitch

The smallest applicable pitch is desirable for quiet operation and for high speed. The horsepower capacity of the chain, however, varies with the chain pitch. Short pitch with a high working load can often be obtained by the use of multiple strand Roller Chain (Figs. 5-3 and 5-4). The horsepower capacity of such a chain is equal to that of a single-strand chain of corresponding pitch, multiplied by the appropriate multiple strand factor as shown in Table 5-5.

Number of Sprocket Teeth

The sprocket selected must be large enough to accommodate the shaft. Table 4-47 gives maximum bore and hub diameters consistent with commercial practice for Roller Chain sprockets having up to 25 teeth.

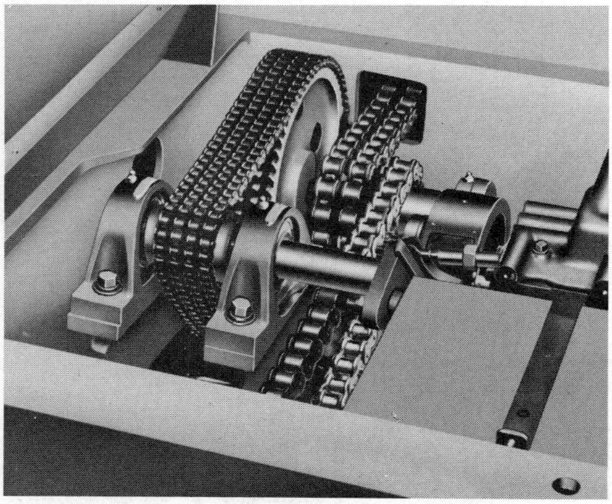

Figure 5-3 Typical drive with multiple strand Roller Chain.

Figure 5-4 Drive with multiple strand Roller Chain.

Table 5-5 Multiple Strand Factors

NUMBER OF STRANDS	MULTIPLE STRAND FACTOR
2	1.7
3	2.5
4	3.3

For a given chain pitch and shaft speed to transmit a given horsepower, the effect of increasing the number of teeth in the sprocket is to increase the chain linear speed and decrease the chain pull proportionally, and to decrease the chordal action. This results in a quieter drive with less impact.

With high speed ratios, the size of the small sprocket is limited by the maximum desirable size of the large sprocket. The size of the large sprocket will limit the allowable wear elongation of the chain. In percent elongation, this is equal to approximately 200 divided by the number of teeth on the large sprocket. Usually, large sprockets should not exceed 120 teeth, although many successful drives use sprockets with 150 teeth and more. The available space may also limit the sprocket size.

Selection of Large Sprocket. After the small sprocket has been selected, the number of teeth in the large sprocket is determined by the specified ratio of the shaft speeds.

Overemphasis on exactness in the speed ratio may result in a cumbersome and expensive installation. In most cases, satisfactory operation can be obtained with a minor change in the speed of one or both of the shafts. The availability of sprocket sizes should be given careful consideration.

The speed ratio for a single drive in which Roller Chain is employed should normally not exceed 10 : 1, although larger ratios are possible. If a greater ratio is required, two drives in series should be used: a drive to a counter shaft and another drive from the counter shaft to the driven shaft.

Shaft Center Distance

Although the distance between shaft centers can be fixed within wide limits, experience dictates certain desirable center distance limitations.

Preferred Center Distance. In general, a center distance of 30 to 50 chain pitches is most desirable (Fig. 5-6) for Roller Chain drives having an adjustable center distance. (Drives with fixed centers should be limited to approximately 30 pitches.) Although the basic reasons for these ranges are mainly technical, the ranges also will be found convenient and economical.

Minimum Center Distance. The minimum center distance must be sufficient to provide clearance between the teeth of the two sprockets (Fig. 5-7). Exclusive of this clearance requirement, for best results the arc of chain engagement should be not less than 120°.

For ratios of 3 : 1 or less, there will be 120° or more wrap. For ratios greater than

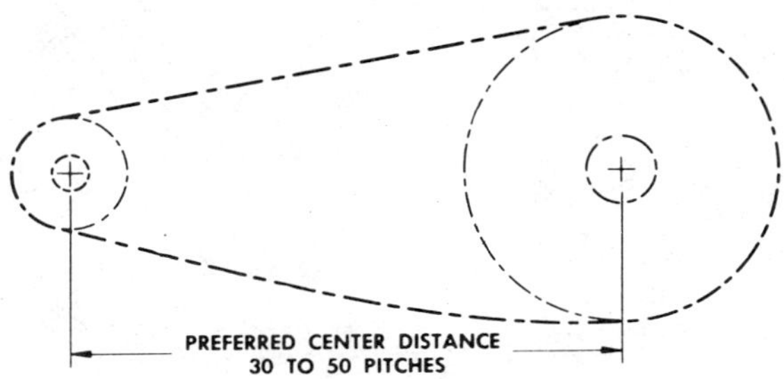

Figure 5-6 Shaft center distance.

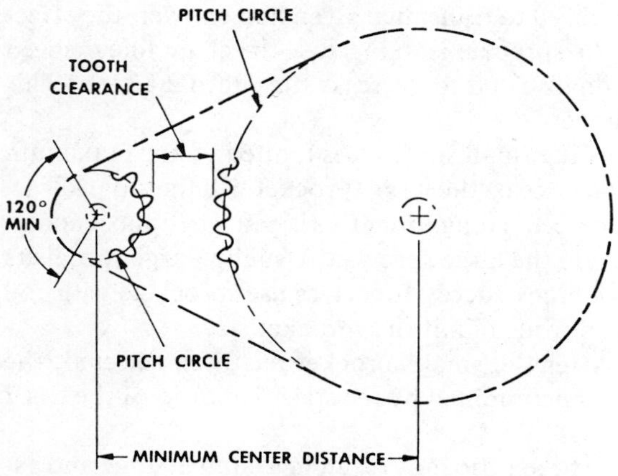

Figure 5-7 Minimum shaft center distance.

3 : 1, the center distance should be no less than the difference in sprocket diameters to provide at least 120° wrap.

Maximum Center Distance. A center distance equivalent to 80 pitches may be considered an approved maximum. Very long center distances result in excessive catenary tension in the chain and may cause the chain to jump teeth. In such cases the chain should be supported by guides (Fig. 5-8) or by idlers. For extremely long center distances, it is recommended that two or more drives in series be substituted for a single drive, or that double-pitch chain be considered in conditions of slow speed and moderate loading.

Adjusting Centers. Drives may be installed with either adjustable or fixed center distances. Adjustable centers simplify the control of chain slack. When provisions are made for the adjustment of the shaft centers, good practice dictates that the range of center adjustment should be at least 1½ pitches. For drives on fixed centers, an idler or shoe may be used to control the backlash that results from normal chain wear.

To obtain full chain life, sufficient housing clearance must always be provided for the chain slack.

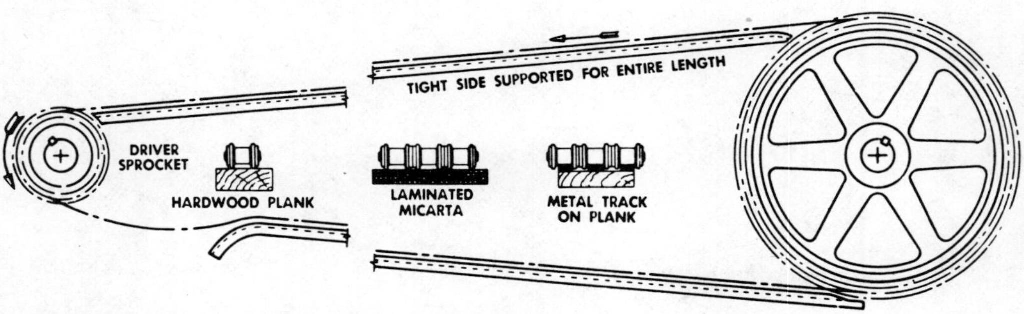

Figure 5-8 Long shaft center distance with guides.

Chain Length and Center Distance

The chain length is a function of the number of teeth in both sprockets and of the center distance. In addition, the chain must consist of an integral number of pitches, with an even number preferably to avoid the use of an offset link.

The most convenient way to obtain this information is by use of center distance tables. If such tables are not available, the values must be calculated.

Chain Length Calculation. For simplicity, it is customary to compute the chain length in terms of chain pitches, and then to multiply the result by the chain pitch to obtain the length in inches.

The pitch circles of the two sprockets and the chain are indicated in Figure 5-9. Referring to equation (4-1), the pitch radii of the two sprockets expressed in chain pitches are

$$DE = \frac{1}{2\sin(180°/N)} \quad \text{and} \quad AB = \frac{1}{2\sin(180°/n)} \tag{5-4}$$

where N and n are the number of teeth in the large and small sprockets.

The length of chain BE between tangent points on the sprockets is

$$BE = AD\cos a = C\cos a$$

where C is the center distance, expressed in chain pitches.

From Figure 5-6, angle a, in degrees, is

$$a = \sin^{-1}\frac{DF}{AD} = \sin^{-1}\frac{DE - AB}{AD} \tag{5-5}$$

$$a = \sin^{-1}\frac{[1/\sin(180°/N)] - [1/\sin(180°/n)]}{2C} \tag{5-6}$$

One-half the length of the chain wrapped on the large sprockets, arc ME, measured in chain pitches, is

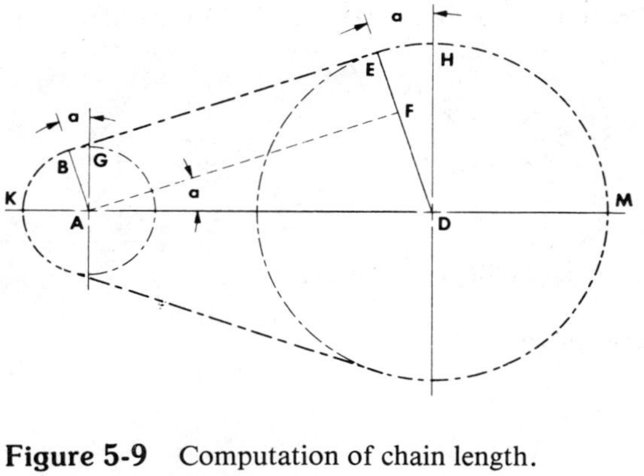

Figure 5-9 Computation of chain length.

$$ME = MH + HE = \frac{N}{4} + N\frac{a}{360°} \tag{5-7}$$

Similarly, one-half the chain wrap on the small sprocket is

$$KB = KG - BG = \frac{n}{4} - n\frac{a}{360°} \tag{5-8}$$

Then if L is the entire length of chain in pitches,

$$L = 2(BE + ME + KB)$$

$$L = 2\left[C\cos a + \frac{N+n}{4} + \frac{a}{360°}(N-n)\right] \tag{5-9}$$

Equation (5-9) gives the exact, accurate answer for chain length. Equation (5-9) can be rearranged to provide a simpler "approximation" formula for use where the center distance is adjustable by plus or minus one-half pitch.

$$L = 2C + \frac{N+n}{2} + \frac{C}{N-n}\left[\frac{a}{180°} - \frac{2C(1-\cos a)}{N-n}\right]\frac{(N-n)^2}{C} \tag{5-10}$$

The term

$$\frac{C}{N-n}\left[\frac{a}{180°} - \frac{2C(1-\cos a)}{N-n}\right]$$

of this formula may be replaced by $1/4\pi^2$, which results in

$$L = 2C + \frac{N+n}{2} + \frac{(N-n)^2}{4\pi^2 C} \tag{5-11}$$

Final Center Distance Calculation. The calculated length, if fractional, must now be modified to a whole (preferably even) number of pitches. The center distance may be corrected for revised chain length by solving equation (5-11) for C. This is also an approximation, and is not for use with fixed center drives or for those with a high ratio.

$$C = \frac{L - \frac{N+n}{2} + \sqrt{\left(L - \frac{N+n}{2}\right)^2 - 8\frac{(N-n)^2}{4\pi^2}}}{4} \tag{5-12}$$

A more exact center distance C can be calculated from Figure 5-9:

$$C' = \frac{L - n[(90-a)/180] - N[(90+a)/180]}{2\cos a} \tag{5-13}$$

In designing a Roller Chain drive, it is desirable that the theoretical center distance (as obtained from the center distance tables) should not be exceeded. This implies that center distances based on table dimensions require negative tolerances. When a plus tolerance is permitted, there is some risk of the new chain being too tight around the sprockets, which could result in chain damage either in installation or during the break-in period.

When extreme accuracy of chain length or center distance is required, consult a chain manufacturer.

Horsepower Chart and Rating Tables

The horsepower rating chart (Fig. 5-10) and the horsepower rating tables (Tables 5-20 to 5-33 at the end of the chapter) are based on the horsepower formulas developed for ANSI Roller Chain earlier in this chapter.

The horsepower rating chart provides a quick means of determining the probable chain requirement without the need for a random search through the horsepower rating tables.

The horsepower rating tables cover a wide range of smaller sprocket sizes, and show more precise horsepower values than the chart (Fig. 5-10).

The tabular ratings apply to single-strand standard and heavy series Roller Chains and, based on roller-bushing, impact values reflect a service factor of 1, a chain length of approximately 100 pitches, use of recommended lubrication methods, and a drive arrangement where two aligned sprockets are mounted on parallel, horizontal shafts. Under these conditions, approximately 15,000 hr of service life at full-load operation may be expected.

The horsepower ratings relate to the speed of the smaller sprocket and drive selections are made on this basis, whether the drive is speed reducing or speed increasing.

Drives with more than two sprockets, idlers, composite duty cycles, or other unusual conditions often require special consideration. It is advisable to consult chain manufacturers for selections of this nature.

Horsepower Rating Chart

When making drive selections, consideration is given to the loads imposed on the chain by the type of input power and the type of equipment to be driven. Service factors (Table 5-1) are used to determine the design horsepower, or the additional chain capacity required to compensate for these loads.

To use the chart, first determine the design horsepower from the following equation:

$$\text{Design HP} = \text{HP to be transmitted} \times \text{service factor} \tag{5-14}$$

Now refer to Figure 5-10 and follow the horizontal line for the design horsepower to the vertical line for the specified speed of the smaller sprocket. The area in which the two lines intersect indicates the probable chain number for the speed and horsepower requirement.

While the single- and multiple strand Roller Chains indicated in the chart satisfy

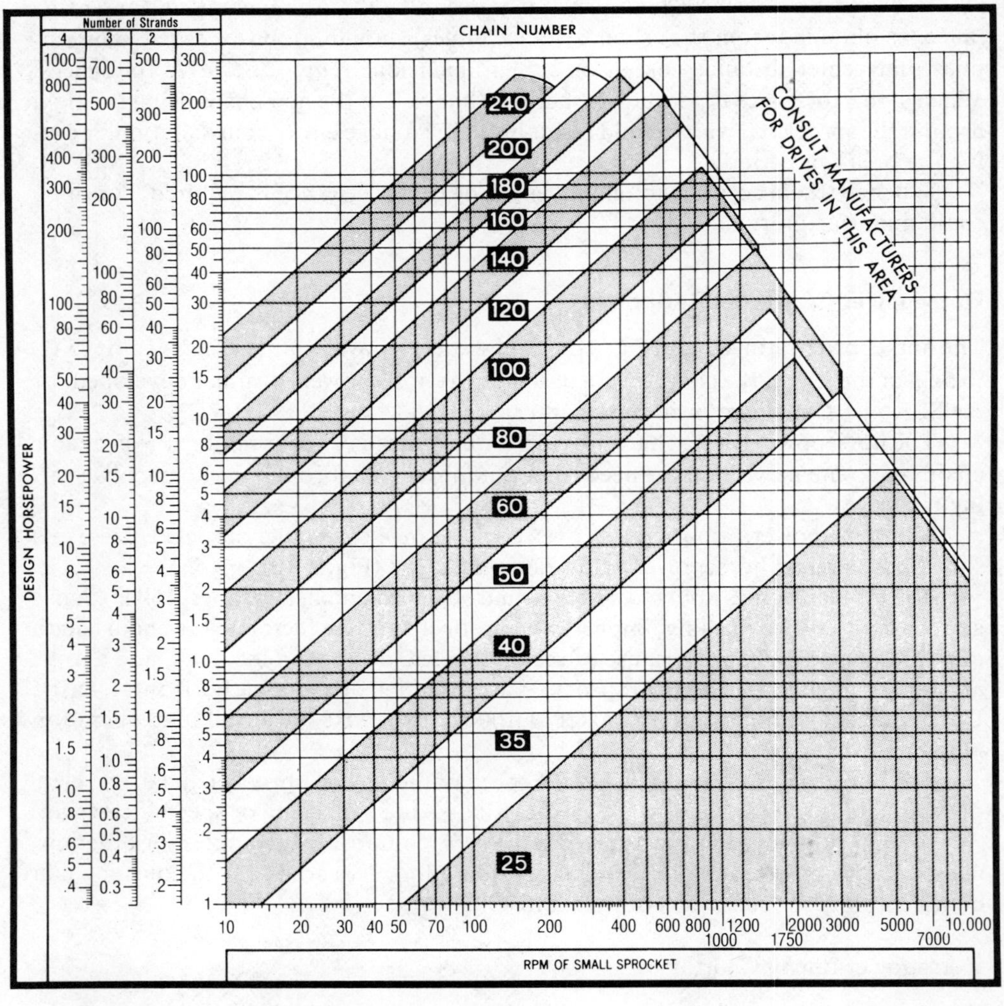

Note: The maximum horsepower rating specified in each of the strand columns in Figure 5-10 is not limiting for Roller Chain drives. See Chapter 6 or consult Roller Chain manufacturers on those applications which are above the horsepower range of the chart.

Figure 5-10 Roller Chain pitch selection chart.

most drive requirements, multiple strand chains of greater width are available. Consult chain manufacturers for selection of this type.

Horsepower Rating Tables

The horsepower rating tables are used to make a precise selection and to determine the number of teeth required for the small sprocket.

The ratings in the horsepower tables are for single-strand chains. Consequently, in order to use the tables for multiple strand chain selections, the required horsepower table rating per strand must be calculated from the following equation:

$$\text{Required HP table rating} = \frac{\text{design HP}}{\text{multiple strand factor}} \tag{5-15}$$

(Multiple strand factors are listed in Table 5-5.)

After tentatively choosing the chain number from the chart (Figure 5-10), refer to the rating table for the selected chain. Follow down the column headed by the speed of the small sprocket and find the nearest value to the design horsepower if single-strand chain has been selected, or the nearest value to the required horsepower table rating if multiple strand chain has been selected. Follow this line horizontally to the left to find the required number of teeth for the small sprocket.

For intermediate speeds or sprocket sizes not tabulated, interpolate between the appropriate columns or lines.

It is often desirable to evaluate additional selections based on the next larger or next smaller chain number. A comparison of two or more selections may also be desirable because of weight, size, noise, or other considerations. Where quietness or extra-smooth operation are of special importance, a small pitch chain operating over large-diameter sprockets will minimize noise and vibration.

Lubrication for Roller Chain drives is covered in Chapter 12. The limiting rpm for each lubrication type is read from the column to the right of the boundary lines shown in the rating tables. Deviation to less than the recommended lubricant or the recommended method of lubrication can result in shorter drive life, whereas using a superior method can improve life.

Roller Chain Drives for Specific Applications

Required Data

Intelligent design of a chain drive is possible only if complete specifications for the drive are available. The required items are discussed in succeeding paragraphs.

Horsepower. The horsepower to be transmitted is a fundamental item in the design of a chain drive. It is either the rated horsepower of an electric motor or prime mover, such as an engine, or the power absorption of the driven equipment, as required by the driven shaft, whichever governs.

Shaft Speeds. The speed of the faster shaft affects the selection of the small sprocket and the chain, while the shaft speed ratio affects the choice of the large sprocket. Any allowable variation in shaft speed or in the speed ratio should be known.

Load and Power Source. Load and power source characteristics are essential data for determination of the service factors. Failure to consider them may lead to misapplications.

Shaft Diameters. Shaft diameters determine the sprocket bore diameters, and therefore limit the minimum sprocket size. In most cases, the large sprocket will be of ample size to accommodate the specified shaft.

Center Distance. The specification of a center distance should always state whether or not adjustment of shaft positions is permissible. If the driving and driven sprockets are on separate units, the center distance can usually be left to the discretion of the designer.

Space Limitations. For most applications, the available space will not be so small as to affect drive design. Where important limitations exist, they should be specified.

Surrounding Conditions. Unusual or adverse surrounding conditions must always be considered. A polluted atmosphere, high or low ambient temperature, or similar conditions may require the use of special drive materials, special lubrication provisions, or chain casings. Chain casings, in addition to ensuring proper lubrication, provide safety for personnel.

Roller Chain and Sprocket Selection

The step-by-step procedure for selection of Roller Chain and sprockets is outlined in the following paragraphs:

Step 1. Service factor. Find the load classification in Table 5-2. If the specified type of load is not listed, classify it by its similarity to a listed item. Then select the service factor from Table 5-1.

Step 2. Computation of design horsepower. Multiply the service factor by the horsepower to be transmitted to obtain the design horsepower.

Step 3. Tentative chain selection. Tentatively select the chain pitch size required from the pitch selection chart, Figure 5-10. Follow the horizontal line representing the design horsepower to the vertical line for the specified speed of the smaller sprocket. The area in which the two lines intersect indicates the probable chain required for the speed and horsepower requirement.

It may also be desirable to evaluate selections based on the next larger or next smaller chain size.

Step 4. Final selection of roller chain and small sprocket. If a single-strand Roller Chain has been selected, refer directly to the horsepower rating table. Follow down the column headed by the speed of the small sprocket and find the value nearest the design horsepower. Follow this line horizontally to the left to find the required number of teeth for the small sprocket.

For intermediate speeds or sprocket sizes not tabulated, interpolate between the appropriate columns or lines.

Check the maximum bore for the selected sprocket in Table 4-47. If the sprocket selected will not accommodate the shaft, use a larger sprocket or make a new sprocket and chain selection from the rating table for the next larger chain size.

Step 5. Selection of the large sprocket. Compute the number of teeth for the large sprocket using the selected small sprocket and the specified shaft speed ratio. If the required precision of speed ratio cannot be obtained with an integral number of teeth on the large sprocket, try a larger size for the small sprocket. Recheck the small sprocket size and the chain pitch.

If space limitations are specified, check for clearances. If the large sprocket size must be decreased, recompute the small sprocket size and the chain pitch. A small sprocket with fewer teeth than the horsepower tables indicate for a given speed is not recommended.

Step 6. Drives with higher chain tension or smaller sprocket requirements. Some drives will be found to have horsepower requirements beyond the capacity for a single-strand Roller Chain. These may be handled by either of two different methods using standard chains described in this manual. These are with multiple strand standard Roller Chain or with heavy duty Engineering Steel drive chain, described in Chapter 6.

Two elements of data usually resolve this decision: those of chain speed and the available space. The type of lubrication available is an important interrelated factor.

Usually, multiple strand Roller Chain will be the smaller pitch, compared to the appropriate Engineering Steel drive chain. The former will be selected where speeds are relatively high and radial space limited (with reference to the sprocket sizes needed). Where speed is relatively slow and torque is therefore high, and the space for chain width is limited, the heavy duty Engineering Steel Chain will usually be selected. The same will be true when a protective case to confine adequate continuous lubrication cannot be provided.

If a multiple strand chain has been selected, determine the required horsepower table rating per strand, using equation (5-15) as discussed earlier in the chapter.

If the use of a heavy duty Engineering Steel Chain is required, its selection may be made by the methods outlined in Chapter 6.

Step 7. Chain length. If center distance tables are not available, compute the length of the chain from equation (5-9) or (5-11), using the tentative center distance. The computed length may be fractional, but a Roller Chain consisting of an integral number of pitches must be used. The next even number of pitches is preferable to an odd number since the latter neccessitates the use of an offset link. The center distance must then be corrected.

Step 8. Final center distance calculation. Using the adopted length, recompute the center distance by equation (5-12) or (5-13). If the center distance is fixed and the adopted chain length results in an objectionable slack, an idler may be used, or a sprocket combination selected that will provide a snug fit without objectionable slack.

Additional Design Features

Relative Shaft Positions (Fig. 5-11). The most favorable operating conditions are obtained when the drive approximates the horizontal. Other practical considerations may require a deviation from the optimum. With a horizontal drive, it is usually

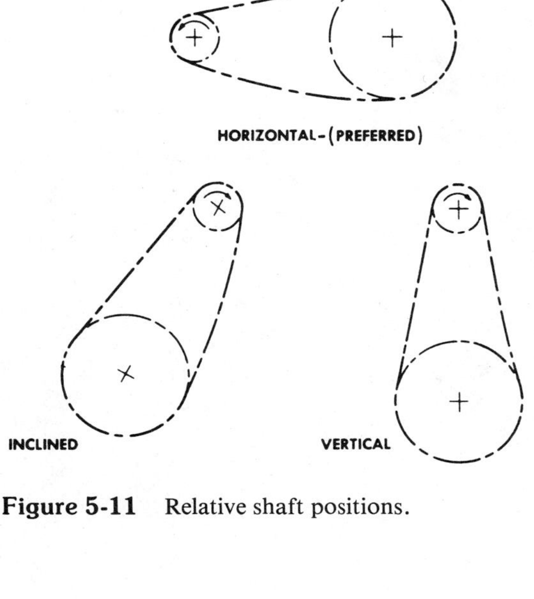

Figure 5-11 Relative shaft positions.

preferable that the tight side of the chain be on top. This results in the slack-side catenary being below, providing ideal conditions for bath lubrication when necessary.

A vertical, or a steeply inclined drive, may necessarily permit less slack than a horizontal drive to keep the chain in proper engagement with the lower sprocket.

Lubrication. The types of lubrication systems for various power capacities are indicated by "stair-step" lines on the rating tables. Details of the different lubrication systems are given in Chapter 12.

Chain Casings. Chain casings (Fig. 5-12) are used to facilitate lubrication and to protect drives from damage by harmful atmospheric dusts. These casings are usually made of sheet metal, stiffened and supported by steel angles, and have access doors for inspection and maintenance of the drive. Proper clearances must be provided to prevent contact with the moving parts of the drive.

Chain guard or case clearance sometimes restricts the useful wear of life of a chain. This happens as normal wear-induced slack collects at the slack or unloaded strand. When clearance is limited, wear-induced slack eventually becomes large enough for the chain to strike the case, damaging both case and chain.

Case clearance for allowable wear may be determined by using the chart information in Figure 5-13.

Detailed instructions for the design of casings and guards are beyond the scope of this manual.

Chain Takeups (Idlers). A chain takeup usually consists of an idler sprocket mounted on a shaft supported by an adjustable base. Its principal function is to provide for adjustment of the chain slack when the drive shaft centers are not adjustable.

Two types of takeups are available. The simpler type (Fig. 5-14) has the idler base bolted to any convenient fixed structure; the bolt holes are slotted to permit manual adjustment. The spring-adjusted type (Fig. 5-15) has the idler sprocket mounted on a pivoted arm, on one arm of a bell crank lever, or on a sliding member. A spring provides for a constant pressure of the idler sprocket on the chain. With this type of takeup, periodic adjustment of the idler sprocket position is not required; however, any

Figure 5-12 Typical oil retaining casing.

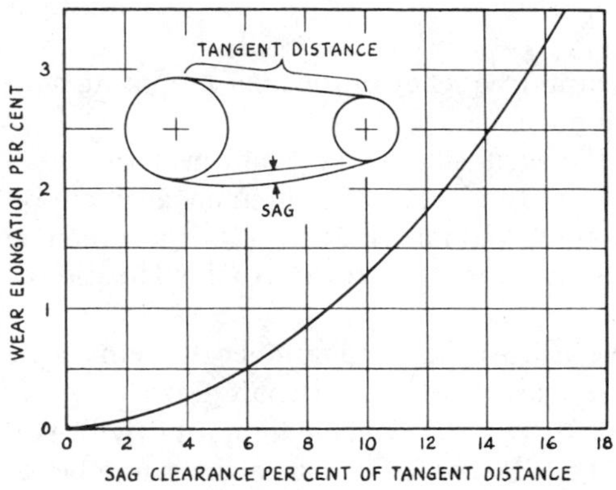

Figure 5-13 Case clearance wear limit.

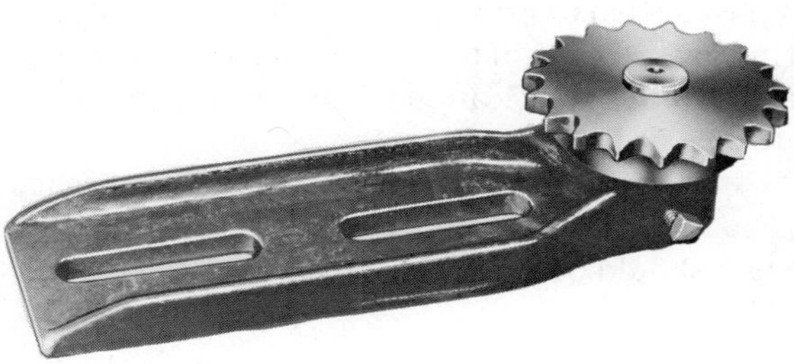

Figure 5-14 Manually adjusted idler mounting.

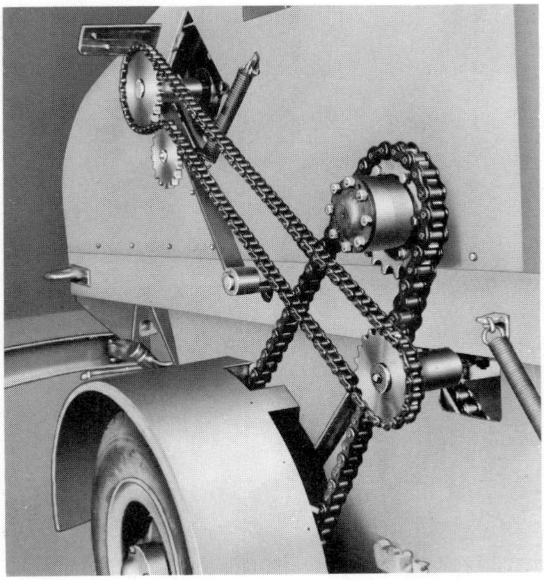

Figure 5-15 Spring-adjusted idler mounting.

chain pulsation may be increased if the natural frequency of the idler mounting happens to synchronize with the chain vibration frequency.

Chain takeups should be located on the slack side of the chain near the small sprocket, and on the outside of the drive so as to increase the chain engagement on that sprocket. Figures 5-16 and 5-17 show preferred idler locations. For a reversible drive, two automatic takeups are required to shift the idler pressure to the slack side as the drive reverses.

In general, idler sprockets should have at least the same number of teeth as the smaller drive sprocket and should engage at least three or four chain links.

Multiple Shaft Drives. A Roller Chain drive is the ideal medium for driving two or more shafts from a single driving shaft (Fig. 5-18), or for synchronizing a number of shafts, such as the print roller shafts on a color print machine. For such drives, the location, the angular speed, and the direction of rotation of each shaft are usually fixed by the design of the machine.

The chain pitch and smallest sprocket are selected, tentatively, as for a two-shaft drive. Then the other sprockets must be selected to provide the required speed ratios. Because each sprocket must have an integral number of teeth, these sprockets can only be selected by trial and error, and may require modification of the selected chain and another selection of the smallest sprocket size.

Next, the chain path must be laid out to provide for adequate chain wrap on each sprocket. To meet this requirement, and to provide for chain adjustment, one or more idlers may be required. The chain length should be determined by means of a graphic layout, preferably to an enlarged scale.

Roller Chain Drive Design Examples

Problem: Select an electric-motor-driven Roller Chain drive to transmit 10 hp from a countershaft to the main shaft of a wire drawing machine. The countershaft is 1 15/16 in. in diameter and operates at 1200 rpm. The main shaft is also 1-15/16 in. in diameter

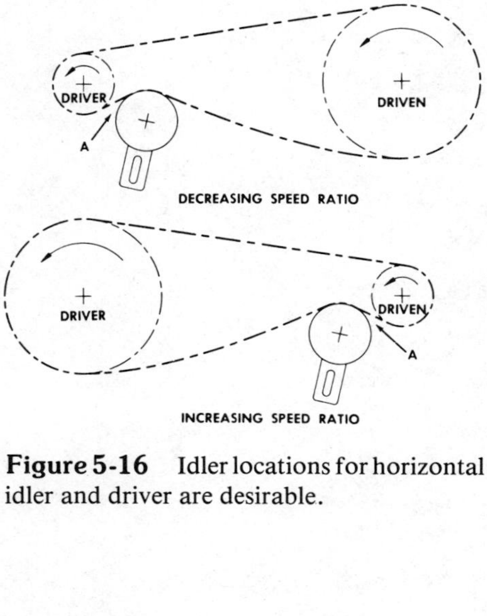

Figure 5-16 Idler locations for horizontal drives. Three or more pitches of chain (A) between idler and driver are desirable.

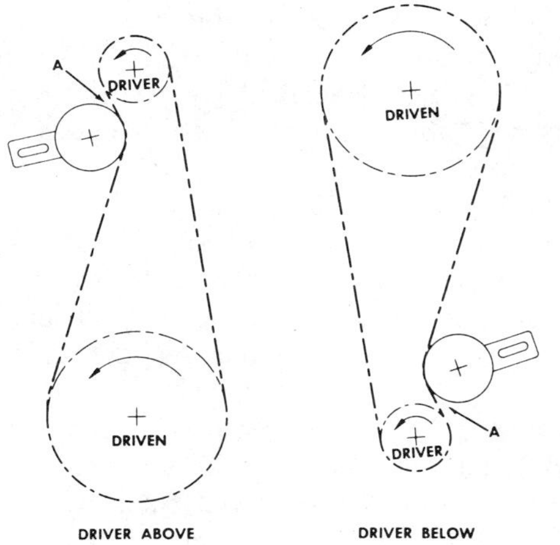

Figure 5-17 Idler locations for vertical drives. Three or more pitches of chain between idler and driver are desirable.

Figure 5-18 Roller Chain driving a series of shafts.

and must operate between 378 and 382 rpm. Shaft centers, once established, are fixed and by initial calculations must be approximately 22-1/2 in. The load on the main shaft is uneven and presents "peaks" which places it in the heavy-shock-load category. The driving head is fully enclosed and all parts are pressure lubricated from a central system; therefore, the drive will receive copious lubrication.

Step 1. Service factor. Classification for this drive is listed in Table 5-2 as heavy shock load. The corresponding service factor from Table 5-1 for heavy shock load and electric motor is 1.5.

Step 2. Design horsepower. The design horsepower is $10 \times 1.5 = 15$ hp.

Step 3. Tentative chain selection. On the horsepower rating chart (Fig. 5-10), locate 15 hp under the single-strand column at the left, and then follow the horizontal "horsepower" line across the chart to where it intersects with the vertical 1200-rpm line for the small sprocket. This intersection lies in the clear diagonal area identified with use of No. 50 (5/8-in. pitch) Roller Chain.

Step 4. Final selection of chain and small sprocket. On the horsepower rating table for No. 50 chain at 1200 rpm, the computed design horsepower of 15 hp is realized with a 20-tooth sprocket. Table 4-47 shows that this sprocket will accept the specified shaft.

Step 5. Selection of the large sprocket. Since the driver is to operate at 1200 rpm and the driven at a minimum of 378 rpm, the speed ratio $= 1200/378 = 3.175$ minimum. Therefore, the large sprocket should have 20×3.175 teeth $= 63.50$ teeth (use 63).

This combination of 20 and 63 teeth will produce a main drive shaft speed of 381 rpm, which is within the limitation of 378 to 382 rpm established in the original specification.

Step 6. Possible alternate. Sometimes space for a sprocket is limited, or higher chain tension capacity is needed in the same format.

Investigate the use of multiple strand chain in the selection chart, Figure 5-10, as a possible alternate drive. Using the double-strand horsepower column, this suggests that a 40-2 chain would transmit 15 design horsepower at 1200 rpm. Obtaining the multiple strand factor of 1.7 from Table 5-5, the required horsepower table rating $= 15/1.7 = 8.83$ hp. The No. 40 chain horsepower rating table (Table 5-22) indicates that 23 teeth are needed to transmit 8.83 hp. Reference to Table 4-47 indicates that the 23-tooth 40-2 sprocket will accept the driving shaft, which is 1-15/16 in. in diameter. This would make an acceptable drive but is more costly than use of the single-strand No. 50 components. It also requires an undesirably long center distance for a 1/2-in. pitch chain.

Step 7. Chain length. Since 20 and 63 tooth sprockets are to be placed in 22-1/2-in. centers, calculations are as follows to determine chain length. From equation (5-11) we have

$$L = 2C + \frac{N+n}{2} + \frac{(N-n)^2}{4\pi^2 C}$$

where

L = chain length, pitches
C = shaft centers, pitches
N = number of teeth in large sprocket
n = number of teeth in small sprocket

Substituting values for C, N, and n, we get

$$L = 2 \times 36 + \frac{63 + 20}{2} + \frac{(63 - 20)^2}{39.48 \times 36}$$

$$= 114.8 \text{ in.}$$

Step 8. Correction of center distance. Since chain is to couple at an even number of pitches, we will use 114 pitches and recompute the centers in pitches using equation (5-12).

$$C = \frac{L - \frac{N+n}{2} + \sqrt{\left(L - \frac{N+n}{2}\right)^2 - 8\frac{(N-n)^2}{4\pi^2}}}{4}$$

$$= \frac{114 - \frac{63+20}{2} + \sqrt{\left(114 - \frac{63+20}{2}\right)^2 - 8\frac{(63-20)^2}{39.48}}}{4}$$

$$= 35.6 \text{ pitches or } 22.25 \text{ in.}$$

Summary. Since the 40-2 drive considered would have a rather long center distance for the chain pitch and is not required for clearance, our final solution is 20 and 63 tooth sprockets mounted on 22.25-in. centers using No. 50, 5/8-in.-pitch Roller Chain.

In the horsepower rating table for No. 50 chain (Table 5-22), type B lubrication is indicated and the existing central system of lubrication more than meets this requirement.

In the section of Chapter 12 covering lubrication, design of cases for chain drives is included. For drives needing type C lubrication, the sizing of the case is important to maintain adequate heat dispersion and to distribute lubricant properly. The size of the case may also be a factor in requiring a double-strand chain with smaller sprockets.

Multiple Shaft and Rack Drives

Most Roller Chain drives assume a rather standard form, as shown elsewhere in this book. Figures 5-3 and 5-4 show typical multiple strand Roller Chains in use and Figures 5-6 to 5-8 present this conventional form.

However, it should not be assumed that use of Roller Chain is limited to just two shafts. Use of a single chain strand to drive numerous shafts, as shown in the example of Figure 5-18, is quite common. Variation in speed of the individual shafts can be achieved by using different sizes of sprockets. Such an arrangement is quite efficient, since sprockets of this nature do not need much "wrap" to operate satisfactorily, but adequate chain tension must be maintained.

Another example of the use of single-strand Roller Chain to drive several shafts simultaneously is the *rack drive*, shown in Figure 5-19. In this configuration, the pinions are arranged with contact surfaces in a straight line, and a straight portion of the strand of Roller Chain serves as a rack.

In the ordinary rack-and-pinion drive, movement of the rack in one direction is

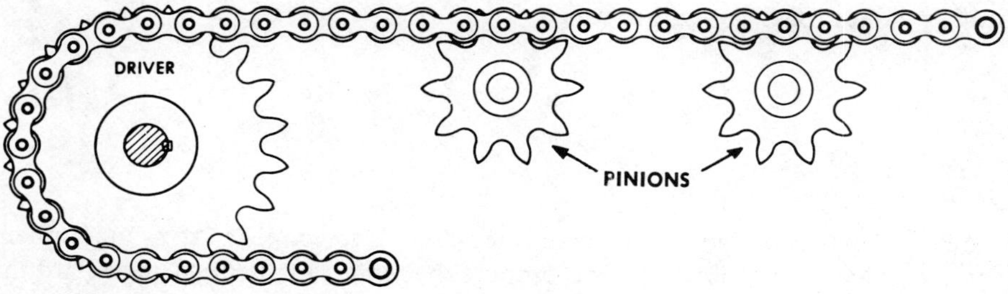

Figure 5-19 Roller Chain rack-and-pinion drive.

limited by its length. By using an endless roller chain running over sprockets at the end of the rack, the equivalent of a rack of infinite length is obtained.

For the pinion to articulate properly with the chain when the pinion is in single tooth contact, the recommended tooth is of cycloidal form.

It is recommended that the chain manufacturer be consulted for exact recommendations for both chain and sprockets.

Table 5-20 Horsepower Ratings, Standard Single-Strand Roller Chain, No. 25, 1/4-in. Pitch

No. of Teeth Small Spkt.	REVOLUTIONS PER MINUTE—SMALL SPROCKET																							
	50	100	300	500	700	900	1200	1500	1800	2100	2500	3000	3500	4000	4500	5000	5500	6000	6500	7000	7500	8000	8500	9000	10000
9	0.02	0.04	0.12	0.18	0.25	0.31	0.41	0.50	0.58	0.67	0.79	0.93	1.06	1.02	0.86	0.73	0.63	0.56	0.49	0.44	0.40	0.36	0.33	0.30	0.26
10	0.03	0.05	0.13	0.21	0.28	0.35	0.45	0.56	0.65	0.75	0.88	1.04	1.19	1.20	1.00	0.86	0.74	0.65	0.58	0.52	0.47	0.42	0.39	0.35	0.30
11	0.03	0.05	0.14	0.23	0.31	0.39	0.50	0.62	0.73	0.83	0.98	1.15	1.32	1.38	1.16	0.99	0.86	0.75	0.67	0.60	0.54	0.49	0.45	0.41	0.35
12	0.03	0.06	0.16	0.25	0.34	0.43	0.55	0.68	0.80	0.92	1.07	1.26	1.45	1.57	1.32	1.12	0.97	0.86	0.76	0.68	0.61	0.56	0.51	0.47	0.40
13	0.04	0.06	0.17	0.27	0.37	0.47	0.60	0.74	0.87	1.00	1.17	1.38	1.58	1.77	1.49	1.27	1.10	0.96	0.86	0.77	0.69	0.63	0.57	0.53	0.45
14	0.04	0.07	0.19	0.30	0.40	0.50	0.65	0.80	0.94	1.08	1.27	1.49	1.71	1.93	1.66	1.42	1.23	1.08	0.96	0.86	0.77	0.70	0.64	0.59	0.50
15	0.04	0.07	0.20	0.32	0.43	0.54	0.70	0.86	1.01	1.17	1.36	1.61	1.85	2.08	1.84	1.57	1.36	1.20	1.06	0.95	0.86	0.78	0.71	0.65	0.56
16	0.04	0.08	0.22	0.34	0.47	0.58	0.76	0.92	1.09	1.25	1.46	1.72	1.98	2.23	2.03	1.73	1.50	1.32	1.17	1.05	0.94	0.86	0.78	0.72	0.61
17	0.05	0.08	0.23	0.37	0.50	0.62	0.81	0.99	1.16	1.33	1.56	1.84	2.11	2.38	2.22	1.90	1.64	1.44	1.28	1.14	1.03	0.94	0.86	0.79	0.67
18	0.05	0.09	0.25	0.39	0.53	0.66	0.86	1.05	1.24	1.42	1.66	1.96	2.25	2.53	2.42	2.07	1.79	1.57	1.39	1.25	1.12	1.02	0.93	0.86	0.73
19	0.05	0.09	0.26	0.41	0.56	0.70	0.91	1.11	1.31	1.50	1.76	2.07	2.38	2.69	2.62	2.24	1.94	1.70	1.51	1.35	1.22	1.11	1.01	0.93	0.79
20	0.06	0.10	0.28	0.44	0.59	0.74	0.96	1.17	1.38	1.59	1.86	2.19	2.52	2.84	2.83	2.42	2.10	1.84	1.63	1.46	1.32	1.20	1.09	1.00	0.86
21	0.06	0.11	0.29	0.46	0.62	0.78	1.01	1.24	1.46	1.68	1.96	2.31	2.66	2.99	3.05	2.60	2.26	1.98	1.76	1.57	1.42	1.29	1.17	1.08	0.92
22	0.06	0.11	0.31	0.48	0.66	0.82	1.07	1.30	1.53	1.76	2.06	2.43	2.79	3.15	3.27	2.79	2.42	2.12	1.88	1.69	1.52	1.38	1.26	1.16	0.99
23	0.06	0.12	0.32	0.51	0.69	0.86	1.12	1.37	1.61	1.85	2.16	2.55	2.93	3.30	3.50	2.98	2.59	2.27	2.01	1.80	1.62	1.47	1.35	1.24	1.06
24	0.07	0.13	0.34	0.53	0.72	0.90	1.17	1.43	1.69	1.94	2.27	2.67	3.07	3.46	3.73	3.18	2.76	2.42	2.15	1.92	1.73	1.57	1.44	1.32	1.12
25	0.07	0.13	0.35	0.56	0.75	0.94	1.22	1.50	1.76	2.02	2.37	2.79	3.21	3.61	3.96	3.38	2.93	2.57	2.28	2.04	1.84	1.67	1.53	1.40	1.20
26	0.07	0.14	0.37	0.58	0.79	0.98	1.28	1.56	1.84	2.11	2.47	2.91	3.34	3.77	4.19	3.59	3.11	2.73	2.42	2.17	1.95	1.77	1.62	1.49	1.27
28	0.08	0.15	0.40	0.63	0.85	1.07	1.38	1.69	1.99	2.29	2.68	3.15	3.62	4.09	4.54	4.01	3.47	3.05	2.70	2.42	2.18	1.98	1.81	1.66	1.42
30	0.08	0.16	0.43	0.68	0.92	1.15	1.49	1.82	2.15	2.46	2.88	3.40	3.90	4.40	4.89	4.45	3.85	3.38	3.00	2.68	2.42	2.20	2.01	1.84	1.57
32	0.09	0.17	0.46	0.73	0.98	1.23	1.60	1.95	2.30	2.64	3.09	3.64	4.18	4.72	5.25	4.90	4.25	3.73	3.30	2.96	2.67	2.42	2.21	2.03	1.73
35	0.10	0.19	0.51	0.80	1.08	1.36	1.76	2.15	2.53	2.91	3.41	4.01	4.61	5.20	5.78	5.60	4.86	4.26	3.78	3.38	3.05	2.77	2.53	2.32	1.98
40	0.12	0.22	0.58	0.92	1.25	1.57	2.03	2.48	2.93	3.36	3.93	4.64	5.32	6.00	6.68	6.85	5.93	5.21	4.62	4.13	3.73	3.38	3.09	2.83	2.42
45	0.13	0.25	0.66	1.05	1.42	1.78	2.31	2.82	3.32	3.82	4.47	5.26	6.05	6.82	7.58	8.17	7.08	6.21	5.51	4.93	4.45	4.04	3.69	3.38	2.89

Type I Type II Type III

Table 5-21 Horsepower Ratings, Standard Single-Strand Roller Chain, No. 35, 3/8-in. Pitch

No. of Teeth Small Spkt.	REVOLUTIONS PER MINUTE—SMALL SPROCKET																							
	50	100	300	500	700	900	1200	1500	1800	2100	2500	3000	3500	4000	4500	5000	5500	6000	6500	7000	7500	8000	8500	9000	10000
9	0.08	0.15	0.39	0.62	0.84	1.06	1.37	1.68	1.98	2.27	2.65	2.17	1.73	1.41	1.18	1.01	0.88	0.77	0.68	0.61	0.55	0.50	0.46	0.42	0.35
10	0.09	0.16	0.44	0.70	0.95	1.19	1.54	1.88	2.21	2.54	2.97	2.55	2.02	1.65	1.39	1.18	1.03	0.90	0.80	0.71	0.64	0.58	0.53	0.49	0.42
11	0.10	0.18	0.49	0.77	1.05	1.31	1.70	2.08	2.45	2.82	3.30	2.94	2.33	1.91	1.60	1.37	1.18	1.04	0.92	0.82	0.74	0.67	0.62	0.57	0.48
12	0.11	0.20	0.54	0.85	1.15	1.44	1.87	2.29	2.70	3.10	3.62	3.35	2.66	2.17	1.82	1.56	1.35	1.18	1.05	0.94	0.85	0.77	0.70	0.64	0.55
13	0.12	0.22	0.59	0.93	1.26	1.57	2.04	2.49	2.94	3.38	3.95	3.77	3.00	2.45	2.05	1.75	1.52	1.33	1.18	1.06	0.95	0.87	0.79	0.73	0.62
14	0.13	0.24	0.63	1.01	1.36	1.71	2.21	2.70	3.18	3.66	4.28	4.22	3.35	2.74	2.30	1.96	1.70	1.49	1.32	1.18	1.07	0.97	0.88	0.81	0.69
15	0.14	0.25	0.68	1.08	1.47	1.84	2.38	2.91	3.43	3.94	4.61	4.68	3.71	3.04	2.55	2.17	1.88	1.65	1.47	1.31	1.18	1.07	0.98	0.90	0.77
16	0.15	0.27	0.73	1.16	1.57	1.97	2.55	3.12	3.68	4.22	4.94	5.15	4.09	3.35	2.81	2.40	2.08	1.82	1.62	1.45	1.30	1.18	1.08	0.99	0.85
17	0.16	0 29	0.78	1.24	1.68	2.10	2.73	3.33	3.93	4.51	5.28	5.64	4.48	3.67	3.07	2.62	2.27	2.00	1.77	1.58	1.43	1.30	1.18	1.09	0.93
18	0.17	0.31	0.83	1.32	1.78	2.24	2.90	3.54	4.18	4.80	5.61	6.15	4.88	3.99	3.35	2.86	2.48	2.17	1.93	1.73	1.56	1.41	1.29	1.18	1.01
19	0.18	0.33	0.88	1.40	1.89	2.37	3.07	3.76	4.43	5.09	5.95	6.67	5.29	4.33	3.63	3.10	2.69	2.36	2.09	1.87	1.69	1.53	1.40	1.28	1.10
20	0.19	0.35	0.93	1.48	2.00	2.51	3.25	3.97	4.68	5.38	6.29	7.20	5.72	4.68	3.92	3.35	2.90	2.55	2.26	2.02	1.82	1.65	1.51	1.39	1.18
21	0.20	0.37	0.98	1.56	2.11	2.64	3.42	4.19	4.93	5.67	6.63	7.75	6.15	5.03	4.22	3.60	3.12	2.74	2.43	2.17	1.96	1.78	1.62	1.49	1.27
22	0.21	0.38	1.03	1.64	2.22	2.78	3.60	4.40	5.19	5.96	6.97	8.21	6.59	5.40	4.52	3.86	3.35	2.94	2.61	2.33	2.10	1.91	1.74	1.60	1.37
23	0.22	0.40	1.08	1.72	2.33	2.92	3.78	4.62	5.44	6.25	7.31	8.62	7.05	5.77	4.83	4.13	3.58	3.14	2.79	2.49	2.25	2.04	1.86	1.71	1.46
24	0.23	0.42	1.14	1.80	2.44	3.05	3.96	4.84	5.70	6.55	7.66	9.02	7.51	6.15	5.15	4.40	3.81	3.35	2.97	2.66	2.40	2.17	1.99	1.82	1.56
25	0.24	0.44	1.19	1.88	2.55	3.19	4.13	5.05	5.95	6.84	8.00	9.43	7.99	6.54	5.48	4.68	4.05	3.56	3.16	2.82	2.55	2.31	2.11	1.94	1.65
26	0.25	0.46	1.24	1.96	2.66	3.33	4.31	5.27	6.21	7.14	8.35	9.84	8.47	6.93	5.81	4.96	4.30	3.77	3.35	3.00	2.70	2.45	2.24	2.05	1.75
28	0.27	0.50	1.34	2.12	2.88	3.61	4.67	5.71	6.73	7.73	9.05	10.7	9.47	7.75	6.49	5.55	4.81	4.22	3.74	3.35	3.02	2.74	2.50	2.30	1.96
30	0.29	0.54	1.45	2.29	3.10	3.89	5.03	6.15	7.25	8.33	9.74	11.5	10.5	8.59	7.20	6.15	5.33	4.68	4.15	3.71	3.35	3.04	2.77	2.55	2.17
32	0.31	0.58	1.55	2.45	3.32	4.17	5.40	6.60	7.77	8.93	10.4	12.3	11.6	9.47	7.93	6.77	5.87	5.15	4.57	4.09	3.69	3.35	3.06	2.81	0
35	0.34	0.64	1.71	2.70	3.66	4.59	5.95	7.27	8.56	9.84	11.5	13.6	13.2	10.8	9.08	7.75	6.72	5.90	5.23	4.68	4.22	3.83	3.50	3.21	0
40	0.39	0.73	1.97	3.12	4.23	5.30	6.87	8.40	9.89	11.4	13.3	15.7	16.2	13.2	11.1	9.47	8.21	7.20	6.39	5.72	5.15	4.68	0		
45	0.45	0.83	2.24	3.55	4.80	6.02	7.80	9.53	11.2	12.9	15.1	17.8	19.3	15.8	13.2	11.3	9.79	8.59	7.62	6.82	0				

Type I Type II Type III

TYPE I: Manual or Drip Lubrication
TYPE II: Bath or Disc Lubrication
TYPE III: Oil Stream Lubrication
See Chapter 12, 'Lubrication' for detailed description of lubrication types.

The limiting RPM for each lubrication type is read from the column to the left of the boundary line shown.

The Horsepower ratings of Multiple Strand Chains are greater than those for Single Strand Chain: see Table 5-5 for 'Multiple Strand Factors'.

Table 5-22 Horsepower Ratings, Standard Single-Strand Roller Chain, No. 40, 1/2-in. Pitch

No. of Teeth Small Spkt.	REVOLUTIONS PER MINUTE—SMALL SPROCKET																							
	10	25	50	100	200	300	400	500	700	900	1000	1200	1400	1600	1800	2100	2400	2700	3000	3500	4000	5000	6000	7000	8000
9	0.04	0.10	0.19	0.35	0.65	0.93	1.21	1.48	2.00	2.51	2.75	3.25	3.73	4.12	3.45	2.74	2.24	1.88	1.60	1.27	1.04	0.75	0.57	0.45	0.37
10	0.05	0.11	0.21	0.39	0.73	1.04	1.35	1.65	2.24	2.81	3.09	3.64	4.18	4.71	4.04	3.21	2.63	2.20	1.88	1.49	1.22	0.87	0.66	0.53	0.43
11	0.05	0.12	0.23	0.43	0.80	1.16	1.50	1.83	2.48	3.11	3.42	4.03	4.63	5.22	4.66	3.70	3.03	2.54-	2.17	1.72	1.41	1.01	0.77	0.61	0.50
12	0.06	0.14	0.25	0.47	0.88	1.27	1.65	2.01	2.73	3.42	3.76	4.43	5.09	5.74	5.31	4.22	3.45	2.89	2.47	1.96	1.60	1.15	0.87	0.69	0.57
13	0.06	0.15	0.28	0.52	0.96	1.39	1.80	2.20	2.97	3.73	4.10	4.83	5.55	6.26	5.99	4.76	3.89	3.26	2.79	2.21	1.81	1.29	0.98	0.78	0.64
14	0.07	0.16	0.30	0.56	1.04	1.50	1.95	2.38	3.22	4.04	4.44	5.23	6.01	6.78	6.70	5.31	4.35	3.65	3.11	2.47	2.02	1.45	1.10	0.87	0.71
15	0.07	0.17	0.32	0.60	1.12	1.62	2.10	2.56	3.47	4.35	4.78	5.64	6.47	7.30	7.43	5.89	4.82	4.04	3.45	2.74	2.24	1.60	1.22	0.97	0.79
16	0.08	0.19	0.35	0.65	1.20	1.74	2.25	2.75	3.72	4.66	5.13	6.04	6.94	7.83	8.18	6.49	5.31	4.45	3.80	3.02	2.47	1.77	1.34	1.07	0.87
17	0.08	0.20	0.37	0.69	1.29	1.85	2.40	2.93	3.97	4.98	5.48	6.45	7.41	8.36	8.96	7.11	5.82	4.88	4.17	3.31	2.71	1.94	1.47	1.17	0.96
18	0.09	0.21	0.39	0.73	1.37	1.97	2.55	3.12	4.22	5.30	5.82	6.86	7.88	8.89	9.76	7.75	6.34	5.31	4.54	3.60	2.95	2.11·	1.60	1.27	0
19	0.09	0.22	0.42	0.78	1.45	2.09	2.71	3.31	4.48	5.62	6.17	7.27	8.36	9.42	10.5	8.40	6.88	5.76	4.92	3.91	3.20	2.29	1.74	1.38	0
20	0.10	0.24	0.44	0.82	1.53	2.21	2.86	3.50	4.73	5.94	6.53	7.69	8.83	9.96	11.1	9.07	7.43	6.22	5.31	4.22	3.45	2.47	1.88	1.49	0
21	0.11	0.25	0.46	0.87	1.62	2.33	3.02	3.69	4.99	6.26	6.88	8.11	9.31	10.5	11.7	9.76	7.99	6.70	5.72	4.54	3.71	2.66	2.02	1.60	0
22	0.11	0.26	0.49	0.91	1.70	2.45	3.17	3.88	5.25	6.58	7.23	8.52	9.79	11.0	12.3	10.5	8.57	7.18	6.13	4.87	3.98	2.85	2.17	1.72	0
23	0.12	0.27	0.51	0.96	1.78	2.57	3.33	4.07	5.51	6.90	7.59	8.94	10.3	11.6	12.9	11.2	9.16	7.68	6.55	5.20	4.26	3.05	2.32	1.84	0
24	0.13	0.29	0.54	1.00	1.87	2.69	3.48	4.26	5.76	7.23	7.95	9.36	10.8	12.1	13.5	11.9	9.76	8.18	6.99	5.54	4.54	3.25	2.47	1.96	0
25	0.13	0.30	0.56	1.05	1.95	2.81	3.64	4.45	6.02	7.55	8.30	9.78	11.2	12.7	14.1	12.7	10.4	8.70	7.43	5.89	4.82	3.45	2.63	0	
26	0.14	0.31	0.58	1.09	2.04	2.93	3.80	4.64	6.28	7.88	8.66	10.2	11.7	13.2	14.7	13.5	11.0	9.23	7.88	6.25	5.12	3.66	2.79	0	
28	0.15	0.34	0.63	1.18	2.20	3.18	4.11	5.03	6.81	8.54	9.39	11.1	12.7	14.3	15.9	15.0	12.3	10.3	8.80	6.99	5.72	4.09	3.11	0	
30	0.16	0.37	0.68	1.27	2.38	3.42	4.43	5.42	7.33	9.20	10.1	11.9	13.7	15.4	17.2	16.7	13.6	11.4	9.76	7.75	6.34	4.54	3.45	0	
32	0.17	0.39	0.73	1.36	2.55	3.67	4.75	5.81	7.86	9.86	10.8	12.8	14.7	16.5	18.4	18.4	15.0	12.6	10.8	8.64	6.99	5.00	0		
35	0.19	0.43	0.81	1.50	2.81	4.04	5.24	6.40	8.66	10.9	11.9	14.1	16.2	18.2	20.3	21.0	17.2	14.4	12.3	9.76	7.99	5.72	0		
40	0.22	0.50	0.93	1.74	3.24	4.67	6.05	7.39	10.0	12.5	13.8	16.3	18.7	21.1	23.4	25.7	21.0	17.6	15.0	11.9	9.76	6.99	0		
45	0.25	0.57	1.06	1.97	3.68	5.30	6.87	8.40	11.4	14.2	15.7	18.5	21.2	23.9	26.6	30.5	25.1	21.0	17.9	14.2	11.7	0			

Type I Type II Type III

Table 5-23 Horsepower Ratings, Standard Single-Strand Roller Chain, No. 41, 1/2-in. Pitch Light Machinery Roller Chain

No. of Teeth Small Spkt.	REVOLUTIONS PER MINUTE—SMALL SPROCKET																								
	10	25	50	100	200	300	400	500	700	900	1000	1200	1400	1600	1800	2100	2400	2700	3000	3500	4000	5000	6000	7000	8000
9	0.02	0.05	0.10	0.19	0.36	0.51	0.66	0.81	1.10	1.38	1.52	1.27	1.01	0.82	0.69	0.55	0.45	0.38	0.32	0.25	0.21	0.15	0.11	0.09	0.07
10	0.03	0.06	0.11	0.21	0.40	0.57	0.74	0.91	1.23	1.54	1.70	1.49	1.18	0.96	0.81	0.64	0.53	0.44	0.38	0.30	0.24	0.17	0.13	0.11	0.08
11	0.03	0.07	0.13	0.24	0.44	0.64	0.82	1.01	1.37	1.71	1.88	1.71	1.36	1.11	0.93	0.74	0.61	0.51	0.43	0.34	0.28	0.20	0.15	0.12	0.10
12	0.03	0.07	0.14	0.26	0.49	0.70	0.91	1.11	1.50	1.88	2.07	1.95	1.55	1.27	1.06	0.84	0.69	0.58	0.49	0.39	0.32	0.23	0.17	0.14	0.11
13	0.04	0.08	0.15	0.28	0.53	0.76	0.99	1.21	1.63	2.05	2.25	2.20	1.75	1.43	1.20	0.95	0.78	0.65	0.56	0.44	0.36	0.26	0.20	0.16	0.13
14	0.04	0.09	0.16	0.31	0.57	0.83	1.07	1.31	1.77	2.22	2.44	2.46	1.95	1.60	1.34	1.06	0.87	0.73	0.62	0.49	0.40	0.29	0.22	0.17	0.14
15	0.04	0.09	0.18	0.33	0.62	0.89	1.15	1.41	1.91	2.39	2.63	2.73	2.17	1.77	1.49	1.18	0.96	0.81	0.69	0.55	0.45	0.32	0.24	0.19	0.16
16	0.04	0.10	0.19	0.36	0.66	0.95	1.24	1.51	2.05	2.57	2.82	3.01	2.39	1.95	1.64	1.30	1.06	0.89	0.76	0.60	0.49	0.35	0.27	0.21	0.17
17	0.05	0.11	0.20	0.38	0.71	1.02	1.32	1.61	2.18	2.74	3.01	3.29	2.61	2.14	1.79	1.42	1.16	0.98	0.83	0.66	0.54	0.39	0.29	0.23	0.19
18	0.05	0.12	0.22	0.40	0.75	1.08	1.40	1.72	2.32	2.91	3.20	3.59	2.85	2.33	1.95	1.55	1.27	1.06	0.91	0.72	0.59	0.42	0.32	0.25	0
19	0.05	0.12	0.23	0.43	0.80	1.15	1.49	1.82	2.46	3.09	3.40	3.89	3.09	2.53	2.12	1.68	1.38	1.15	0.98	0.78	0.64	0.46	0.35	0.28	0
20	0.06	0.13	0.24	0.45	0.84	1.21	1.57	1.92	2.60	3.26	3.59	4.20	3.33	2.73	2.29	1.81	1.49	1.24	1.06	0.84	0.69	0.49	0.38	0.30	0
21	0.06	0.14	0.26	0.48	0.89	1.28	1.66	2.03	2.74	3.44	3.78	4.46	3.59	2.94	2.46	1.95	1.60	1.34	1.14	0.91	0.74	0.53	0.40	0.32	0
22	0.06	0.14	0.27	0.50	0.93	1.35	1.74	2.13	2.89	3.62	3.98	4.69	3.85	3.15	2.64	2.09	1.71	1.44	1.23	0.97	0.80	0.57	0.43	0.34	0
23	0.06	0.15	0.28	0.53	0.98	1.41	1.83	2.24	3.03	3.80	4.17	4.92	4.11	3.37	2.82	2.24	1.83	1.54	1.31	1.04	0.85	0.61	0.46	0.37	0
24	0.07	0.16	0.29	0.55	1.03	1.48	1.92	2.34	3.17	3.97	4.37	5.15	4.38	3.59	3.01	2.39	1.95	1.64	1.40	1.11	0.91	0.65	0.49	0.39	0
25	0.07	0.17	0.31	0.57	1.07	1.55	2.00	2.45	3.31	4.15	4.57	5.38	4.66	3.81	3.20	2.54	2.08	1.74	1.49	1.18	0.96	0.69	0.53	0	
26	0.07	0.17	0.32	0.60	1.12	1.61	2.09	2.55	3.46	4.33	4.76	5.61	4.94	4.05	3.39	2.69	2.20	1.85	1.58	1.25	1.02	0.73	0.56	0	
28	0.08	0.19	0.35	0.65	1.21	1.75	2.26	2.77	3.74	4.69	5.16	6.08	5.52	4.52	3.79	3.01	2.46	2.06	1.76	1.40	1.14	0.82	0.62	0	
30	0.08	0.20	0.38	0.70	1.31	1.88	2.44	2.98	4.03	5.06	5.56	6.55	6.13	5.01	4.20	3.33	2.73	2.29	1.95	1.55	1.27	0.91	0.69	0	
32	0.09	0.22	0.40	0.75	1.40	2.02	2.61	3.20	4.33	5.42	5.96	7.03	6.75	5.52	4.63	3.67	3.01	2.52	2.15	1.71	1.40	1.00	0		
35	0.10	0.24	0.44	0.83	1.54	2.22	2.88	3.52	4.76	5.97	6.57	7.74	7.72	6.32	5.29	4.20	3.44	2.88	2.46	1.95	1.60	1.14	0		
40	0.12	0.27	0.51	0.96	1.78	2.57	3.33	4.07	5.50	6.90	7.59	8.94	9.43	7.72	6.47	5.13	4.20	3.52	3.01	2.39	1.95	1.40	0		
45	0.14	0.31	0.58	1.08	2.02	2.92	3.78	4.62	6.25	7.84	8.62	10.2	11.3	9.21	7.72	6.13	5.01	4.20	3.59	2.85	2.33	0			

Type I — Type II — Type III

Type I: Manual or Drip Lubrication
Type II: Bath or Disc Lubrication
Type III: Oil Stream Lubrication
See Chapter 12, 'Lubrication for detailed description of lubrication types.

The limiting RPM for each lubrication type is read from the column to the left of the boundary line shown.

The Horsepower ratings of Multiple Strand Chains are greater than those for Single Strand Chain: see Table 5-5 for 'Multiple Strand Factors'.

Table 5-24 Horsepower Ratings, Standard Single-Strand Roller Chain, No. 50, 5/8-in. Pitch

No. of Teeth Small Spkt.	REVOLUTIONS PER MINUTE—SMALL SPROCKET																								
	10	25	50	100	200	300	400	500	700	900	1000	1200	1400	1600	1800	2100	2400	2700	3000	3500	4000	4500	5000	5500	6000
9	0.09	0.19	0.36	0.67	1.26	1.81	2.35	2.87	3.89	4.88	5.36	6.32	6.02	4.92	4.13	3.27	2.68	2.25	1.92	1.52	1.25	1.04	0.89	0.77	0.58
10	0.10	0.22	0.41	0.76	1.41	2.03	2.63	3.22	4.36	5.46	6.01	7.08	7.05	5.77	4.83	3.84	3.14	2.63	2.25	1.78	1.46	1.22	1.04	0.90	0.79
11	0.11	0.24	0.45	0.84	1.56	2.25	2.92	3.57	4.83	6.06	6.66	7.85	8.13	6.65	5.58	4.42	3.62	3.04	2.59	2.06	1.68	1.41	1.20	1.04	0.92
12	0.12	0.26	0.49	0.92	1.72	2.47	3.21	3.92	5.31	6.65	7.31	8.62	9.26	7.58	6.35	5.04	4.13	3.46	2.95	2.34	1.92	1.61	1.37	1.19	1.04
13	0.13	0.29	0.54	1.00	1.87	2.70	3.50	4.27	5.78	7.25	7.97	9.40	10.4	8.55	7.16	5.69	4.65	3.90	3.33	2.64	2.16	1.81	1.55	1.34	0
14	0.14	0.31	0.58	1.09	2.03	2.92	3.79	4.63	6.27	7.86	8.64	10.2	11.7	9.55	8.01	6.35	5.20	4.36	3.72	2.95	2.42	2.03	1.73	1.50	0
15	0.15	0.34	0.63	1.17	2.19	3.15	4.08	4.99	6.75	8.47	9.31	11.0	12.6	10.6	8.88	7.05	5.77	4.83	4.13	3.27	2.68	2.25	1.92	1.66	0
16	0.16	0.36	0.67	1.26	2.34	3.38	4.37	5.35	7.24	9.08	9.98	11.8	13.5	11.7	9.78	7.76	6.35	5.32	4.55	3.61	2.95	2.47	2.11	1.83	0
17	0.17	0.39	0.72	1.34	2.50	3.61	4.67	5.71	7.73	9.69	10.7	12.6	14.4	12.8	10.7	8.50	6.96	5.83	4.98	3.95	3.23	2.71	2.31	2.01	0
18	0.18	0.41	0.76	1.43	2.66	3.83	4.97	6.07	8.22	10.3	11.3	13.4	15.3	13.9	11.7	9.26	7.58	6.35	5.42	4.30	3.52	2.95	2.52	0	
19	0.19	0.43	0.81	1.51	2.82	4.07	5.27	6.44	8.72	10.9	12.0	14.2	16.3	15.1	12.7	10.0	8.22	6.89	5.88	4.67	3.82	3.20	2.73	0	
20	0.20	0.46	0.86	1.60	2.98	4.30	5.57	6.80	9.21	11.5	12.7	15.0	17.2	16.3	13.7	10.8	8.88	7.44	6.35	5.04	4.13	3.46	2.95	0	
21	0.21	0.48	0.90	1.69	3.14	4.53	5.87	7.17	9.71	12.2	13.4	15.8	18.1	17.6	14.7	11.7	9.55	8.01	6.84	5.42	4.44	3.72	3.18	0	
22	0.22	0.51	0.95	1.77	3.31	4.76	6.17	7.54	10.2	12.8	14.1	16.6	19.1	18.8	15.8	12.5	10.2	8.59	7.33	5.82	4.76	3.99	3.41	0	
23	0.23	0.53	1.00	1.86	3.47	5.00	6.47	7.91	10.7	13.4	14.8	17.4	20.0	20.1	16.9	13.4	11.0	9.18	7.84	6.22	5.09	4.27	0		
24	0.25	0.56	1.04	1.95	3.63	5.23	6.78	8.29	11.2	14.1	15.5	18.2	20.9	21.4	18.0	14.3	11.7	9.78	8.35	6.63	5.42	4.55	0		
25	0.26	0.58	1.09	2.03	3.80	5.47	7.08	8.66	11.7	14.7	16.2	19.0	21.9	22.8	19.1	15.2	12.4	10.4	8.88	7.05	5.77	4.83	0		
26	0.27	0.61	1.14	2.12	3.96	5.70	7.39	9.03	12.2	15.3	16.9	19.9	22.8	24.2	20.3	16.1	13.2	11.0	9.42	7.47	6.12	5.13	0		
28	0.29	0.66	1.23	2.30	4.29	6.18	8.01	9.79	13.2	16.6	18.3	21.5	24.7	27.0	22.6	18.0	14.7	12.3	10.5	8.35	6.84	5.73	0		
30	0.31	0.71	1.33	2.49	4.62	6.66	8.63	10.5	14.3	17.9	19.7	23.2	26.6	30.0	25.1	19.9	16.3	13.7	11.7	9.26	7.58	0			
32	0.33	0.76	1.42	2.66	4.96	7.14	9.25	11.3	15.3	19.2	21.1	24.9	28.6	32.2	27.7	22.0	18.0	15.1	12.9	10.2	8.35	0			
35	0.37	0.84	1.57	2.93	5.46	7.86	10.2	12.5	16.9	21.1	23.2	27.4	31.5	35.5	31.6	25.1	20.6	17.2	14.7	11.7	9.55	0			
40	0.43	0.97	1.81	3.38	6.31	9.08	11.8	14.4	19.5	24.4	26.8	31.6	36.3	41.0	38.7	30.7	25.1	21.0	18.0	14.3	0				
45	0.48	1.10	2.06	3.84	7.16	10.3	13.4	16.3	22.1	27.7	30.5	35.9	41.3	46.5	46.1	36.6	30.0	25.1	21.4	0					

Type I — Type II — Type III

Table 5-25 Horsepower Ratings, Standard Single-Strand Roller Chain, No. 60, 3/4-in. Pitch

No. of Teeth Small Spkt.	REVOLUTIONS PER MINUTE—SMALL SPROCKET																								
	10	25	50	100	150	200	300	400	500	600	700	800	900	1000	1100	1200	1400	1600	1800	2000	2500	3000	3500	4000	4500
9	0.15	0.33	0.62	1.16	1.67	2.16	3.12	4.04	4.94	5.82	6.68	7.54	8.38	9.21	9.99	8.77	6.96	5.70	4.77	4.08	2.92	2.22	1.76	1.44	1.21
10	0.16	0.37	0.70	1.30	1.87	2.43	3.49	4.53	5.53	6.52	7.49	8.44	9.39	10.3	11.2	10.3	8.15	6.67	5.59	4.77	3.42	2.60	2.06	1.69	1.41
11	0.18	0.41	0.77	1.44	2.07	2.69	3.87	5.02	6.13	7.23	8.30	9.36	10.4	11.4	12.5	11.9	9.41	7.70	6.45	5.51	3.94	3.00	2.38	1.95	1.63
12	0.20	0.45	0.85	1.58	2.28	2.95	4.25	5.51	6.74	7.94	9.12	10.3	11.4	12.6	13.7	13.5	10.7	8.77	7.35	6.28	4.49	3.42	2.71	2.22	1.86
13	0.22	0.50	0.92	1.73	2.49	3.22	4.64	6.01	7.34	8.65	9.94	11.2	12.5	13.7	14.9	15.2	12.1	9.89	8.29	7.08	5.06	3.85	3.06	2.50	0
14	0.24	0.54	1.00	1.87	2.69	3.49	5.02	6.51	7.96	9.37	10.8	12.1	13.5	14.8	16.2	17.0	13.5	11.1	9.26	7.91	5.66	4.31	3.42	2.80	0
15	0.25	0.58	1.08	2.01	2.90	3.76	5.41	7.01	8.57	10.1	11.6	13.1	14.5	16.0	17.4	18.8	15.0	12.3	10.3	8.77	6.28	4.77	3.79	3.10	0
16	0.27	0.62	1.16	2.16	3.11	4.03	5.80	7.52	9.19	10.8	12.4	14.0	15.6	17.1	18.7	20.2	16.5	13.5	11.3	9.66	6.91	5.26	4.17	3.42	0
17	0.29	0.66	1.24	2.31	3.32	4.30	6.20	8.03	9.81	11.6	13.3	15.0	16.7	18.3	19.9	21.6	18.1	14.8	12.4	10.6	7.57	5.76	4.57	3.74	0
18	0.31	0.70	1.31	2.45	3.53	4.58	6.59	8.54	10.4	12.3	14.1	15.9	17.7	19.5	21.2	22.9	19.7	16.1	13.5	11.5	8.25	6.28	4.98	4.08	0
19	0.33	0.75	1.39	2.60	3.74	4.85	6.99	9.05	11.1	13.0	15.0	16.9	18.8	20.6	22.5	24.3	21.4	17.5	14.6	12.5	8.95	6.81	5.40	4.42	0
20	0.35	0.79	1.47	2.75	3.96	5.13	7.38	9.57	11.7	13.8	15.8	17.9	19.8	21.8	23.8	25.7	23.1	18.9	15.8	13.5	9.66	7.35	5.83	0	
21	0.36	0.83	1.55	2.90	4.17	5.40	7.78	10.1	12.3	14.5	16.7	18.8	20.9	23.0	25.1	27.1	24.8	20.3	17.0	14.5	10.4	7.91	6.28	0	
22	0.38	0.87	1.63	3.05	4.39	5.68	8.19	10.6	13.0	15.3	17.5	19.8	22.0	24.2	26.4	28.5	26.6	21.8	18.2	15.6	11.1	8.48	6.73	0	
23	0.40	0.92	1.71	3.19	4.60	5.96	8.59	11.1	13.6	16.0	18.4	20.8	23.1	25.4	27.7	29.9	28.4	23.3	19.5	16.7	11.9	9.07	7.19	0	
24	0.42	0.96	1.79	3.35	4.82	6.24	8.99	11.6	14.2	16.8	19.3	21.7	24.2	26.6	29.0	31.3	30.3	24.8	20.8	17.8	12.7	9.66	7.67	0	
25	0.44	1.00	1.87	3.50	5.04	6.52	9.40	12.2	14.9	17.5	20.1	22.7	25.3	27.8	30.3	32.7	32.2	26.4	22.1	18.9	13.5	10.3	8.15	0	
26	0.46	1.05	1.95	3.65	5.25	6.81	9.80	12.7	15.5	18.3	21.0	23.7	26.4	29.0	31.6	34.1	34.2	28.0	23.4	20.0	14.3	10.9	8.65	0	
28	0.50	1.13	2.12	3.95	5.69	7.37	10.6	13.8	16.8	19.8	22.8	25.7	28.5	31.4	34.2	37.0	38.2	31.3	26.2	22.4	16.0	12.2	0		
30	0.54	1.22	2.28	4.26	6.13	7.94	11.4	14.8	18.1	21.4	24.5	27.7	30.8	33.8	36.8	39.8	42.4	34.7	29.1	24.8	17.8	13.5	0		
32	0.57	1.31	2.45	4.56	6.57	8.52	12.3	15.9	19.4	22.9	26.3	29.7	33.0	36.3	39.5	42.7	46.7	38.2	32.0	27.3	19.6	14.9	0		
35	0.63	1.44	2.69	5.03	7.24	9.38	13.5	17.5	21.4	25.2	29.0	32.7	36.3	39.9	43.5	47.1	53.4	43.7	36.6	31.3	22.4	17.0	0		
40	0.73	1.67	3.11	5.81	8.37	10.8	15.6	20.2	24.7	29.1	33.5	37.7	42.0	46.1	50.3	54.4	62.5	53.4	44.7	38.2	27.3	0			
45	0.83	1.89	3.53	6.60	9.50	12.3	17.7	23.0	28.1	33.1	38.0	42.0	47.7	52.4	57.1	61.7	70.9	63.7	53.4	45.6	32.6	0			

Type I Type II Type III

TYPE I: Manual or Drip Lubrication
TYPE II: Bath or Disc Lubrication
TYPE III: Oil Stream Lubrication
See Chapter 12, 'Lubrication' for detailed description of lubrication types.

The limiting RPM for each lubrication type is read from the column to the left of the boundary line shown.

The Horsepower ratings of Multiple Strand Chains are greater than those for Single Strand Chain: see Table 5-5 for 'Multiple Strand Factors'.

Table 5-26 Horsepower Ratings, Standard Single-Strand Roller Chain, No. 80, 1-in. Pitch

No. of Teeth Small Spkt.	REVOLUTIONS PER MINUTE—SMALL SPROCKET																								
	10	25	50	100	150	200	300	400	500	600	700	800	900	1000	1100	1200	1400	1600	1800	2000	2200	2400	2700	3000	3400
9	0.34	0.78	1.45	2.71	3.90	5.05	7.28	9.43	11.5	13.6	15.6	17.6	17.0	14.5	12.6	11.0	8.76	7.17	6.01	5.13	4.45	3.90	3.27	2.79	2.32
10	0.38	0.87	1.63	3.03	4.37	5.66	8.16	10.6	12.9	15.2	17.5	19.7	19.9	17.0	14.7	12.9	10.3	8.40	7.04	6.01	5.21	4.57	3.83	3.27	2.71
11	0.42	0.97	1.80	3.36	4.84	6.28	9.04	11.7	14.3	16.9	19.4	21.9	23.0	19.6	17.0	14.9	11.8	9.69	8.12	6.93	6.01	5.27	4.42	3.77	1.70
12	0.47	1.06	1.98	3.69	5.32	6.89	9.93	12.9	15.7	18.5	21.3	24.0	26.2	22.3	19.4	17.0	13.5	11.0	9.25	7.90	6.85	6.01	5.04	4.30	0
13	0.51	1.16	2.16	4.03	5.80	7.52	10.8	14.0	17.1	20.2	23.2	26.2	29.1	25.2	21.8	19.2	15.2	12.5	10.4	8.91	7.72	6.78	5.68	4.85	0
14	0.55	1.25	2.34	4.36	6.29	8.14	11.7	15.2	18.6	21.9	25.1	28.4	31.5	28.2	24.4	21.4	17.0	13.9	11.7	9.96	8.63	7.57	6.35	5.42	0
15	0.59	1.35	2.52	4.70	6.77	8.77	12.6	16.4	20.0	23.6	27.1	30.6	34.0	31.2	27.1	23.8	18.9	15.4	12.9	11.0	9.57	8.40	7.04	6.01	0
16	0.63	1.45	2.70	5.04	7.26	9.41	13.5	17.6	21.5	25.3	29.0	32.8	36.4	34.4	29.8	26.2	20.8	17.0	14.2	12.2	10.5	9.25	7.76	6.62	0
17	0.68	1.55	2.88	5.38	7.75	10.0	14.5	18.7	22.9	27.0	31.0	35.0	38.9	37.7	32.7	28.7	22.7	18.6	15.6	13.3	11.5	10.1	8.49	7.25	0
18	0.72	1.64	3.07	5.72	8.25	10.7	15.4	19.9	24.4	28.7	33.0	37.2	41.4	41.1	35.5	31.2	24.8	20.3	17.0	14.5	12.6	11.0	9.25	7.90	0
19	0.76	1.74	3.25	6.07	8.74	11.3	16.3	21.1	25.8	30.4	35.0	39.4	43.8	44.5	38.6	33.9	26.9	22.0	18.4	15.7	13.6	12.0	10.0	8.57	0
20	0.81	1.84	3.44	6.41	9.24	12.0	17.2	22.3	27.3	32.2	37.0	41.7	46.3	48.1	41.7	36.6	29.0	23.8	19.9	17.0	14.7	12.9	10.8	0	
21	0.85	1.94	3.62	6.76	9.74	12.6	18.2	23.5	28.8	33.9	39.0	43.9	48.9	51.7	44.8	39.4	31.2	25.6	21.4	18.3	15.9	13.9	11.7	0	
22	0.90	2.04	3.81	7.11	10.2	13.3	19.1	24.8	30.3	35.7	41.0	46.2	51.4	55.5	48.1	42.2	33.5	27.4	23.0	19.6	17.0	14.9	12.5	0	
23	0.94	2.14	4.00	7.46	10.7	13.9	20.1	26.0	31.8	37.4	43.0	48.5	53.9	59.3	51.4	45.1	35.8	29.3	24.6	21.0	18.2	15.9	13.4	0	
24	0.98	2.24	4.19	7.81	11.3	14.6	21.0	27.2	33.2	39.2	45.0	50.8	56.4	62.0	54.8	48.1	38.2	31.2	26.2	22.3	19.4	17.0	14.2	0	
25	1.03	2.34	4.37	8.16	11.8	15.2	21.9	28.4	34.7	40.9	47.0	53.0	59.0	64.8	58.2	51.1	40.6	33.2	27.8	23.8	20.6	18.1	15.1	0	
26	1.07	2.45	4.56	8.52	12.3	15.9	22.9	29.7	36.2	42.7	49.1	55.3	61.5	67.6	61.8	54.2	43.0	35.2	29.5	25.2	21.8	19.2	16.1	0	
28	1.16	2.65	4.94	9.23	13.3	17.2	24.8	32.1	39.3	46.3	53.2	59.9	66.7	73.3	69.0	60.6	48.1	39.4	33.0	28.2	24.4	21.4	0		
30	1.25	2.85	5.33	9.94	14.3	18.5	26.7	34.6	42.3	49.9	57.3	64.6	71.8	78.9	76.6	67.2	53.3	43.6	36.6	31.2	27.1	23.8	0		
32	1.34	3.06	5.71	10.7	15.3	19.9	28.6	37.1	45.4	53.5	61.4	69.2	77.0	84.6	84.3	74.0	58.7	48.1	40.3	34.4	29.8	26.2	0		
35	1.48	3.37	6.29	11.7	16.9	21.9	31.6	40.9	50.0	58.9	67.6	76.3	84.8	93.3	96.5	84.7	67.2	55.0	46.1	39.4	34.1	0			
40	1.71	3.89	7.27	13.6	19.5	25.3	36.4	47.2	57.7	68.0	78.1	88.1	98.0	108	117	103	82.1	67.2	56.3	48.1	20.0	0			
45	1.94	4.42	8.25	15.4	22.2	28.7	41.4	53.6	65.6	77.2	88.7	100	111	122	133	123	98.0	80.2	67.2	54.1	0				

Type I Type II Type III

Table 5-27 Horsepower Ratings, Standard Single-Strand Roller Chain, No. 100, 1-1/4-in. Pitch

No. of Teeth Small Spkt.	REVOLUTIONS PER MINUTE—SMALL SPROCKET																								
	10	25	50	100	150	200	300	400	500	600	700	800	900	1000	1100	1200	1300	1400	1600	1800	2000	2200	2400	2600	2700
9	0.65	1.49	2.78	5.19	7.47	9.68	13.9	18.1	22.1	26.0	29.6	24.2	20.3	17.4	15.0	13.2	11.7	10.5	8.57	7.19	6.13	5.32	4.67	4.14	0
10	0.73	1.67	3.11	5.81	8.37	10.8	15.6	20.2	24.7	29.2	33.5	28.4	23.8	20.3	17.6	15.5	13.7	12.3	10.0	8.42	7.19	6.23	5.47	4.85	0
11	0.81	1.85	3.45	6.44	9.28	12.0	17.3	22.4	27.4	32.3	37.1	32.8	27.5	23.4	20.3	17.8	15.8	14.2	11.6	9.71	8.29	7.19	6.31	1.29	0
12	0.89	2.03	3.79	7.08	10.2	13.2	19.0	24.6	30.1	35.5	40.8	37.3	31.3	26.7	23.2	20.3	18.0	16.1	13.2	11.1	9.45	8.19	7.19	0	
13	0.97	2.22	4.13	7.72	11.1	14.4	20.7	26.9	32.8	38.7	44.5	42.1	35.3	30.1	26.1	22.9	20.3	18.2	14.9	12.5	10.6	9.23	8.10	0	
14	1.05	2.40	4.48	8.36	12.0	15.6	22.5	29.1	35.6	41.9	48.2	47.0	39.4	33.7	29.2	25.6	22.7	20.3	16.6	13.9	11.9	10.3	9.05	0	
15	1.13	2.59	4.83	9.01	13.0	16.8	24.2	31.4	38.3	45.2	51.9	52.2	43.7	37.3	32.4	28.4	25.2	22.5	18.4	15.5	13.2	11.4	10.0	0	
16	1.22	2.77	5.17	9.66	13.9	18.0	26.0	33.6	41.1	48.4	55.6	57.5	48.2	41.1	35.7	31.3	27.7	24.8	20.3	17.0	14.5	12.6	11.1	0	
17	1.30	2.96	5.52	10.3	14.8	19.2	27.7	35.9	43.9	51.7	59.4	63.0	52.8	45.0	39.0	34.3	30.4	27.2	22.3	18.7	15.9	13.8	0.79	0	
18	1.38	3.15	5.88	11.0	15.8	20.5	29.5	38.2	46.7	55.0	63.2	68.6	57.5	49.1	42.5	37.3	33.1	29.6	24.2	20.3	17.4	15.0	0		
19	1.46	3.34	6.23	11.6	16.7	21.7	31.2	40.5	49.5	58.3	67.0	74.4	62.3	53.2	46.1	40.5	35.9	32.1	26.3	22.0	18.8	16.3	0		
20	1.55	3.53	6.58	12.3	17.7	22.9	33.0	42.8	52.3	61.6	70.8	79.8	67.3	57.5	49.8	43.7	38.8	34.7	28.4	23.8	20.3	17.6	0		
21	1.63	3.72	6.94	13.0	18.7	24.2	34.8	45.1	55.1	65.0	74.6	84.2	72.4	61.8	53.6	47.0	41.7	37.3	30.6	25.6	21.9	19.0	0		
22	1.71	3.91	7.30	13.6	19.6	25.4	36.6	47.4	58.0	68.3	78.5	88.5	77.7	66.3	57.5	50.4	44.7	40.0	32.8	27.5	23.4	20.3	0		
23	1.80	4.10	7.66	14.3	20.6	26.7	38.4	49.8	60.8	71.7	82.3	92.8	83.0	70.9	61.4	53.9	47.8	42.8	35.0	29.4	25.1	7.74	0		
24	1.88	4.30	8.02	15.0	21.5	27.9	40.2	52.1	63.7	75.0	86.2	97.2	88.5	75.6	65.5	57.5	51.0	45.6	37.3	31.3	26.7	0			
25	1.97	4.49	8.38	15.6	22.5	29.2	42.0	54.4	66.6	78.4	90.1	102	94.1	80.3	69.6	61.1	54.2	48.5	39.7	33.3	28.4	0			
26	2.05	4.68	8.74	16.3	23.5	30.4	43.8	56.8	69.4	81.8	94.0	106	99.8	85.2	73.8	64.8	57.5	51.4	42.1	35.3	30.1	0			
28	2.22	5.07	9.47	17.7	25.5	33.0	47.5	61.5	75.2	88.6	102	115	112	95.2	82.5	72.4	64.2	57.5	47.0	39.4	33.7	0			
30	2.40	5.47	10.2	19.0	27.4	35.5	51.2	66.3	81.0	95.5	110	124	124	106	91.5	80.3	71.2	63.7	52.2	43.7	10.0	0			
32	2.57	5.86	10.9	20.4	29.4	38.1	54.9	71.1	86.9	102	118	133	136	116	101	88.5	78.5	70.2	57.5	48.2	0				
35	2.83	6.46	12.0	22.5	32.4	42.0	60.4	78.3	95.7	113	130	146	156	133	115	101	89.8	80.3	65.8	55.1	0				
40	3.27	7.46	13.9	26.0	37.4	48.5	69.8	90.4	111	130	150	169	188	163	141	124	110	98.1	80.3	0					
45	3.71	8.47	15.8	29.5	42.5	55.0	79.3	103	126	148	170	192	213	194	168	148	131	117	45.3	0					

Type I Type II Type III

TYPE I: Manual or Drip Lubrication
TYPE II: Bath or Disc Lubrication
TYPE III: Oil Stream Lubrication
See Chapter 12, 'Lubrication' for detailed description of lubrication types.

The limiting RPM for each lubrication type is read from the column to the left of the boundary line shown. For optimum results, it is recommended that the Roller Chain manufacturer be given the opportunity to evaluate the conditions of operation of chains in the shaded (galling range) speed area. The Horsepower ratings of Multiple Strand Chains are greater than those for Single Strand Chain: see Table 5-5 for 'Multiple Strand Factors'.

Table 5-28 Horsepower Ratings, Standard Single-Strand Roller Chain, No. 120, 1-1/2 in. Pitch

No. of Teeth Small Spkt.	REVOLUTIONS PER MINUTE—SMALL SPROCKET																								
	10	25	50	100	150	200	300	400	500	600	700	800	900	1000	1100	1200	1300	1400	1500	1600	1700	1800	1900	2000	2100
9	1.10	2.52	4.69	8.76	12.6	16.3	23.5	30.5	37.3	43.2	34.3	28.1	23.5	20.1	17.4	15.3	13.5	12.1	10.9	9.92	9.06	8.31	7.67	7.10	6.60
10	1.24	2.82	5.26	9.81	14.1	18.3	26.4	34.2	41.8	49.2	40.1	32.9	27.5	23.5	20.4	17.9	15.9	14.2	12.8	11.6	10.6	9.74	8.98	8.31	7.73
11	1.37	3.12	5.83	10.9	15.7	20.3	29.2	37.9	46.3	54.6	46.3	37.9	31.8	27.1	23.5	20.6	18.3	16.4	14.8	13.4	12.2	11.2	10.4	9.59	0
12	1.50	3.43	6.40	11.9	17.2	22.3	32.1	41.6	50.9	59.9	52.8	43.2	36.2	30.9	26.8	23.5	20.9	18.7	16.8	15.3	13.9	12.8	11.8	10.9	0
13	1.64	3.74	6.98	13.0	18.8	24.3	35.0	45.4	55.5	65.3	59.5	48.7	40.8	34.9	30.2	26.5	23.5	21.0	19.0	17.2	15.7	14.4	13.3	12.3	0
14	1.78	4.05	7.56	14.1	20.3	26.3	37.9	49.1	60.1	70.8	66.5	54.4	45.6	39.0	33.8	29.6	26.3	23.5	21.2	19.2	17.6	16.1	14.9	8.94	0
15	1.91	4.37	8.15	15.2	21.9	28.4	40.9	53.0	64.7	76.3	73.8	60.4	50.6	43.2	37.4	32.9	29.1	26.1	23.5	21.3	19.5	17.0	16.5	0	
16	2.05	4.68	8.74	16.3	23.5	30.4	43.8	56.8	69.4	81.8	81.3	66.5	55.7	47.6	41.2	36.2	32.1	28.7	25.9	23.5	21.5	19.7	18.2	0	
17	2.19	5.00	9.33	17.4	25.1	32.5	46.8	60.6	74.1	87.3	89.0	72.8	61.0	52.1	45.2	39.6	35.2	31.5	28.4	25.8	23.5	21.6	19.9	0	
18	2.33	5.32	9.92	18.5	26.7	34.6	49.8	64.5	78.8	92.9	97.0	79.4	66.5	56.8	49.2	43.2	38.3	34.3	30.9	28.1	25.6	23.5	11.3	0	
19	2.47	5.64	10.5	19.6	28.3	36.6	52.8	68.4	83.6	98.5	105	86.1	72.1	61.6	53.4	46.8	41.5	37.2	33.5	30.4	27.8	25.5	0		
20	2.61	5.96	11.1	20.7	29.9	38.7	55.8	72.2	88.3	104	114	92.9	77.9	66.5	57.6	50.6	44.9	40.1	36.2	32.9	30.0	27.5	0		
21	2.75	6.28	11.7	21.9	31.5	40.8	58.8	76.2	93.1	110	122	100	83.8	71.6	62.0	54.4	48.3	43.2	39.0	35.4	32.3	29.6	0		
22	2.90	6.60	12.3	23.0	33.1	42.9	61.8	80.1	97.9	115	131	107	89.9	76.7	66.5	58.4	51.8	46.3	41.8	37.9	34.6	16.6	0		
23	3.04	6.93	12.9	24.1	34.8	45.0	64.9	84.0	103	121	139	115	96.1	82.0	71.1	62.4	55.3	49.5	44.6	40.5	37.0	0			
24	3.18	7.25	13.5	25.3	36.4	47.1	67.9	88.0	108	127	146	122	102	87.4	75.8	66.5	59.0	52.8	47.6	43.2	39.4	0			
25	3.32	7.58	14.1	26.4	38.0	49.3	71.0	91.9	112	132	152	130	109	92.9	80.6	70.7	62.7	56.1	50.6	45.9	41.3	0			
26	3.47	7.91	14.8	27.5	39.7	51.4	74.0	95.9	117	138	159	138	115	98.6	85.4	75.0	66.5	59.5	53.7	48.7	26.6	0			
28	3.76	8.57	16.0	29.8	43.0	55.7	80.2	104	127	150	172	154	129	110	95.5	83.8	74.3	66.5	60.0	54.4	0				
30	4.05	9.23	17.2	32.1	46.3	60.0	86.4	112	137	161	185	171	143	122	106	92.9	82.4	73.8	66.5	42.4	0				
32	4.34	9.90	18.5	34.5	49.6	64.3	92.6	120	147	173	199	188	158	135	117	102	90.8	81.3	73.3	0					
35	4.78	10.9	20.3	38.0	54.7	70.9	102	132	162	190	219	215	180	154	133	117	104	92.9	47.7	0					
40	5.52	12.6	23.5	43.9	63.2	81.8	118	153	187	220	253	263	220	188	163	143	127	59.5	0						
45	6.27	14.3	26.7	49.8	71.7	92.9	134	173	212	250	287	314	263	224	195	171	80.0	0							

Type I Type II Type III

Table 5-29 Horsepower Ratings, Standard Single-Strand Roller Chain, No. 140, 1-3/4-in. Pitch

NO. OF TEETH SMALL SPKT.	REVOLUTIONS PER MINUTE—SMALL SPROCKET 10	25	50	100	150	200	250	300	350	400	450	500	550	600	700	800	900	1000	1100	1200	1300	1400	1500	1600	1700
9	1.71	3.89	7.26	13.6	19.5	25.3	30.9	36.4	41.8	47.2	52.5	57.7	55.7	48.9	38.8	31.7	26.6	22.7	19.7	17.3	15.3	13.7	12.4	11.2	10.2
10	1.91	4.36	8.14	15.2	21.9	28.3	34.6	40.8	46.9	52.9	58.8	64.6	65.2	57.2	45.4	37.2	31.2	26.6	23.1	20.2	17.9	16.1	14.5	13.1	0
11	2.12	4.83	9.02	16.8	24.2	31.4	38.4	45.2	52.0	58.6	65.2	71.6	75.2	66.0	52.4	42.9	35.9	30.7	26.6	23.3	20.7	18.5	16.7	15.2	0
12	2.33	5.31	9.91	18.5	26.6	34.5	42.2	49.7	57.1	64.4	71.6	78.7	85.7	75.2	59.7	48.9	41.0	35.0	30.3	26.6	23.6	21.1	19.0	17.3	0
13	2.54	5.79	10.8	20.2	29.0	37.6	46.0	54.2	62.2	70.2	78.0	85.8	93.5	84.8	67.3	55.1	46.2	39.4	34.2	30.0	26.5	23.8	21.5	19.5	0
14	2.75	6.27	11.7	21.8	31.5	40.8	49.8	58.7	67.4	76.0	84.5	93.0	101	94.8	75.2	61.6	51.6	44.1	38.2	33.5	29.7	26.6	24.0	21.8	0
15	2.96	6.76	12.6	23.5	33.9	43.9	53.7	63.2	72.7	81.9	91.1	100	109	105	83.4	68.3	57.2	48.9	42.4	37.2	33.0	29.5	26.6	0	
16	3.18	7.24	13.5	25.2	36.3	47.1	57.5	67.8	77.9	87.8	97.7	107	117	116	91.9	75.2	63.1	53.8	46.7	41.0	36.3	32.5	29.3	0	
17	3.39	7.73	14.4	26.9	38.8	50.3	61.4	72.4	83.2	93.8	104	115	125	127	101	82.4	69.1	59.0	51.1	44.9	39.8	35.6	32.1	0	
18	3.61	8.23	15.4	28.6	41.3	53.5	65.3	77.0	88.5	99.8	111	122	133	138	110	89.8	75.2	64.2	55.7	48.9	43.3	38.8	35.0	0	
19	3.82	8.72	16.3	30.4	43.7	56.7	60.3	81.6	93.8	106	118	129	141	150	119	97.4	81.6	69.7	60.4	53.0	47.0	42.1	37.9	0	
20	4.04	9.22	17.2	32.1	46.2	59.9	73.2	86.3	99.1	112	124	137	149	161	128	105	88.1	75.2	65.2	57.2	50.8	45.4	0		
21	4.26	9.72	18.1	33.8	48.7	63.1	77.2	91.0	104	118	131	144	157	170	138	113	94.8	80.9	70.2	61.6	54.6	48.9	0		
22	4.48	10.2	19.1	35.6	51.3	66.4	81.2	95.6	110	124	138	151	165	178	148	121	102	86.8	75.2	66.0	58.6	52.4	0		
23	4.70	10.7	20.0	37.3	53.8	69.7	85.2	100	115	130	145	159	173	187	158	130	109	92.8	80.4	70.6	62.6	56.0	0		
24	4.92	11.2	20.9	39.1	56.3	72.9	89.2	105	121	136	151	166	181	196	169	138	116	98.9	85.7	75.2	66.7	59.7	0		
25	5.14	11.7	21.9	40.8	58.8	76.2	93.2	110	126	142	158	174	189	205	180	147	123	105	91.1	80.0	70.9	63.5	0		
26	5.37	12.2	22.8	42.6	61.4	79.5	97.2	115	132	148	165	181	198	214	190	156	131	112	96.7	84.8	75.2	0			
28	5.81	13.3	24.7	46.2	66.5	86.2	105	124	143	161	179	197	214	232	213	174	146	125	108	94.8	84.1	0			
30	6.26	14.3	26.7	49.7	71.6	92.8	113	134	154	173	193	212	231	249	236	193	162	138	120	105	93.2	0			
32	6.71	15.3	28.6	53.3	76.8	99.5	122	143	165	186	206	227	247	267	260	213	178	152	132	116	0				
35	7.40	16.9	31.5	58.7	84.6	110	134	158	181	205	227	250	272	295	297	243	204	174	151	130	0				
40	8.54	19.5	36.4	67.9	97.7	127	155	182	210	236	263	289	315	340	363	297	249	213	178	0					
45	9.70	22.1	41.3	77.1	111	144	176	207	238	268	298	328	357	387	434	355	297	237	92.7	0					

Type I Type II Type III

TYPE I: Manual or Drip Lubrication
TYPE II: Bath or Disc Lubrication
TYPE III: Oil Stream Lubrication
See Chapter 12, 'Lubrication' for detailed description of lubrication types.

The limiting RPM for each lubrication type is read from the column to the left of the boundary line shown.
For optimum results, it is recommended that the Roller Chain manufacturer be given the opportunity to evaluate the conditions of operation of chains in the shaded (galling range) speed area.
The Horsepower ratings of Multiple Strand Chains are greater than those for Single Strand Chain: see Table 5-5 for 'Multiple Strand Factors'

Table 5-30 Horsepower Ratings, Standard Single-Strand Roller Chain, No. 160, 2-in. Pitch

NO. OF TEETH SMALL SPKT.	REVOLUTIONS PER MINUTE—SMALL SPROCKET 10	25	50	100	150	200	260	300	350	400	450	500	550	600	650	700	750	800	850	900	1000	1100	1200	1300	1400
9	2.48	5.65	10.5	19.7	28.3	36.7	44.8	52.8	60.7	68.5	76.1	71.5	62.0	54.4	48.2	43.2	38.9	35.3	32.2	29.6	25.3	21.9	19.2	17.0	0
10	2.77	6.33	11.8	22.0	31.7	41.1	50.3	59.2	68.0	76.7	85.3	83.7	72.6	63.7	56.5	50.5	45.6	41.4	37.8	34.7	29.6	25.7	22.5	20.0	0
11	3.07	7.01	13.1	24.4	35.2	45.6	55.7	65.6	75.4	85.0	94.5	96.6	83.7	73.5	65.2	58.3	52.6	47.7	43.6	40.0	34.1	29.6	26.0	23.0	0
12	3.38	7.70	14.4	26.8	38.6	50.1	61.2	72.1	82.8	93.4	104	110	95.4	83.7	74.2	66.4	59.9	54.4	49.6	45.6	38.9	33.7	29.6	26.3	0
13	3.68	8.40	15.7	29.2	42.1	54.6	66.7	78.6	90.3	102	113	124	108	94.4	83.7	74.9	67.5	61.3	56.0	51.4	43.9	38.0	33.4	29.6	0
14	3.99	9.10	17.0	31.7	45.6	59.1	72.3	85.2	97.8	110	123	135	120	105	93.6	83.7	75.5	68.5	62.6	57.4	49.0	42.5	37.3	33.1	0
15	4.30	9.80	18.3	34.1	49.2	63.7	77.9	91.7	105	119	132	145	133	117	104	92.8	83.7	76.0	69.4	63.7	54.4	47.1	41.4	0	
16	4.61	10.5	19.6	36.6	52.7	68.3	83.5	98.4	113	127	142	156	147	129	114	102	92.2	83.7	76.4	70.2	59.9	51.9	45.6	0	
17	4.92	11.2	20.9	39.1	56.3	72.9	89.1	105	121	136	151	166	161	141	125	112	101	91.7	83.7	75.8	65.6	56.9	49.9	0	
18	5.23	11.9	22.3	41.6	59.9	77.6	94.8	112	128	145	161	177	175	154	136	122	110	99.9	91.2	83.7	71.5	62.0	54.4	0	
19	5.55	12.7	23.6	44.1	63.5	82.2	101	118	136	153	171	188	190	167	148	132	119	108	98.9	90.8	77.5	67.2	59.0	0	
20	5.86	13.4	25.0	46.6	67.1	86.9	106	125	144	162	180	198	205	180	160	143	129	117	107	98.1	83.7	72.6	63.7	0	
21	6.18	14.1	26.3	49.1	70.7	91.6	112	132	152	171	190	209	221	194	172	154	139	126	115	105	90.1	78.1	68.5	0	
22	6.50	14.8	27.7	51.6	74.4	96.3	118	139	159	180	200	220	237	208	184	165	149	135	123	113	96.6	83.7	0		
23	6.82	15.6	29.0	54.2	78.0	101	124	146	167	189	210	231	251	222	197	176	159	144	132	121	103	98.5	0		
24	7.14	16.3	30.4	56.7	81.7	106	129	152	175	197	220	241	263	237	210	188	169	154	140	129	110	95.4	0		
25	7.46	17.0	31.8	59.3	85.4	111	135	159	183	206	229	252	275	252	223	200	180	164	149	137	117	101	0		
26	7.78	17.8	33.1	61.8	89.1	115	141	166	191	215	239	263	287	267	237	212	191	173	158	145	124	108	0		
28	8.43	19.2	35.9	67.0	96.5	125	153	180	207	233	259	285	311	298	265	237	214	194	177	162	139	120	0		
30	9.08	20.7	38.7	72.2	104	135	165	194	223	251	279	307	335	331	293	263	237	215	196	180	154	0			
32	9.74	22.2	41.5	77.4	111	144	176	208	239	269	300	329	359	365	323	289	261	237	216	198	169	0			
35	10.7	24.5	45.7	85.2	123	159	194	229	263	297	330	363	395	417	370	331	298	271	247	227	180	0			
40	12.4	28.3	52.8	98.5	142	184	225	265	304	343	381	419	457	494	452	404	365	331	302	257	0				
45	14.1	32.1	59.9	112	161	209	255	301	345	389	433	476	519	561	538	482	418	348	271	189	0				

Type I Type II Type III

Table 5-31 Horsepower Ratings, Standard Single-Strand Roller Chain, No. 180, 2-1/4-in. Pitch

NO. OF TEETH SMALL SPKT.	10	25	50	100	150	200	250	300	350	400	450	500	550	600	650	700	750	800	850	900	950	1000	1050	1100	1150
	REVOLUTIONS PER MINUTE—SMALL SPROCKET																								
9	3.42	7.80	14.5	27.1	39.1	50.7	61.9	73.0	83.8	94.5	92.0	78.5	68.1	59.7	53.0	47.4	42.8	38.8	35.4	32.5	30.0	27.8	25.8	24.1	0
10	3.83	8.74	16.3	30.4	43.8	56.8	69.4	81.8	93.9	106	108	92.0	79.7	70.0	62.1	55.5	50.1	45.5	41.5	38.1	35.1	32.5	30.2	28.2	0
11	4.24	9.68	18.1	33.7	48.6	62.9	76.9	90.6	104	117	124	106	92.0	80.7	71.6	64.1	57.8	52.4	47.9	43.9	40.5	37.5	34.9	32.5	0
12	4.66	10.6	19.8	37.0	53.4	69.1	84.5	99.6	114	129	142	121	105	92.0	81.6	73.0	65.8	59.7	54.6	50.1	46.2	42.8	39.7	37.1	0
13	5.08	11.6	21.6	40.4	58.2	75.4	92.1	109	125	141	156	136	118	104	92.0	82.3	74.2	67.4	61.5	55.5	52.1	48.2	44.8	0	
14	5.51	12.6	23.4	43.7	63.0	81.6	99.8	118	135	152	169	152	132	116	103	92.0	82.9	75.3	68.7	63.1	58.2	53.9	50.1	0	
15	5.93	13.5	25.3	47.1	67.9	88.0	108	127	146	164	182	169	146	129	114	102	92.0	83.5	76.2	70.0	64.5	59.7	55.5	0	
16	6.36	14.5	27.1	50.5	72.8	94.3	115	136	156	176	196	186	161	142	126	112	101	92.0	84.0	77.1	71.1	65.8	61.2	0	
17	6.79	15.5	28.9	54.0	77.7	101	123	145	167	188	209	204	177	155	138	123	111	101	92.0	84.4	77.9	72.1	0		
18	7.22	16.5	30.8	57.4	82.7	107	131	154	177	200	222	222	193	169	150	134	121	110	100	92.0	84.8	78.5	0		
19	7.66	17.5	32.6	60.8	87.6	114	139	164	188	212	236	241	209	183	163	145	131	119	109	99.8	92.0	85.2	0		
20	8.10	18.5	34.5	64.3	92.6	120	147	173	199	224	249	260	226	198	175	157	142	129	117	108	99.3	92.0	0		
21	8.53	19.5	36.3	67.8	97.6	126	155	182	209	236	262	280	243	213	189	169	152	138	126	116	107	99.0	0		
22	8.97	20.5	38.2	71.3	103	133	163	192	220	248	276	300	260	228	203	181	163	148	135	124	115	0			
23	9.41	21.5	40.1	74.8	108	140	171	201	231	260	290	318	278	244	216	194	175	159	145	133	123	0			
24	9.86	22.5	42.0	78.3	113	146	179	210	242	273	303	333	296	260	231	206	186	169	154	142	131	0			
25	10.3	23.5	43.9	81.8	118	153	187	220	253	285	317	348	315	277	245	220	198	180	164	151	139	0			
26	10.7	24.5	45.7	85.4	123	159	195	229	264	297	331	363	334	293	260	233	210	191	174	160	0				
28	11.6	26.6	49.6	92.5	133	173	211	249	286	322	358	394	374	328	291	260	235	213	194	178	0				
30	12.5	28.6	53.4	99.6	144	186	227	268	308	347	386	424	414	364	322	289	260	236	216	198	0				
32	13.4	30.7	57.2	107	154	199	244	287	330	372	414	455	456	401	355	318	287	260	238	0					
35	14.8	33.8	63.1	118	170	220	268	316	363	410	456	501	522	458	406	364	328	291	220	0					
40	17.1	39.0	72.9	136	196	254	310	365	420	473	526	579	575	524	465	398	324	244	0						
45	19.4	44.3	82.7	154	222	288	352	415	477	538	598	631	578	514	441	360	271	0							

Type I Type II Type III

TYPE I: Manual or Drip Lubrication
TYPE II: Bath or Disc Lubrication
TYPE III: Oil Stream Lubrication
See Chapter 12, 'Lubrication' for detailed description of lubrication types.

The limiting RPM for each lubrication type is read from the column to the left of the boundary line shown.
For optimum results, it is recommended that the Roller Chain manufacturer be given the opportunity to evaluate the conditions of operation of chains in the shaded (galling range) speed area.
The Horsepower ratings of Multiple Strand Chains are greater than those for Single Strand Chain: see Table 5-5 for 'Multiple Strand Factors'.

Table 5-32 Horsepower Ratings, Standard Single-Strand Roller Chain, No. 200, 2-1/2 in. Pitch

No. of Teeth Small Spkt.	10	15	20	30	40	50	70	100	150	200	250	300	350	400	450	500	550	600	650	700
	REVOLUTIONS PER MINUTE—SMALL SPROCKET																			
9	4.54	6.54	8.47	12.2	15.8	19.3	26.1	36.0	51.9	67.3	82.2	96.9	111	119	100	85.4	74.1	65.0	57.6	0
10	5.08	7.32	9.49	13.7	17.7	21.6	29.3	40.4	58.2	75.4	92.1	109	125	140	117	100	86.7	76.1	67.5	0
11	5.64	8.12	10.5	15.1	19.6	24.0	32.5	44.8	64.5	83.5	102	120	138	156	135	115	100	87.8	77.9	0
12	6.19	8.92	11.6	16.6	21.6	26.4	35.7	49.2	70.8	91.8	112	132	152	171	154	132	114	100	0	
13	6.75	9.72	12.6	18.1	23.5	28.7	38.9	53.6	77.2	100	122	144	166	187	174	148	129	113	0	
14	7.31	10.5	13.6	19.7	25.5	31.1	42.1	58.1	83.7	108	132	156	179	202	194	166	144	126	0	
15	7.88	11.3	14.7	21.2	27.4	33.5	45.4	62.6	90.1	117	143	168	193	218	215	184	159	140	0	
16	8.45	12.2	15.8	22.7	29.4	36.0	48.7	67.1	96.6	125	153	180	207	234	237	203	176	154	0	
17	9.02	13.0	16.8	24.2	31.4	38.4	52.0	71.6	103	134	163	193	221	249	260	222	192	169	0	
18	9.59	13.8	17.9	25.8	33.4	40.8	55.3	76.2	110	142	174	205	235	265	283	242	209	184	0	
19	10.2	14.6	19.0	27.3	35.4	43.3	58.6	80.8	116	151	184	217	249	281	307	262	227	199	0	
20	10.7	15.5	20.1	28.9	37.4	45.8	61.9	85.4	123	159	195	229	264	297	331	283	245	0		
21	11.3	16.3	21.1	30.5	39.5	48.2	65.3	90.0	130	168	205	242	278	313	348	305	264	0		
22	11.9	17.2	22.2	32.0	41.5	50.7	68.7	94.6	136	177	216	254	292	330	366	327	283	0		
23	12.5	18.0	23.3	33.6	43.5	53.2	72.0	99.3	143	185	226	267	307	346	384	349	303	0		
24	13.1	18.9	24.4	35.2	45.6	55.7	75.4	104	150	194	237	279	321	362	402	372	323	0		
25	13.7	19.7	25.5	36.8	47.6	58.2	78.8	109	156	203	248	292	335	378	421	396	343	0		
26	14.3	20.6	26.6	38.4	49.7	60.7	82.2	113	163	212	259	305	350	395	439	420	364	0		

Type I Type II Type III

Table 5-33 Horsepower Ratings, Standard Single-Strand Roller Chain, No. 240, 3-in. Pitch

No. of Teeth Small Spkt.	REVOLUTIONS PER MINUTE—SMALL SPROCKET																			
	5	10	15	20	25	30	40	50	60	80	100	125	150	175	200	250	300	350	400	450	500
9	3.92	7.31	10.5	13.6	16.7	19.6	25.4	31.1	36.7	47.5	58.1	71.0	83.6	96.1	108	132	156	169	138	116	0
10	4.39	8.19	11.8	15.3	18.7	22.0	28.5	34.9	41.1	53.2	65.0	79.5	93.7	108	121	148	175	198	162	136	0
11	4.86	9.08	13.1	16.9	20.7	24.4	31.6	38.6	45.5	59.0	72.1	88.1	104	119	135	164	194	223	187	156	0
12	5.34	9.97	14.4	18.6	22.7	26.8	34.7	42.4	50.0	64.8	79.2	96.8	114	131	148	181	213	245	218	0	
13	5.83	10.9	15.7	20.3	24.8	29.2	37.9	46.3	54.5	70.6	86.4	106	124	143	161	197	232	267	240	0	
14	6.31	11.8	17.0	22.0	26.9	31.7	41.0	50.1	59.1	76.5	93.6	114	135	155	175	213	251	289	268	0	
15	6.80	12.7	18.3	23.7	28.9	34.1	44.2	54.0	63.6	82.4	101	123	145	167	188	230	271	311	297	0	
16	7.29	13.6	19.6	25.4	31.0	36.6	47.4	57.9	68.2	88.4	108	132	156	179	202	247	290	334	328	0	
17	7.78	14.5	20.9	27.1	33.1	39.0	50.6	61.8	72.9	94.4	115	141	166	191	215	263	310	356	359	0	
18	8.28	15.4	22.3	28.8	35.2	41.5	53.8	65.8	77.5	100	123	150	177	203	229	280	330	379	377	0	
19	8.78	16.4	23.6	30.6	37.4	44.0	57.0	69.7	82.2	106	130	159	187	215	243	297	350	402	393	0	
20	9.28	17.3	24.9	32.3	39.5	46.5	60.3	73.7	86.8	112	138	168	198	228	257	314	370	423	407	0	
21	9.78	18.2	26.3	34.1	41.6	49.0	63.5	77.7	91.5	119	145	177	209	240	270	331	390	439	421	0	
22	10.3	19.2	27.6	35.8	43.8	51.6	66.8	81.7	96.2	125	152	186	220	252	284	348	410	454	435	0	
23	10.8	20.1	29.0	37.6	45.9	54.1	70.1	85.7	101	131	160	195	230	265	298	365	430	469	448	0	
24	11.3	21.1	30.4	39.3	48.1	56.7	73.4	89.7	106	137	167	205	241	277	312	382	450	483	0		
25	11.8	22.0	31.7	41.1	50.3	59.2	76.7	93.8	110	143	175	214	252	290	327	399	470	496	0		
26	12.3	23.0	33.1	42.9	52.4	61.8	80.0	97.8	115	149	183	223	263	302	341	416	491	509	0		

Type I — Type II — Type III

TYPE I: Manual or Drip Lubrication
TYPE II: Bath or Disc Lubrication
TYPE III: Oil Stream Lubrication
See Chapter 12 for detailed descriptions of lubrication types.

The limiting RPM for each lubrication type is read from the column to the left of the boundary line shown.

The Horsepower ratings of Multiple Strand Chains are greater than those for Single Strand Chain: see Table 5-5 for 'Multiple Strand Factors'.

6

Engineering Steel Drive Chains

Engineering Steel drive chains are frequently the most economical and efficient means of transmitting power mechanically. Chain can bridge large spaces with a single strand, whereas gearing may be used efficiently only in much smaller spaces. Where drive shafts are parallel, a single strand of chain may replace several sets of gears with great economic advantage. Chains are less sensitive to foreign material and will operate better in rugged environments.

ANSI Standard B29.10—1978 (Heavy Duty Offset Sidebar Power Transmission Roller Chains and Sprocket Teeth, UDS No. 621.855 and 621.851) lists and describes the sizes of Engineering Steel drive chains which are standard in the industry, although it limits the selection to eight of the more popular sizes. Other sizes, even though not listed in the standard, are widely used in power transmission applications and are readily available from chain manufacturers.

As was noted in Chapter 2, Engineering Steel drive chains are usually of the offset sidebar design. This is true of all chains listed in ANSI B29.10. Most chains used in drive applications have rollers. Typical Engineering Steel drive chains are shown in Figures 6-1 and 6-2.

Generally speaking, the heavier the load and the larger the drive chain, the slower

Figure 6-1 Engineering Steel, heavy duty, offset drive chain.

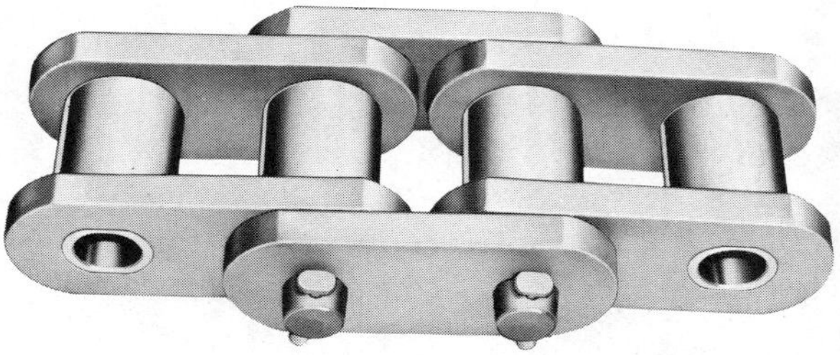

Figure 6-2 Engineering Steel, heavy duty, straight link drive chain.

the speed of operation. However, applications exist today where very heavy chains are being operated successfully at speeds deemed unlikely only a few years ago. These factors are reflected in the horsepower curves applicable to heavy duty offset power transmission chains.

Applications

Selection of the proper chain is made on the basis of the desired operating characteristics of the drive under consideration: shaft speed and reduction, available space, constant or fluctuating load, desired speeds and possible variations, environment, anticipated daily hours of operation, and life expectancy. These factors and many others are considered in selecting chains.

Selection of heavy duty offset drive chains was based on working loads until horsepower curves were developed by the American Chain Association. These working loads, as given in manufacturers' catalogs, were necessarily conservative since they were intended for general usage. Although service and speed factors were furnished for computing design working loads, this method of chain selection was not considered to be sufficiently precise by those responsible for the design of machinery using such chains.

After a great deal of testing by ACA and its member companies, horsepower curves were developed giving more definite chain selections. These curves are easier to use and more accurate than the working load/speed factor combination. However, service factors, which include consideration of shock, lubrication methods, environment and hours of service, must still be applied. Furthermore, in selecting heavy duty Engineering Steel drive chains, it is obviously desirable that the operating speeds fall into the ranges of speeds covered by the curves.

Probably the major application fields for drive chains of the Engineering Steel heavy duty type are on construction equipment and on stationary drives of conveyors and other processing equipment. There are also many other usages, particularly where centers are far apart and operating conditions are rugged.

Among the construction equipment applications, power shovels usually use chain throughout the power trains from engine to propel, and also to the crowd mechanism. Figures 6-3 and 6-4 show typical applications for offset sidebar drive chain. Shovel and crane crawler drives, and transit-truck mixer drives, are typical applications of drive chains in the construction equipment field.

Most chain drives simply consist of a pair of sprockets with shaft centers a short

Figure 6-3 Application of offset sidebar chain on the crowd drive of a power shovel.

Figure 6-4 Application of offset sidebar on the propel drive of the crawlers of a crane.

distance apart. However, the nature of chain makes possible some very ingenious drive arrangements. Usually, the major problem to be solved in arranging a chain drive is the maintenance of the proper amount of slack-side tension. If there is insufficient slack-side tension, the chain rollers may tend to jump the sprocket teeth; too much tension and the chain life may be lessened. Consideration of this problem is particularly important in drives with short shaft centers or vertical arrangement.

Direction of Travel

Offset Sidebar Chain

Standard Engineering Steel drive chains are of the offset type, which is to say, they are not symmetric relative to direction of travel.

The direction of travel of drive chains can have considerable influence on the wear life of the chains when the sprockets are not of equal size. The elongation resulting from pin and bushing wear is an important factor of drive chain life. This wear occurs as the chain articulates in passing over the sprockets. In this situation, the greater the difference in sprocket size, the more wear-life advantage can be secured by using offset sidebar chain traveling in the correct direction. Figure 6-5 shows the basic idea of travel direction. The general rule for direction of chain travel for offset drive chains is as follows. The narrow or roller end of the link in the tight-side strand should always face the smaller sprocket, regardless of whether this is a driver or a driven wheel.

The usual drive is a step-down arrangement—that is, the drive sprocket is smaller and turns faster than the driven sprocket—but there are many step-up drive arrangements in existence also. Therefore, a rule of thumb based on driver or driven sprockets to determine direction of travel of chain cannot be used.

When the narrow end of the offset link faces the sprocket (whether entering or

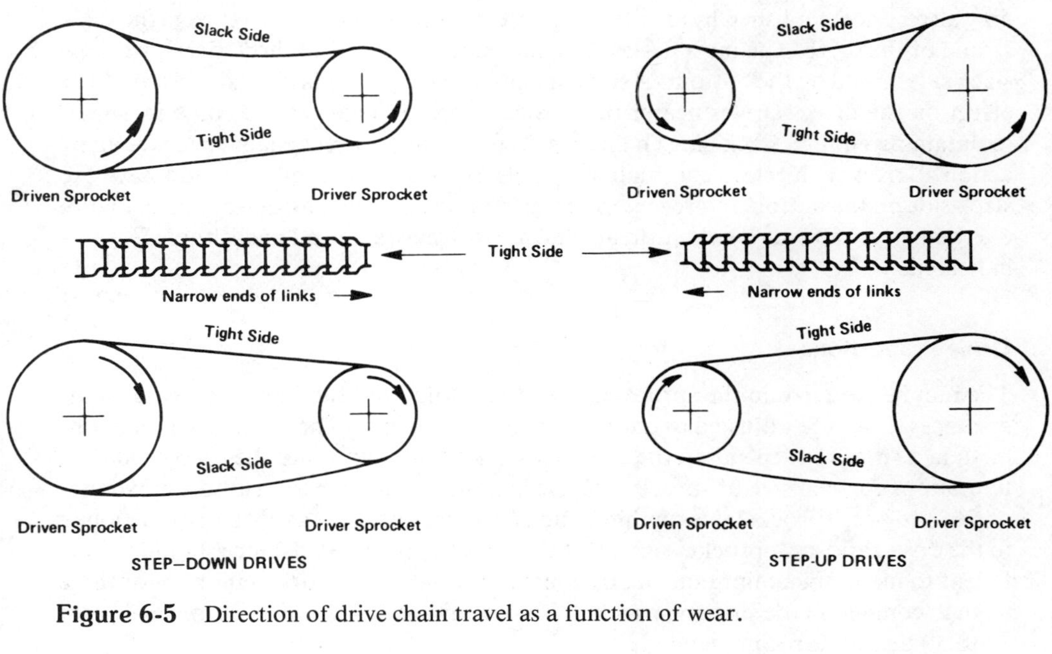

Figure 6-5 Direction of drive chain travel as a function of wear.

leaving), the sliding that produces wear takes place primarily between the roller and bushing. Articulation between pin and bushing bore also occurs, but the load quickly decreases between these parts as articulation starts and the full load between the sprocket tooth and the tight-side strand is transferred from roller to bushing.

When the wide or pin end of the link faces the sprocket tooth (entering or leaving), the articulation is entirely between the pin and the bushing bore. Wear between the pin and the bushing bore causes the chain to elongate in pitch. This elongation is probably the major factor for chain replacement. Thus the direction of travel that gives the least amount of wear between pin and bushing bore should be the correct direction for the chain to travel. Therefore, with other than a one-to-one drive ratio, the narrow or roller end of the link in the tight-side strand should face the smaller sprocket because of the greater angle of articulation.

Straight Sidebar Chain

When straight sidebar chains are used as drive chains, it will usually be noted that wear patterns will be different for trailing and leading ends of the roller links. The reason for this is exactly the same as has been noted between running offset sidebar chain with roller end forward as opposed to open end forward. In straight sidebar chain, one end of the block link operates as if it were an offset link operated open end forward; the other end acts as if it were an offset link with roller end forward.

Drive Arrangements

The arrangement or relative position of the driver and driven sprockets and the tight- and slack-side strands can have a great deal of influence on the life of a drive. It is recommended that means be provided for adjusting slack in a chain, preferably by having one shaft adjustable so that shaft center-to-center distance may be varied. This may also be accomplished by an idler sprocket on a moveable shaft to engage the slack strand of the drive. The most desirable arrangement of a drive is horizontal, with the slack side strand on the bottom. A small inclination from the horizontal will have little effect on the drive. Drives nearer the vertical than 45° may require some means of maintaining enough slack-side chain tension to provide correct chain-sprocket interaction. Drives with relatively small sprockets or long shaft centers should have the slack side on the bottom to prevent the chain strands from hitting one another. Figure 6-6 shows a group of drive chain arrangements with desirable and undesirable features indicated.

Drive Proportions

The number of teeth in the sprockets, the drive ratio, and the center-to-center shaft distances should be adjusted to one another for best results. The minimum number of teeth in a sprocket depend on the chain clearance limitations and the speed and load limitations for that size of sprocket. Drive ratios are usually best held to a maximum of 6 to 1 to keep sprocket sizes within bounds. Center distance should be proportioned to the drive ratio and sprocket sizes. If the ratio is large, center distance should be sufficient to maintain a minimum arc of contact of at least 135° or a minimum of three teeth in contact on the smaller sprocket. For normal design, a center distance of from 30 to 50 pitches is recommended.

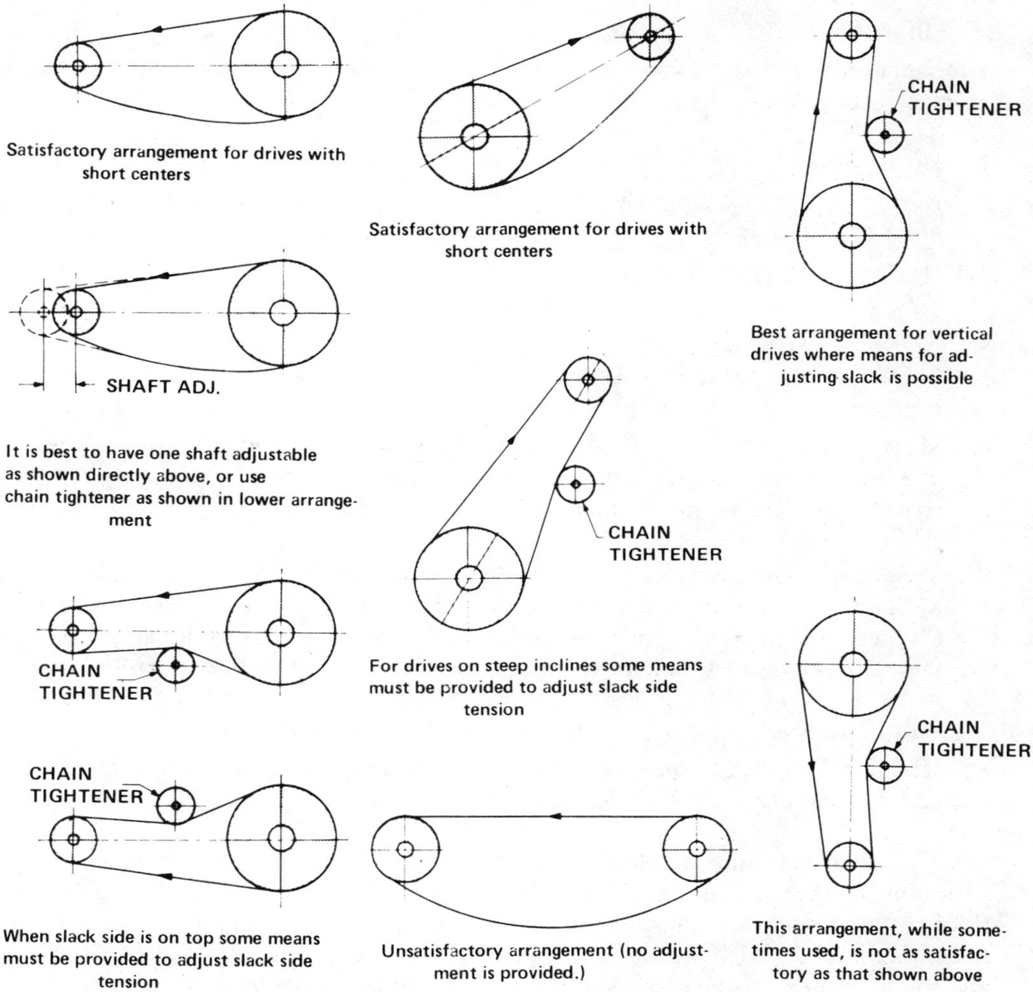

Figure 6-6 Drive chain arrangements.

Chain Selection

In any application, certain specifications of the drive will be established and others will be more or less at the designer's discretion. For example, the horsepower to be transmitted and the desired rotative speed of the driven sprocket may be known. The shaft size of the driven sprocket may be established by the known dimensions of the equipment and certain dimensional limitations. Examples of these are: maximum size of the driven sprocket, minimum center-to-center distance of the drive, and the width of the chain and sprocket hub lengths. These factors are established or at least limited by the dimensions and arrangement of other equipment. It is usually necessary that all remaining optional specifications be manipulated by the designer to the best advantage; on these will depend the efficiency and ultimate economics of the drive.

Data Required for Selection

The data needed for selection of the drive equipment—chain, sprockets, shafting and bearings, and so on—is included in the following list:

1. Horsepower to be transmitted.
2. Revolutions per minute of the driven shaft. If variable, then maximum and minimum of the range.
3. Desired speed ratio or revolutions per minute of the drive shaft. If variable, maximum and range.
4. Source and type of power.
5. Available space for the drive. For smoothest operation with the least pulsation, use the shortest practicable pitch consistent with sprocket diameter requirements.
6. Shafting diameters. This is usually a function of torques and types of loading, and must be determined in conjunction with sprocket sizes or after these are selected. The sizes of the smaller sprocket and fastest turning shaft should be selected first, if possible.
7. Type of equipment to be driven and operating environment. Also load patterns, if available, such as starting characteristics, whether intermittent or constant operation, whether fluctuating or constant loads and speeds, and possible shock loadings. Consideration should be given to conditions such as possible corrosion, elevated temperature, ease and method of lubrication, presence of abrasive materials, rigidity and accuracy of alignment of the drive, and accessibility.
8. Desired life of the installation, that is, hours of running time per week or per day and whether drive will be used seasonably.

The actual mechanics of selecting a drive sometimes involves some trial and error, depending on the method used. This, in turn, depends somewhat on which specifications are given and which must be developed. In any case, the first requirement is the determination of service factors. These factors are to be applied to the loads, speed, and so on, desired for the drive and thus provide for greater or lesser severity of the operating conditions of the application. Table 6-7 gives some typical applications and their load classifications.

The fact that Table 6-7 is similar to Table 5-2 in the chapter on Roller Chain drives is not exactly a coincidence. The similarities, duplications, and also the differences simply reflect the fact that heavy-duty Engineering Steel drive chains may be considered an extension of the Roller Chain drive field. Usually, Engineering Steel drive chains will carry heavier loads but at lower speeds than Roller Chain. The former are often operated in less friendly surroundings than the latter relative to dirt, moisture, or other foreign materials.

Service factors are listed in Tables 6-8 to 6-10, and will be found similar to those factors given for Engineering Steel Chains used for conveyors and bucket elevators. Applications of these service factors are somewhat more complex than those given in Chapter 5 for Roller Chain drives, for the reasons discussed in the preceding paragraph.

Chain selection may be based on either of two methods. The preferred method is

Table 6-7 Load Classifications

SMOOTH LOAD	MODERATE SHOCK LOAD	HEAVY SHOCK LOAD
Agitators – Pure liquid	Clay working machinery – Pug mills	Clay working machinery – Brick press, briquetting machinery
Conveyors – Uniformly loaded or fed (apron, assembly, belt, flight, oven, screw)	Conveyors – Heavy duty and NOT uniformly loaded (apron, assembly, belt, bucket, flight, oven, screw)	Conveyors – Reciprocating and shaker
Fans – Centrifugal and light, small diameter	Cranes and hoists – Medium duty, skip hoists (travel motion and trolley motion)	Cranes and hoists – Heavy duty, including logging, lumbering, and rotary drilling rigs
Line shafts – Light	Dredges – Cable, reel and conveyor	Dredges – Cutter head drives, jig drives, screen drives
Machines – All types with uniform non-reversing loads	Food industry – Beet slicers, dough mixers, meat grinders	Hammer mills
Screens – Rotary (uniformly fed), traveling water intake	Grinders	Machine tools – Punch press, shears, plate planers
Sewage disposal equipment – Inside service (uniformly fed)	Laundry industry – Washers, tumblers	Machines – All types with severe impact shock loads or speed variations, and reversing service
	Line shafts – Heavy service	Metal mills – Draw bench, forming machines, slitters, small rolling mill drives, wire drawing or flattening
	Machine tools – Main and auxiliary drives	Mills – Rotary type (ball, cement kilns, rod mills, tumbling mills)
	Machine – All types with moderate shock and non-reversing loads	Paper industry – Mixers, calendars, rubber mill sheeters
	Screens – Rotary (stone or gravel)	Textile industry – Carding machinery
	Textile industry – Calenders, dyeing machinery, mangles, nappers, soapers, spinners, tenter frames	

to use horsepower ratings when available. The other method is based on rated working loads. To determine the design working load of the chain in either case, service factors must be applied: in the former, to power to be delivered, and in the other, to the working load required. Usually, the torque and speed requirements to the input shaft are known and the service factor may be determined. These factors produce a horsepower figure. From this a drive unit or at least the speed of the drive shaft or the speed ratio may be determined. Power to be delivered, service conditions, and speeds of the drive and driven shafts are usually known. When using the horsepower ratings, the first thing to do is to calculate the horsepower. The process may then follow on a step-by-step basis.

Step 1. Determine the combined service factor from the different service conditions listed in Tables 6-8 to 6-10. The individual factors are combined by multiplying them together.

$$SF = (SF_1)(SF_2)(SF_3) \tag{6-1}$$

Step 2. Calculate the design horsepower by applying the combined service factor to the required horsepower of the system.

Table 6-8 Service Factors for Chain Loading, SF_1

TYPE OF DRIVEN LOAD	TYPE OF INPUT POWER		
	Internal Combustion Engine with Hydraulic Drive	Electric Motor or Turbine	Internal Combustion Engine with Mechanical Drive
SMOOTH	1.0	1.0	1.2
MODERATE SHOCK	1.2	1.3	1.4
HEAVY SHOCK	1.4	1.5	1.7

Table 6-9 Service Factors for Environment, SF_2

ATMOSPHERIC CONDITIONS	Relatively clean and moderate temperature	1.0
	Moderately dirty and moderate temperature	1.2
	Exposed to weather, very dirty, abrasive, midly corrosive and reasonably high temperatures	1.4

Table 6-10 Service Factors for Operating Time, SF_3

DAILY OPERATING RANGE	8–10 Hours	1.0
	10–24 Hours	1.4

$$\text{Chain design HP} = \text{HP} \times \text{SF} \tag{6-2}$$

Step 3. With design HP from step 2 and rpm of the smaller sprocket, use the quick selector chart (Fig. 6-11) to determine which specific chains to consider. This chart covers the eight ANSI B29.10 standard offset drive chains ranging from 2½ to 7 in. in pitch. These standard sizes as well as additional drive chain sizes are available from chain manufacturers. The design horsepower and small sprocket speed values are traced on the chart. The point of intersection determines the chain or chains to be considered.

Step 4. After the preliminary selection of the chain or chains to be considered, the specific horsepower ratings, either the curves (Figs. 6-12 to 6-19) or the tabulated form (Tables 6-20 to 6-27), should be consulted. The selection data and ratings that appear in this book do not cover the many unusual or special application conditions that can be encountered, such as:

Higher-than-listed speeds
Inadequate lubrication
Pulsating or shock loads
Abrasive or corrosive conditions
More than two sprockets or idlers required
Any other unusual application conditions
Sprockets other than precision cut, such as flame-cut or cast tooth

When the conditions listed above exist, or lack of experience makes a satisfactory selection difficult, a chain manufacturer should be consulted for the proper drive selection.

The horsepower and speed values are intersected to give the correct point for sprocket tooth selections. The value for the curve or table listing that corresponds to the number of sprocket teeth should equal or exceed the calculated design horsepower for the system. Both the tables and curves are included herein so that the user may exercise his or her preference. Results will be the same regardless of whether curves or tables are used.

Step 5. The tables and curves are set up so that the type of lubrication needed can be readily determined. The HP-RPM intersection will fall within one of the designated lubrication areas. All the factors of the application will have to be considered when determining the type of lubrication to use. It should be noted, however, that in the horsepower rating tables, for continuous operation in the stippled area, galling of the live bearing surfaces of the chain joints may be expected even though lubrication is as recommended.

Step 6. The number of teeth in the driven sprocket is determined by calculating the drive ratio. This is done by dividing the higher speed by the lower. Multiply the number of teeth of the drive sprocket by the calculated ratio to obtain the number of teeth in the driven sprocket. Since a sprocket cannot have a fractional tooth, good judgment will have to be used to change the calculated number of teeth either up or down. Data concerning sprockets for Engineering Steel drive chains are given in Chapter 4. For information on shafting selection, see Figures 4-50 to 4-54.

Step 7. The system or application will usually dictate the distance between sprock-

(Text continued on page 177)

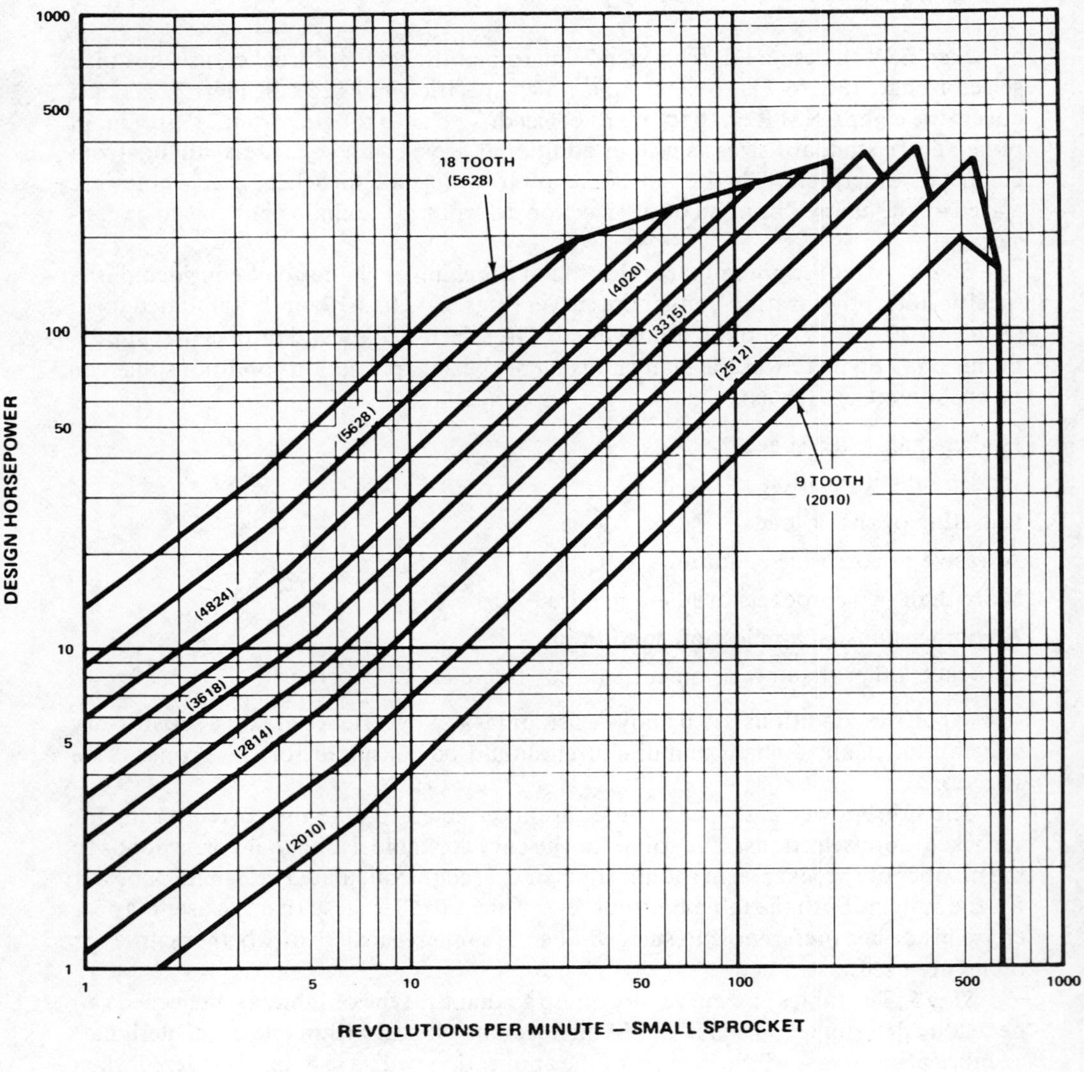

NOTES:
Lower line is a 9-tooth 2010. Top line is for an 18-tooth 5628. Intermediate lines are approximate midpoints for sprocket-tooth range shown on charts. Where the HORSEPOWER-RPM intersection takes place near a line, both chains on each side should be checked on the charts.

The table to the right shows the corresponding pitch and pin diameter for each chain. The first two, or left-hand digits in the chain designation denote the number 1/8" in the pitch. The last two, or right-hand digits, denote the number of 1/16" in the pin diameter.

ANSI B29.10 STANDARD CHAINS		
CHAIN NO.	PITCH	PIN DIA.
2010	2.5"	0.625"
2512	3.067"	0.750"
2814	3.500"	0.875"
3315	4.073"	0.938"
3618	4.500"	1.100"
4020	5.000"	1.250"
4824	6.000"	1.500"
5628	7.000"	1.750"

Figure 6-11 Quick selector chart for Engineering Steel drive chains.

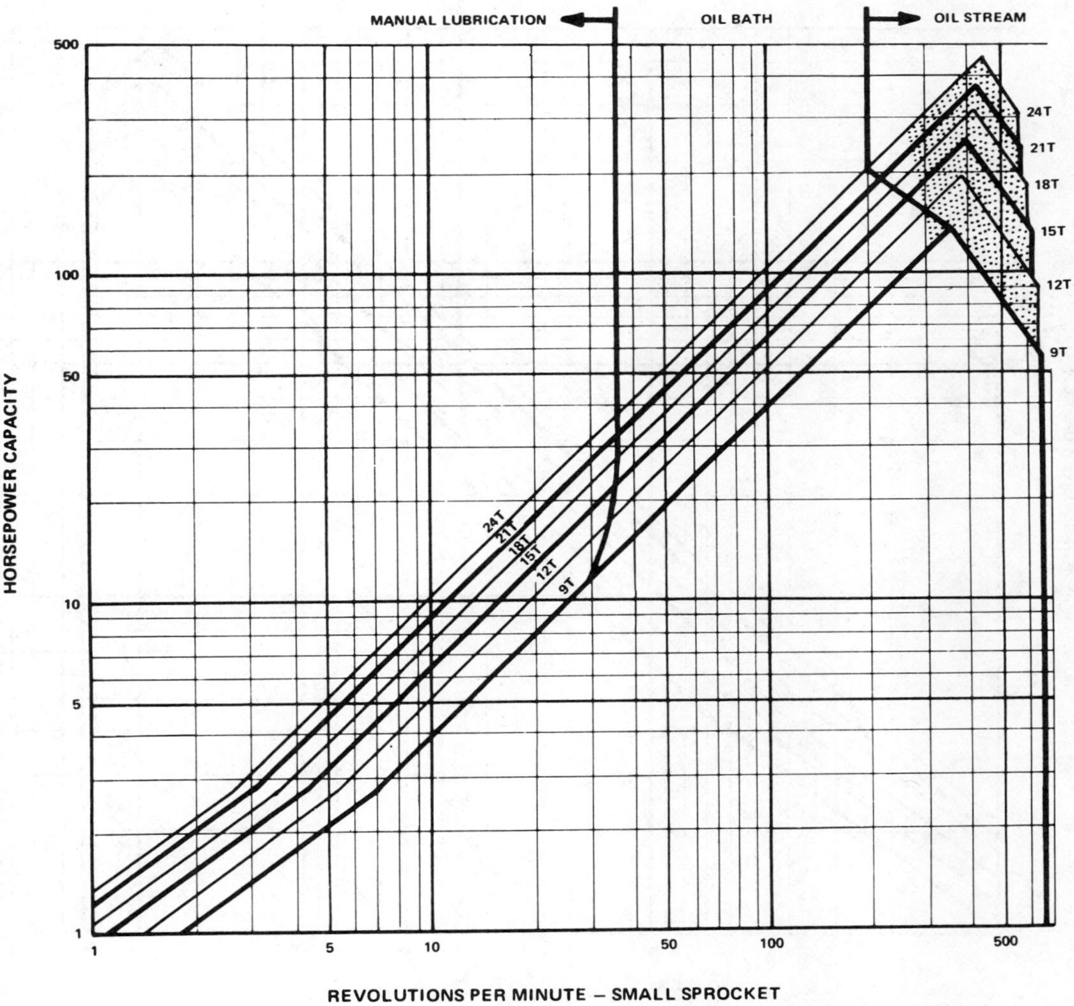

For continuous operation in the stippled area, some galling of the live bearing surfaces of the chain joints may be expected even though lubrication is as recommended.

The ratings shown on these charts are based on chain which operates over machine-cut tooth sprockets.

Figure 6-12 Horsepower rating curves for No. 2010, 2.500-in. pitch heavy duty offset power transmission chain.

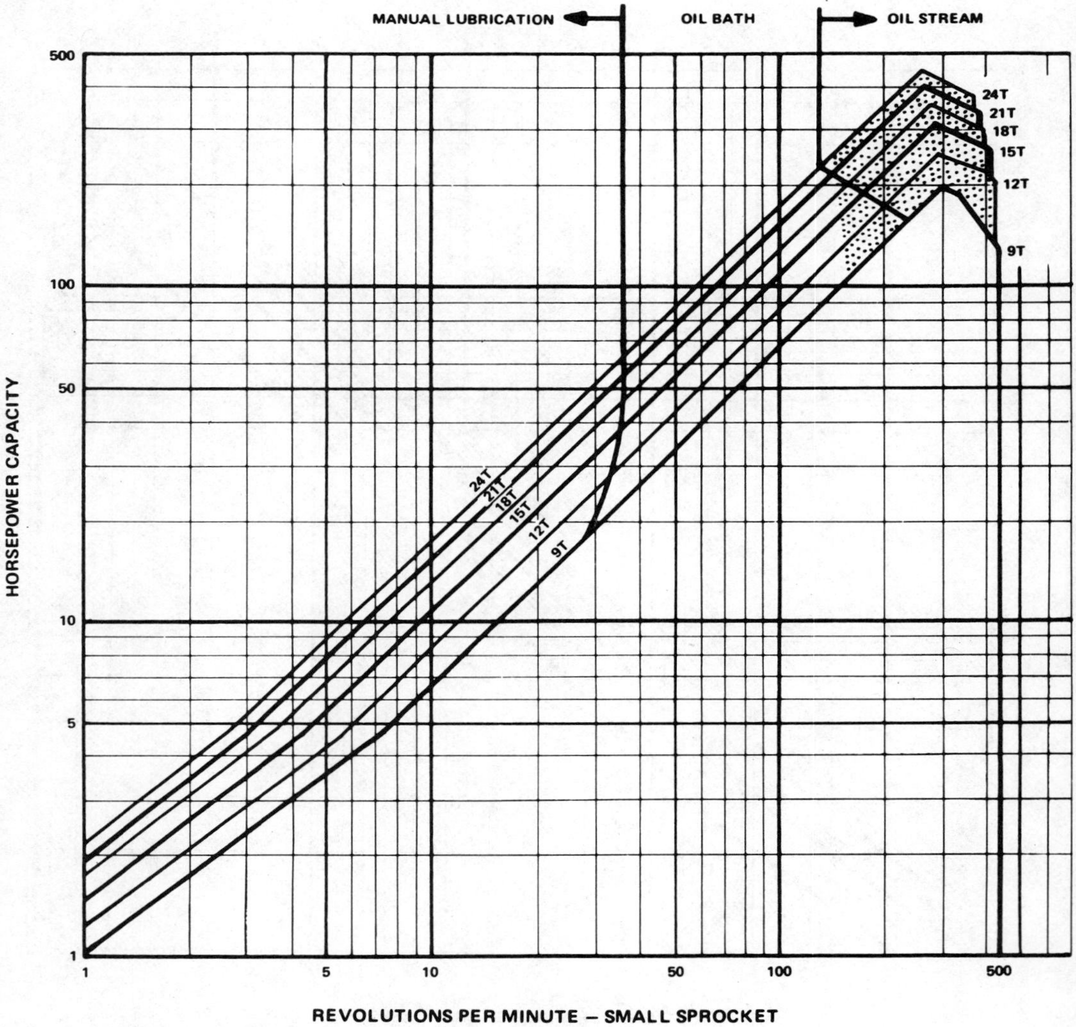

For continuous operation in the stippled area, some galling of the live bearing surfaces of the chain joints may be expected even though lubrication is as recommended.

The ratings shown on these charts are based on chain which operates over machine-cut tooth sprockets.

Figure 6-13 Horsepower rating curves for No. 2512, 3.067-in.-pitch heavy duty offset side-bar power transmission chain.

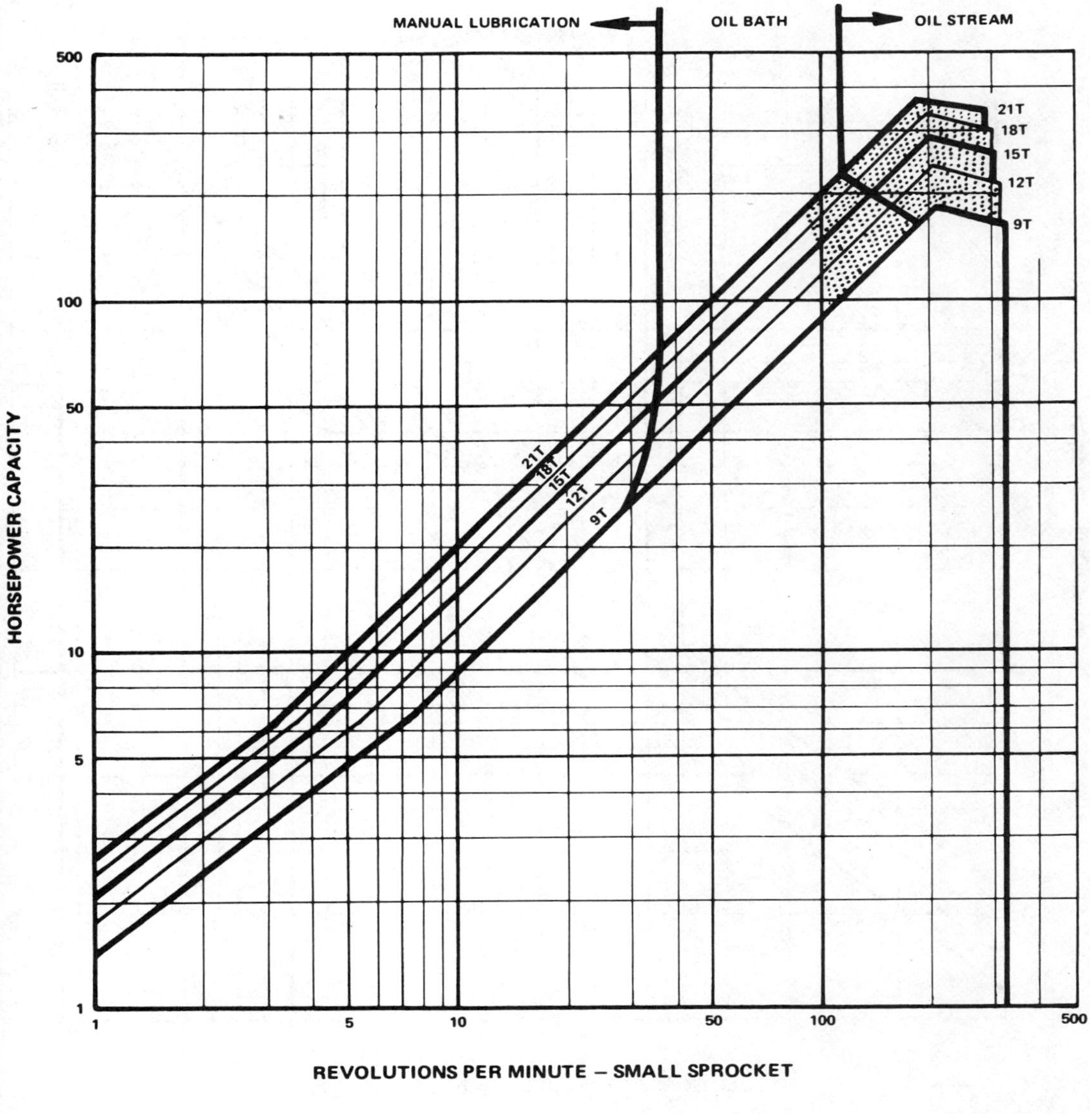

For continuous operation in the stippled area, some galling of the live bearing surfaces of the chain joints may be expected even though lubrication is as recommended.

The ratings shown on these charts are based on chain which operates over machine-cut tooth sprockets.

Figure 6-14 Horsepower rating curves for No. 2814, 3.500-in.-pitch heavy duty offset sidebar power transmission chain.

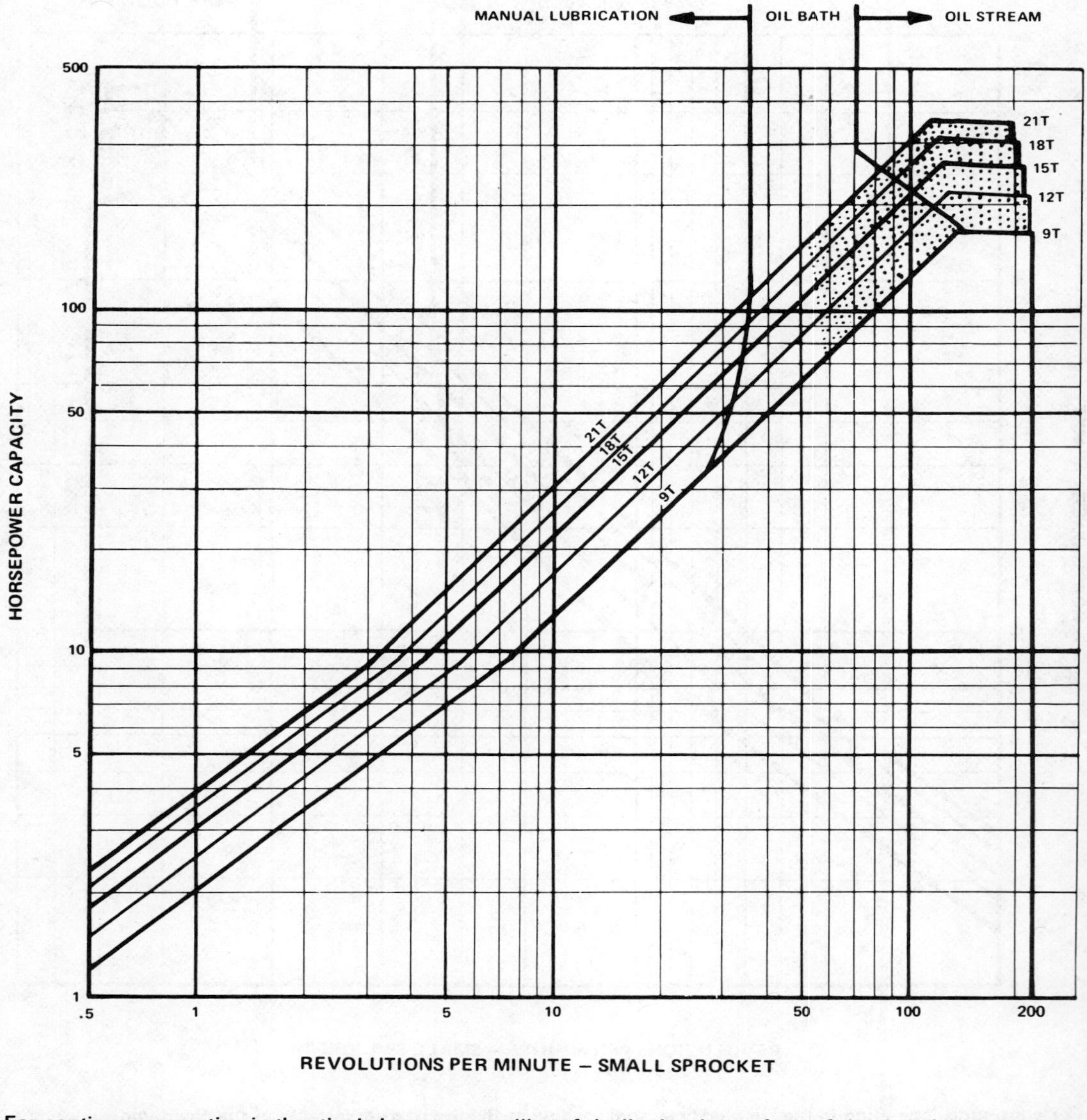

For continuous operation in the stippled area, some galling of the live bearing surfaces of the chain joints may be expected even though lubrication is as recommended.

The ratings shown on these charts are based on chain which operates over machine-cut tooth sprockets.

Figure 6-15 Horsepower rating curves for No. 3315, 4.073-in.-pitch heavy duty offset side-bar power transmission chain.

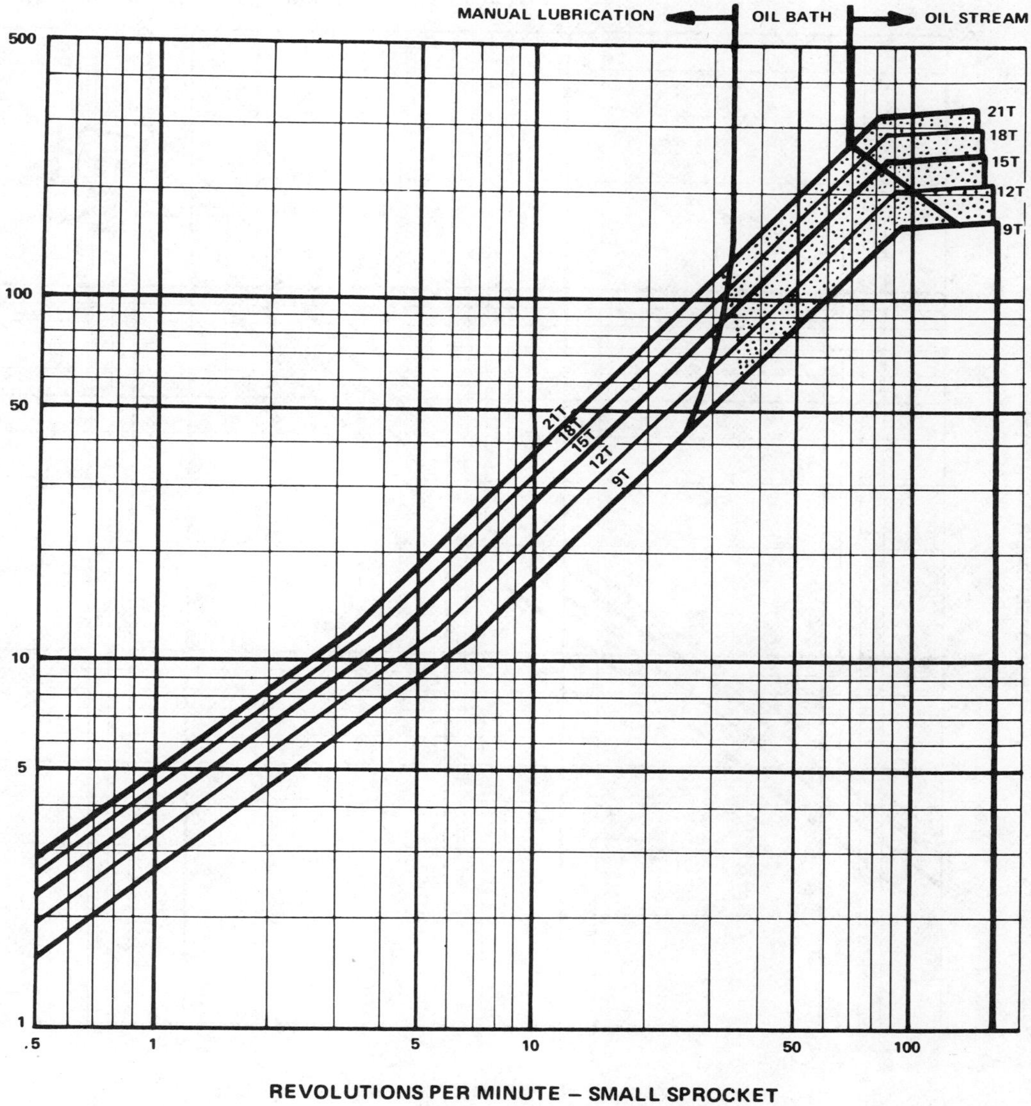

For continuous operation in the stippled area, some galling of the live bearing surfaces of the chain joints may be expected even though lubrication is as recommended.

The ratings shown on these charts are based on chain which operates over machine-cut tooth sprockets.

Figure 6-16 Horsepower rating curves for No. 3618, 4.500-in.-pitch heavy duty offset side-bar power transmission chain.

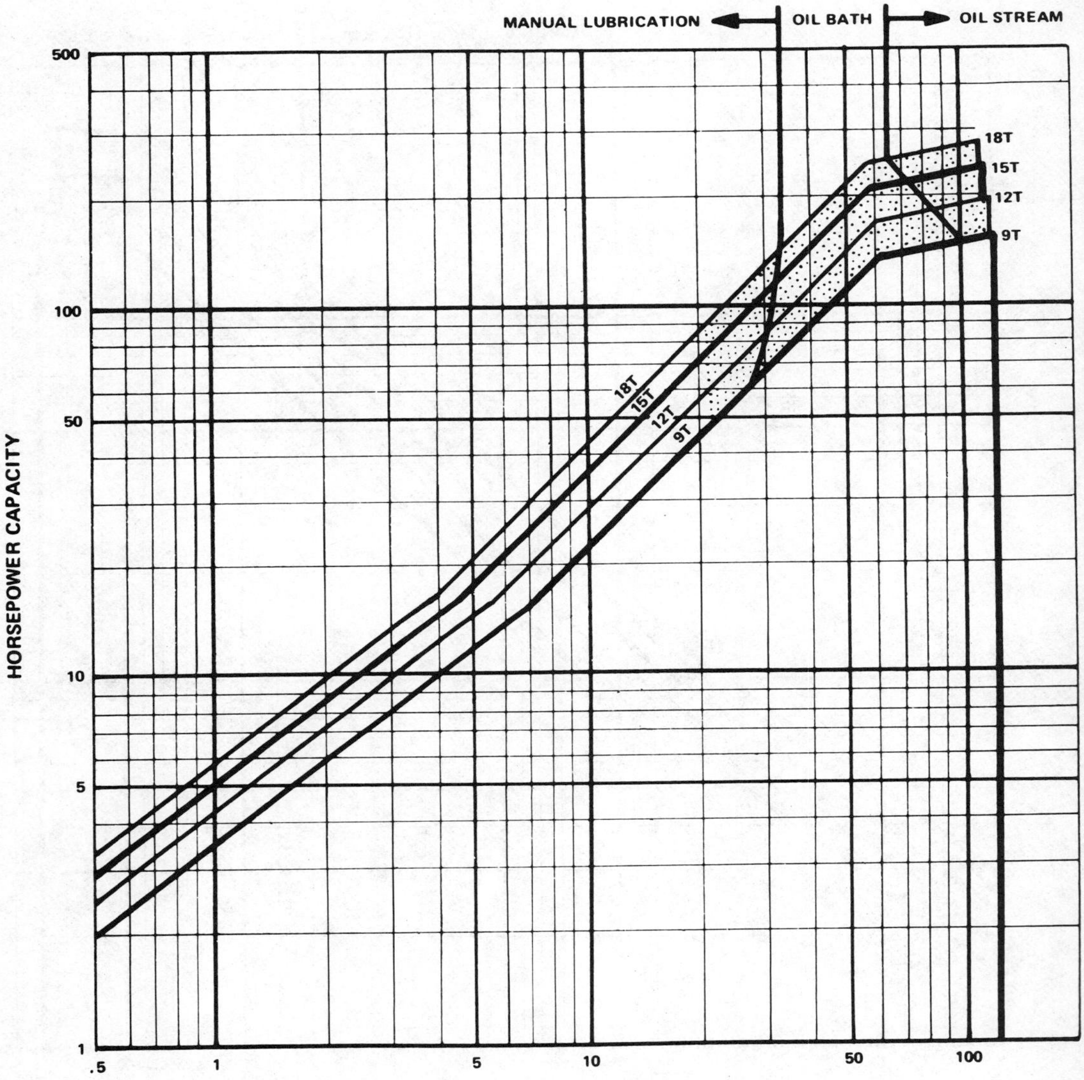

For continuous operation in the stippled area, some galling of the live bearing surfaces of the chain joints may be expected even though lubrication is as recommended.

The ratings shown on these charts are based on chain which operates over machine-cut tooth sprockets.

Figure 6-17 Horsepower rating curves for No. 4020, 5.000-in.-pitch heavy duty offset side-bar power transmission chain.

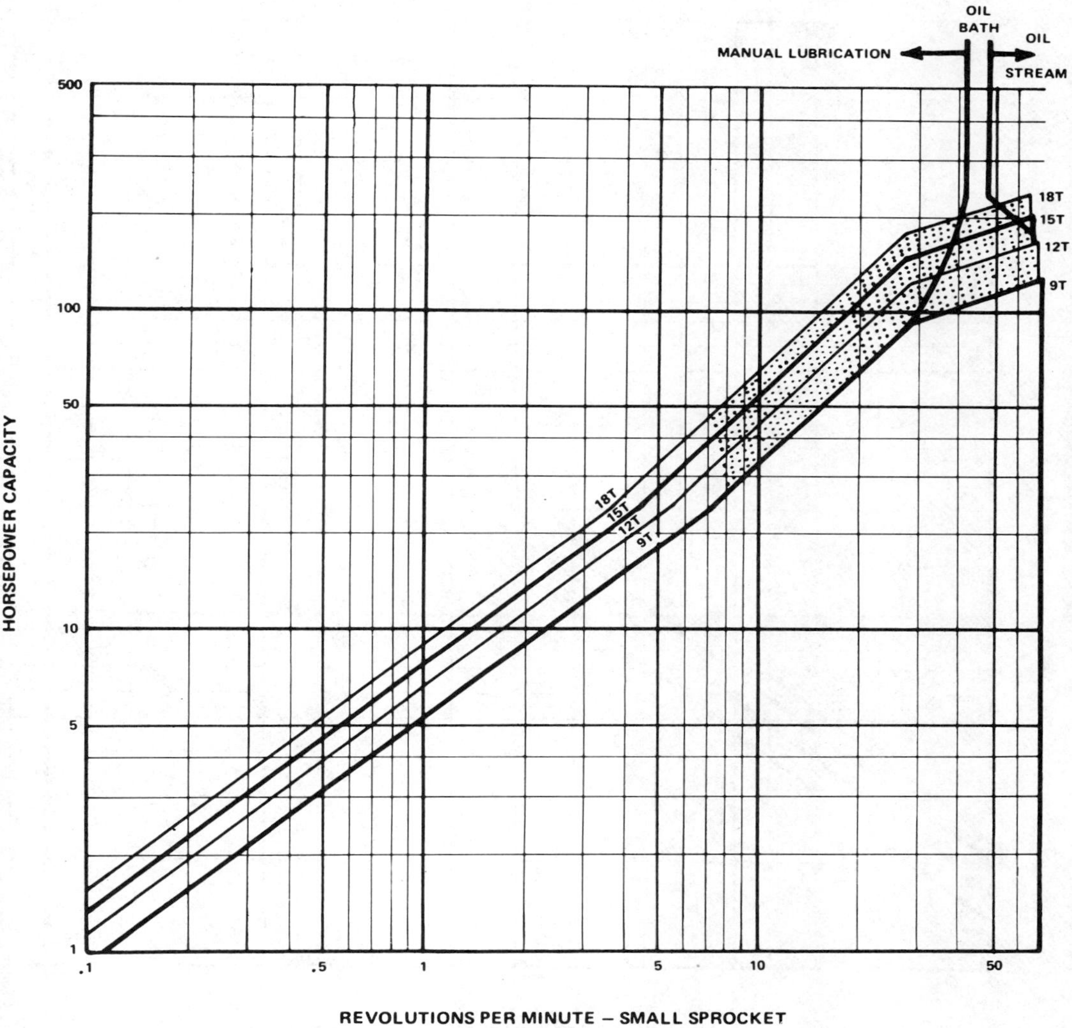

For continuous operation in the stippled area, some galling of the live bearing surfaces of the chain joints may be expected even though lubrication is as recommended.

The ratings shown on these charts are based on chain which operates over machine-cut tooth sprockets.

Figure 6.18 Horsepower rating curves for No. 4824, 6.000-in.-pitch heavy duty offset side-bar power transmission chain.

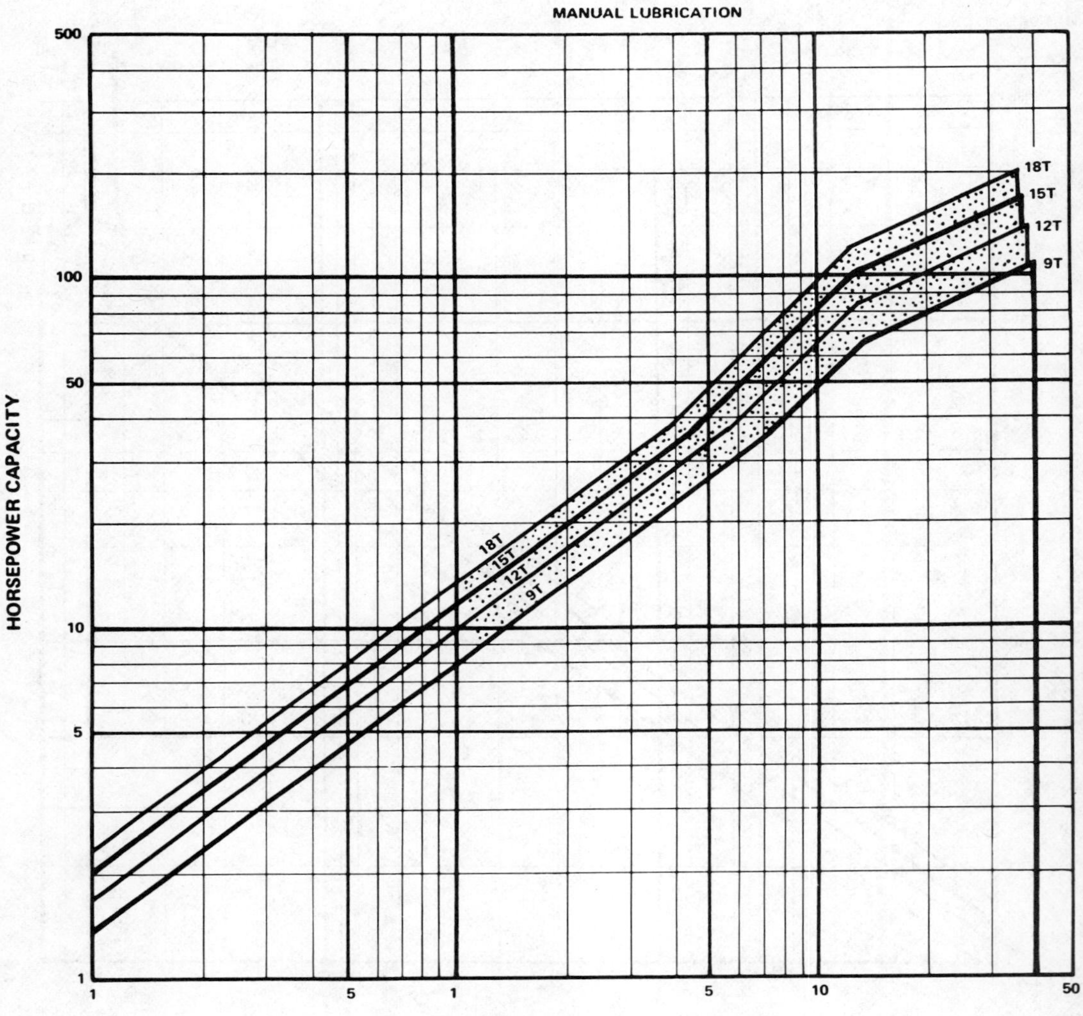

For continuous operation in the stippled area, some galling of the live bearing surfaces of the chain joints may be expected even though lubrication is as recommended.

The ratings shown on these charts are based on chain which operates over machine-cut tooth sprockets.

Figure 6-19 Horsepower rating curves for No. 5628, 7.000-in.-pitch heavy duty offset sidebar power transmission chain.

Table 6-20 Horsepower Ratings for No. 2010 Chain, 2.500-In.-Pitch Heavy Duty Offset Sidebar Power Transmission Chain

TEETH	HORSEPOWER CAPACITY													
	RPM													
	2	3	7	10	20	30	40	100	200	250	350	450	600	700
9	1.1	1.4	2.7	3.9	7.7	11.6	15.4	38.6	77.2	96.5	135.1	100.1	65.0	.0
10	1.1	1.5	3.0	4.3	8.6	12.9	17.2	42.9	85.8	107.3	150.2	117.2	76.1	.0
11	1.2	1.7	3.3	4.7	9.4	14.2	18.9	47.2	94.4	118.0	165.2	135.2	87.8	.0
12	1.3	1.8	3.6	5.1	10.3	15.4	20.6	51.5	103.0	128.7	180.2	154.1	100.1	.0
13	1.4	1.9	3.9	5.6	11.2	16.7	22.3	55.8	111.5	139.4	195.2	173.7	112.8	.0
14	1.5	2.0	4.2	6.0	12.0	18.0	24.0	60.1	120.1	150.2	210.2	194.2	126.1	.0
15	1.5	2.1	4.5	6.4	12.9	19.3	25.7	64.4	128.7	160.9	225.2	215.3	139.9	.0
16	1.6	2.2	4.8	6.9	13.7	20.6	27.5	68.6	137.3	171.6	240.3	237.2	154.1	.0
17	1.7	2.3	5.1	7.3	14.6	21.9	29.2	72.9	145.9	182.3	255.3	259.8	168.8	.0
18	1.8	2.4	5.4	7.7	15.4	23.2	30.9	77.2	154.5	193.1	270.3	283.1	183.9	.0
19	1.9	2.5	5.7	8.2	16.3	24.5	32.6	81.5	163.0	203.8	285.3	307.0	.0	.0
20	1.9	2.6	6.0	8.6	17.2	25.7	34.3	85.8	171.6	214.5	300.3	331.5	.0	.0
21	2.0	2.7	6.3	9.0	18.0	27.0	36.0	90.1	180.2	225.2	315.3	356.7	.0	.0
22	2.1	2.8	6.6	9.4	18.9	28.3	37.8	94.4	188.8	236.0	330.4	382.5	.0	.0
23	2.1	3.0	6.9	9.9	19.7	29.6	39.5	98.7	197.4	246.7	345.4	405.3	.0	.0
24	2.2	3.1	7.2	10.3	20.6	30.9	41.2	103.0	205.9	257.4	360.4	414.4	.0	.0
	MANUAL LUBRICATION						OIL BATH		OIL STREAM LUBRICATION					

Table 6-21 Horsepower Ratings for No. 2512 Chain, 3.067-In.-Pitch Heavy Duty Offset Sidebar Power Transmission Chain

TEETH	HORSEPOWER CAPACITY													
	RPM													
	1	3	6	10	20	40	100	150	200	250	300	350	400	450
9	1.0	2.4	4.0	6.4	12.7	25.5	63.7	95.6	127.4	159.3	191.1	171.8	140.6	.0
10	1.1	2.6	4.3	7.1	14.2	28.3	70.8	106.2	141.6	177.0	212.4	198.9	164.7	.0
11	1.2	2.7	4.7	7.8	15.6	31.1	77.9	116.8	155.7	194.7	231.3	215.5	190.0	.0
12	1.3	2.9	5.1	8.5	17.0	34.0	85.0	127.4	169.9	212.4	248.6	231.5	216.5	.0
13	1.4	3.1	5.5	9.2	18.4	36.8	92.0	138.0	184.1	230.1	265.3	247.0	232.3	.0
14	1.4	3.3	5.9	9.9	19.8	39.6	99.1	148.7	198.2	247.8	281.4	262.1	246.4	.0
15	1.5	3.5	6.4	10.6	21.2	42.5	106.2	159.3	212.4	265.5	296.9	276.6	260.0	.0
16	1.6	3.7	6.8	11.3	22.7	45.3	113.3	169.9	226.5	283.2	312.0	290.6	273.2	.0
17	1.7	3.8	7.2	12.0	24.1	48.1	120.3	180.5	240.7	300.9	326.5	304.1	285.9	.0
18	1.7	4.0	7.6	12.7	25.5	51.0	127.4	191.1	245.9	318.6	340.5	317.1	.0	.0
19	1.8	4.2	8.1	13.5	26.9	53.8	134.5	201.8	269.0	336.3	354.0	329.7	.0	.0
20	1.9	4.3	8.5	14.2	28.3	56.6	141.6	212.4	283.2	354.0	367.1	341.9	.0	.0
21	1.9	4.5	8.9	14.9	29.7	59.5	148.7	223.0	297.3	371.7	379.6	353.6	.0	.0
22	2.0	4.7	9.3	15.6	31.1	62.3	155.7	233.6	311.5	389.4	391.7	364.8	.0	.0
23	2.1	4.9	9.8	16.3	32.6	65.1	162.8	244.2	325.6	407.1	403.4	375.7	.0	.0
24	2.2	5.1	10.2	17.0	34.0	68.0	169.9	254.9	339.8	424.8	414.6	386.1	.0	.0
	MANUAL LUBRICATION					OIL BATH		OIL STREAM LUBRICATION						

For continuous operation in the stippled area, some galling of the live bearing surfaces of the chain joints may be expected even though lubrication is as recommended.

The ratings shown on these charts are based on chain which operates over machine-cut tooth sprockets.

Table 6-22 Horsepower Ratings for No. 2814 Chain, 3.500-In.-Pitch Heavy Duty Offset Sidebar Power Transmission Chain

TEETH	HORSEPOWER CAPACITY RPM													
	1	3	6	10	20	35	60	100	125	150	200	250	300	325
9	1.4	3.3	5.5	8.8	17.6	30.8	52.8	88.1	110.1	132.1	176.1	178.7	170.8	.0
10	1.5	3.5	6.0	9.8	19.6	34.2	58.7	97.8	122.3	146.8	195.7	196.1	187.4	.0
11	1.6	3.8	6.5	10.8	21.5	37.7	64.6	107.6	134.5	161.4	215.2	213.0	203.6	.0
12	1.8	4.1	7.0	11.7	23.5	41.1	70.4	117.4	146.8	176.1	234.8	229.5	219.4	.0
13	1,9	4,3	7,6	12.7	25.4	44.5	76.3	127.2	159.0	190.8	254.4	245.6	234.7	.0
14	2.0	4.6	8.2	13.7	27.4	47.9	82.2	137.0	171.2	205.5	273.9	261.2	249.6	.0
15	2.1	4.8	8.8	14.7	29.4	51.4	88.1	146.8	183.4	220.1	292.1	276.3	264.1	.0
16	2.2	5.1	9.4	15.7	31.3	54.8	93.9	156.5	195.7	234.8	307.7	291.1	278.2	.0
17	2.3	5.3	10.0	16.6	33.3	58.2	99.8	166.3	207.9	249.5	322.8	305.5	.0	.0
18	2.4	5.5	10.6	17.6	35.2	61.6	105.7	176.1	220.1	264.2	337.6	319.4	.0	.0
19	2.5	5.8	11.2	18.6	37.2	65.1	111.5	185.9	232.4	278.8	351.9	333.0	.0	.0
20	2.6	6.0	11.7	19.6	39.1	68.5	117.4	195.7	244.6	293.5	365.8	346.1	.0	.0
21	2.7	6.2	12.3	20.5	41.1	71.9	123.3	205.5	256.8	308.2	379.3	358.9	.0	.0
	MANUAL LUBRICATION						OIL BATH		OIL STREAM LUBRICATION					

Table 6-23 Horsepower Ratings for No. 3315 Chain, 4.073-In.-Pitch Heavy Duty Offset Sidebar Power Transmission Chain

TEETH	HORSEPOWER CAPACITY RPM													
	1	3	6	10	20	30	40	65	80	100	125	150	200	225
9	2.0	4.7	8.0	12.8	25.5	38.3	51.1	83.0	102.1	127.7	159.6	168.2	166.3	.0
10	2.2	5.1	8.7	14.2	28.4	42.6	56.7	92.2	113.5	141.8	177.3	185.0	182.9	.0
11	2.4	5.5	9.4	15.6	31.2	46.8	62.4	101.4	124.8	156.0	195.0	201.5	199.2	.0
12	2.5	5.9	10.2	17.0	34.0	51.1	68.1	110.6	136.2	170.2	212.8	217.6	215.1	.0
13	2.7	6.3	11.1	18.4	36.9	55.3	73.8	119.9	147.5	184.4	230.5	233.4	230.7	.0
14	2.9	6.6	11.9	19.9	39.7	59.6	79.4	129.1	158.9	198.6	248.2	248.8	246.0	.0
15	3.0	7.0	12.8	21.3	42.6	63.8	85.1	138.3	170.2	212.8	265.9	263.9	261.0	.0
16	3.2	7.3	13.6	22.7	45.4	68.1	90.8	147.5	181.6	227.0	280.7	278.7	275.6	.0
17	3.3	7.7	14.5	24.1	48.2	72.3	96.5	156.7	192.9	241.1	295.3	293.2	289.9	.0
18	3.5	8.0	15.3	25.5	51.1	76.6	102.1	166.0	204.3	255.3	309.6	307.3	303.9	.0
19	3.6	8.4	16.2	27.0	53.9	80.9	107.8	175.2	215.6	269.5	323.5	321.2	317.6	.0
20	3.8	8.7	17.0	28.4	56.7	85.1	113.5	184.4	227.0	283.7	337.1	334.7	.0	.0
21	3.9	9.0	17.9	29.8	59.6	89.4	119.2	193.6	238.3	297.9	350.5	347.9	.0	.0
	MANUAL LUBRICATION						OIL BATH			OIL STREAM LUBRICATION				

For continuous operation in the stippled area, some galling of the live bearing surfaces of the chain joints may be expected even though lubrication is as recommended.

The ratings shown on these charts are based on chain which operates over machine-cut tooth sprockets.

Table 6-24 Horsepower Ratings for No. 3618 Chain, 4.500-In.-Pitch Heavy Duty Offset Sidebar Power Transmission Chain

TEETH	HORSEPOWER CAPACITY													
	RPM													
	1	3	6	10	20	30	35	50	65	80	100	125	150	175
9	2.6	6.0	10.2	16.3	32.6	48.9	57.0	81.5	105.9	130.4	153.8	156.6	158.8	.0
10	2.8	6.5	11.1	18.1	36.2	54.3	63.4	90.5	117.7	144.9	169.5	172.5	175.0	.0
11	3.0	7.0	12.0	19.9	39.8	59.8	69.7	99.6	129.5	159.4	184.8	188.1	190.8	.0
12	3.3	7.5	13.0	21.7	43.5	65.2	76.1	108.7	141.3	173.9	199.8	203.4	206.3	.0
13	3.5	8.0	14.1	23.5	47.1	70.6	82.4	117.7	153.0	188.3	214.6	218.4	221.6	.0
14	3.7	8.5	15.2	25.4	50.7	76.1	88.7	126.8	164.8	202.8	229.1	233.2	236.6	.0
15	3.9	8.9	16.3	27.2	54.3	81.5	95.1	135.8	176.6	217.3	243.4	247.7	251.3	.0
16	4.1	9.4	17.4	29.0	58.0	86.9	101.4	144.9	188.3	231.8	257.4	261.9	265.7	.0
17	4.2	9.8	18.5	30.8	61.6	92.4	107.8	153.9	200.1	246.3	271.1	275.9	279.9	.0
18	4.4	10.2	19.6	32.6	65.2	97.8	114.1	163.0	211.9	260.8	284.6	289.6	293.8	.0
19	4.6	10.7	20.6	34.4	68.8	103.2	120.4	172.0	223.7	275.3	297.8	303.1	307.5	.0
20	4.8	11.1	21.7	36.2	72.4	108.7	126.8	181.1	235.4	289.8	310.7	316.3	320.9	.0
21	5.0	11.5	22.8	38.0	76.1	114.1	133.1	190.1	247.2	304.2	323.5	329.2	334.0	.0
	MANUAL LUBRICATION						OIL BATH			OIL STREAM LUBRICATION				

Table 6-25 Horsepower Ratings for No. 4020 Chain, 5.000-In.-Pitch Heavy Duty Offset Sidebar Power Transmission Chain

TEETH	HORSEPOWER CAPACITY													
	RPM													
	.5	1.0	3	6	10	20	30	35	50	65	80	100	125	130
9	2.0	3.4	7.8	13.3	21.1	42.2	63.3	73.8	105.5	133.9	139.3	145.3	151.6	.0
10	2.2	3.7	8.5	14.4	23.4	46.9	70.3	82.0	117.2	147.6	153.6	160.2	.0	.0
11	2.3	3.9	9.1	15.5	25.8	51.6	77.4	90.3	128.9	161.2	167.7	174.9	.0	.0
12	2.5	4.2	9.7	16.9	28.1	56.3	84.4	98.5	140.7	174.5	181.6	189.4	.0	.0
13	2.6	4.5	10.3	18.3	30.5	61.0	91.4	106.7	152.4	187.7	195.2	203.7	.0	.0
14	2.8	4.7	10.9	19.7	32.8	65.6	98.5	114.9	164.1	200.6	208.7	217.7	.0	.0
15	2.9	5.0	11.5	21.1	35.2	70.3	105.5	123.1	175.8	213.4	222.0	231.6	.0	.0
16	3.1	5.2	12.1	22.5	37.5	75.0	112.5	131.3	187.5	225.9	235.0	245.2	.0	.0
17	3.2	5.5	12.7	23.9	39.9	79.7	119.6	139.5	199.3	238.2	247.8	258.6	.0	.0
18	3.4	5.7	13.3	25.3	42.2	84.4	126.6	147.7	211.0	250.4	260.5	271.7	.0	.0
	MANUAL LUBRICATION							OIL BATH		OIL STREAM LUBRICATION				

For continuous operation in the stippled area, some galling of the live bearing surfaces of the chain joints may be expected even though lubrication is as recommended.

The ratings shown on these charts are based on chain which operates over machine-cut tooth sprockets.

Table 6-26 Horsepower Ratings for No. 4824 Chain, 6.000-In.-Pitch Heavy Duty Offset Sidebar Power Transmission Chain

TEETH	HORSEPOWER CAPACITY													
	RPM													
	.5	1.0	3	6	10	20	30	35	40	45	50	60	70	75
9	3.1	5.3	12.2	20.7	33.0	66.0	96.1	101.5	106.3	110.8	115.0	122.6	129.0	.0
10	3.4	5.7	13.2	22.4	36.6	73.3	106.2	112.1	117.5	122.5	127.1	135.5	.0	.0
11	3.6	6.2	14.2	24.2	40.3	80.6	116.1	122.6	128.5	133.9	139.0	148.2	.0	.0
12	3.9	6.6	15.2	26.4	44.0	87.9	126.0	133.0	139.4	145.3	150.8	160.8	.0	.0
13	4.1	7.0	16.2	28.6	47.6	95.3	135.7	143.2	150.1	156.5	162.4	173.2	.0	.0
14	4.4	7.4	17.1	30.8	51.3	102.6	145.3	153.4	160.8	167.6	173.9	185.4	.0	.0
15	4.6	7.8	18.0	33.0	55.0	109.9	154.8	163.4	171.3	178.5	185.3	197.5	.0	.0
16	4.8	8.2	18.9	35.2	58.6	117.3	164.2	173.3	181.6	189.3	196.5	209.5	.0	.0
17	5.1	8.6	19.8	37.4	62.3	124.6	173.4	183.1	191.9	200.0	207.6	221.3	.0	.0
18	5.3	9.0	20.7	39.6	66.0	131.9	182.6	192.7	202.0	210.6	218.5	233.0	.0	.0
	MANUAL LUBRICATION									OIL BATH			OIL STREAM	

Table 6-27 Horsepower Ratings for No. 5628 Chain, 7.000-In.-Pitch Heavy Duty Offset Sidebar Power Transmission Chain

TEETH	HORSEPOWER CAPACITY													
	RPM													
	.1	.5	1	2	4	6	10	15	20	25	30	35	40	45
9	1.3	4.6	7.7	13.1	22.2	30.2	48.1	67.1	76.7	85.0	92.5	99.4	105.7	.0
10	1.4	4.9	8.4	14.2	24.0	32.7	53.5	74.2	84.8	94.0	102.3	109.9	.0	.0
11	1.6	5.3	9.0	15.2	25.9	35.3	58.8	81.2	92.8	103.0	112.0	120.3	.0	.0
12	1.7	5.7	9.6	16.3	27.6	38.5	64.2	88.2	100.8	111.8	121.7	130.7	.0	.0
13	1.8	6.0	10.2	17.3	29.4	41.7	69.5	95.1	108.7	120.6	131.2	140.9	.0	.0
14	1.9	6.4	10.8	18.3	31.1	44.9	74.8	102.0	116.5	129.2	140.6	151.1	.0	.0
15	2.0	6.7	11.4	19.3	32.7	48.1	80.2	108.8	124.3	137.8	150.0	161.1	.0	.0
16	2.1	7.1	12.0	20.3	34.4	51.3	85.5	115.5	132.0	146.4	159.3	171.1	.0	.0
17	2.2	7.4	12.5	21.2	36.4	54.5	90.9	122.2	139.6	154.8	168.5	180.9	.0	.0
18	2.3	7.7	13.1	22.2	38.5	57.7	96.2	128.8	147.1	163.2	177.5	190.7	.0	.0
	MANUAL LUBRICATION													OIL BATH

For continuous operation in the stippled area, some galling of the live bearing surfaces of the chain joints may be expected even though lubrication is as recommended.

The ratings shown on these charts are based on chain which operates over machine-cut tooth sprockets.

et centers; that is, there are limitations placed in the system by loads, speeds, space, and other special conditions. A center distance of 30 to 50 pitches of the chain represents good practice for the average application. If the shaft centers are required to be close, the center-to-center distance must exceed one-half the sum of the outside diameters of the driving and driven sprockets, plus one chain pitch. The shorter the center distance, the more frequently each link of the chain comes in contact with the sprockets, thus reducing chain life. Too long shaft centers may cause the drive to develop considerable slapping, vibrating, or whipping action.

Step 8. Determine the chain length in pitches. This length L may be calculated from the equation

$$L_p = 2C_p + \frac{N}{2} + \frac{n}{2} + \frac{K(N-n)^2}{100\,C_p} \qquad (6\text{-}3)$$

where

L_p = chain length, pitches
C_p = distance between centers of shafts, pitches
N = number of teeth on larger sprocket
n = number of teeth on smaller sprocket
K = correction factor depending on the

$$\text{values of } \frac{N-n}{C_p} \quad \text{(see Table 6-28)} \qquad (6\text{-}4)$$

Typical Selection Examples

Example 1. Select a chain drive for driving directly to the headshaft of an elevator that is to operate at 20 rpm, requires 12 hp, and runs an average of 20 hr per day. The source of power is the slow-speed shaft of a gear head motor at 64 rpm.

Step 1. Refer to Tables 6-8 to 6-10 for the conditions and service factors for this application, which are:

Moderate shock load—electric motor input	1.3
Moderately dirty	1.2
20 hr of daily service	1.2

Combined service factor = (1.3)(1.2)(1.2) = 1.87

Step 2. Apply service factor to horsepower.

Design horsepower = (1.87)(12.0) = 22.4

Step 3. Refer to the quick selector chart (Fig. 6-11). The 22.4 hp/64 rpm intersection falls below the 9 T sprocket line for No. 2010 chain (Fig. 6-12).

Step 4. The HP-RPM intersection falls slightly below the 9 T sprocket line. Next, looking at Table 6-20, by interpolation, it is seen that the 9 T sprocket for No. 2010

Table 6-28 Values of K for Computing Chain Length

$\frac{N-n}{Cp}$	Factor K	$\frac{N-n}{Cp}$	Factor K	$\frac{N-n}{Cp}$	Factor K	$\frac{N-n}{Cp}$	Factor K
.05	2.5326	1.65	2.5474	3.25	2.5940	4.85	2.6908
.10	2.5327	1.70	2.5483	3.30	2.5961	4.90	2.6950
.15	2.5328	1.75	2.5492	3.35	2.5982	4.95	2.6993
.20	2.5330	1.80	2.5502	3.40	2.6004	5.00	2.7036
.25	2.5331	1.85	2.5513	3.45	2.6026	5.05	2.7081
.30	2.5333	1.90	2.5524	3.50	2.6048	5.10	2.7127
.35	2.5335	1.95	2.5535	3.55	2.6072	5.15	2.7175
.40	2.5337	2.00	2.5546	3.60	2.6096	5.20	2.7224
.45	2.5339	2.05	2.5558	3.65	2.6120	5.25	2.7275
.50	2.5342	2.10	2.5570	3.70	2.6145	5.30	2.7328
.55	2.5345	2.15	2.5582	3.75	2.6170	5.35	2.7382
.60	2.5348	2.20	2.5594	3.80	2.6197	5.40	2.7439
.65	2.5351	2.25	2.5607	3.85	2.6223	5.45	2.7498
.70	2.5355	2.30	2.5621	3.90	2.6251	5.50	2.7560
.75	2.5359	2.35	2.5634	3.95	2.6279	5.55	2.7623
.80	2.5363	2.40	2.5648	4.00	2.6309	5.60	2.7688
.85	2.5367	2.45	2.5663	4.05	2.6338	5.65	2.7755
.90	2.5371	2.50	2.5677	4.10	2.6369	5.70	2.7824
.95	2.5376	2.55	2.5692	4.15	2.6400	5.75	2.7895
1.00	2.5382	2.60	2.5708	4.20	2.6431	5.80	2.7968
1.05	2.5387	2.65	2.5723	4.25	2.6464	5.85	2.8043
1.10	2.5393	2.70	2.5739	4.30	2.6497	5.90	2.8120
1.15	2.5399	2.75	2.5756	4.35	2.6530	5.95	2.8200
1.20	2.5405	2.80	2.5773	4.40	2.6565	6.00	2.8284
1.25	2.5411	2.85	2.5790	4.45	2.6600	6.05	2.8374
1.30	2.5418	2.90	2.5807	4.50	2.6636	6.10	2.8472
1.35	2.5425	2.95	2.5825	4.55	2.6672	6.15	2.8581
1.40	2.5432	3.00	2.5843	4.60	2.6710	6.20	2.8704
1.45	2.5439	3.05	2.5862	4.65	2.6748	6.25	2.8844
1.50	2.5447	3.10	2.5881	4.70	2.6787	6.30	2.9011
1.55	2.5456	3.15	2.5900	4.75	2.6826	6.35	2.9215
1.60	2.5465	3.20	2.5920	4.80	2.6867		

chain is good for 24.7 hp at 64 rpm. Therefore, this chain may be used for the elevator drive, which requires 22.4 hp. The result of using the table or the curve is the same.

Step 5. The table and the curve show that oil bath lubrication is needed.

Step 6. The speeds of both sprockets are known. So the number of teeth in the driven sprocket is calculated by the equation

$$\text{Number of teeth in driver} \times \frac{\text{rpm of driving shaft}}{\text{rpm of driven shaft}}$$

$$= \text{number of teeth in driven sprocket} \qquad (6\text{-}5)$$

or

$$(9)\,\frac{64}{20} = 28$$

Therefore, the 9-tooth driver sprocket will require a 28-tooth driven sprocket to provide for 20 rpm of the driven shaft.

Step 7. For average applications, 30 to 50 pitches of chain is a good center distrance to use. In this example, we arbitrarily use a center distance of 40 pitches, so

Number of pitches (40) × chain pitch (2.5 in.) = 100-in. center distance

Step 8. To calculate the chain length required, use equation (6-3):

$$L_p = 2C_p + \frac{N}{2} + \frac{n}{2} + \frac{K(N-n)^2}{100C_p}$$

From Table 6-28, K = 2.53 and

$$L_p = 2(40) + \frac{28}{2} + \frac{9}{2} + \frac{2.53(28-9)^2}{100(40)} = 98.782 \text{ pitches}$$

Since a fractional link is not available, 99 links of chain are required for the drive.

Example 2. Select a suitable drive for a conveyor that is to operate at 72 ft/min. The head shaft is to turn at 12 rpm. The total load on the conveyor chain is 6500 lb. The drive will be open and exposed to moderately dirty conditions. The power will be supplied by an electric motor and moderate shock loads may occasionally be encountered. The normal operating time will be for an 8-hr shift. The reducer to be used has an output speed of 30 rpm.

Step 1. Referring to Tables 6-8 to 6-10, the service factors are:

Moderate shock load	1.3
Moderately dirty	1.2
8-hr daily service	1.0

Combined service factor = (1.3)(1.2)(1.0) = 1.56

Step 2. The horsepower of the system can be calculated as follows:

$$HP = \frac{(W_p)(S)}{33{,}000}$$

where

W_p = chain load, lb
S = linear conveyor speed, ft/min

so

$$HP = \frac{(6500)(72)}{33{,}000} = 14.18$$

Apply the combined service factors to the calculated horsepower.

Design horsepower = (1.56)(14.18) = 22.12

Step 3. Use the quick selector chart, Figure 6-11, to find the 22.12 hp/30 rpm intersection. This point is slightly above the line for chain No. 2512. Check the curves for chain Nos. 2512 and 2814.

Step 4. The intersection point on the chart for No. 2814 chain, Figure 6-14, falls below the 9-tooth sprocket line. On the chart for No. 2512 chain, Figure 6-13, the intersection point is between the 9-tooth and 12-tooth sprocket lines. The 2512 chain should be selected for this drive. The 11-tooth sprocket meets the horsepower requirement. To check this selection, Table 6-21 can be used to show that the 11-tooth sprocket has a capacity of 23.35 hp.

Step 5. The type of lubrication to be used is on the borderline. In this application, manual lubrication will be sufficient.

Step 6. The speed ratio of this drive is

$$\frac{30 \text{ rpm (driving shaft)}}{12 \text{ rpm (driven shaft)}} = 2.5$$

The number of teeth in the driven sprocket may be calculated as follows:

Number of teeth in driven sprocket = number of teeth in driver × speed ratio

or

(11)(2.5) = 27.5

Therefore, the 11-tooth driver sprocket will require a 27.5-tooth driven sprocket to provide for 12 rpm of the driven shaft. Since a fractional tooth sprocket is not available, a 27-tooth sprocket will provide a driven shaft speed of 12.22 rpm. This sprocket may be used. If the 12 rpm of the driven shaft must be held, then considering the speed ratio of 2.5, a set of sprockets for this drive could consist of a 12-tooth driving and a 30-tooth driven sprocket.

Step 7. Use a center distance of 40 pitches for this drive:

Number of pitches (40) × chain pitch (3.067) = 122.680-in. center distance

Step 8. The length of chain required is

$$L_p - 2(40) + \frac{27}{2} + \frac{11}{2} + \frac{2.53(27-11)^2}{40(100)} = 99.162 \text{ pitches}$$

Therefore, 100 links of No. 2512 chain are required for this drive.

Alternate Method

The alternate method must be used when horsepower curves or tables do not exist. The alternate method uses rated working loads, which are listed in the manufacturers' catalogs. These working loads should not be confused with the ultimate or actual tensile strength of the chains.

Using the design horsepower derived in step 2 of example 1, the number of teeth in the drive sprocket, and the chain pitch in inches, the chain working load may be calculated using the following formula:

$$WL = HP_c \times \frac{396{,}000}{\text{chain pitch} \times \text{number of teeth} \times \text{rpm}} \tag{6-6}$$

where

chain pitch is in inches

WL = working load

HP_c = horsepower input, corrected by service factors (design horsepower)

It should be noted that the chain working load is often directly available from the raw data. This may be used after application of the service factors, Tables 6-8 to 6-10. Referring to the section covering Engineering Steel drive chains in a chain manufacturer's catalog, a chain or chains of the selected pitch may usually be found with a rated working load equal to or slightly in excess of the calculated value. If a chain of the proper value cannot be found, a longer pitch and heavier chain should be selected and the calculations rechecked. If the lightest working load shown for a chain of the selected pitch is greatly in excess of the calculated requirements, a shorter pitch should be selected and the calculation redone.

Table 6-29 lists the ANSI standard drive chains, showing the standardized dimensions necessary to specify these chains, including recommended maximum working loads and weights per foot of the chains.

The figures for recommended working loads were compiled from the catalogs of American Chain Association members, who manufacture and describe these chains in their catalogs. Although not every manufacturer shows exactly the same figure for each chain, variations in no case exceed 2% of the recommended load.

The weights per foot given are ranges, showing maximums and minimums, taken from the manufacturers' catalogs. The dimensions referenced in Table 6-29 are as shown in Figure 6-30.

The remainder of the selection process for selection of chains based on working load is the same as that used when the horsepower charts and tables are the source of the data for selection.

Chain manufacturers provide a relatively large selection of chains in addition to those made in accordance with ANSI standards. Although most manufacturers in-

Table 6-29 Dimensional Data and Working Loads of ANSI Standard Offset Power Transmission Chains

Chain number	Pitch, P	Maximum working load (lb)	Average weight (lb/ft)	Roller diameter, R	Distance between sidebars, W	Maximum chain height, U	Maximum dimensions pin ends to chain centerline from:	
							Cotter ends, J	Head ends, H
2010	2.500	4,650	7.8–8.6	1-1/4	1-1/2	1-7/8	1-7/8	1-11/16
2512	3.067	6,000	12.0–13.2	1-5/8	1-9/16	2-3/8	2-3/16	1-7/8
2814	3.500	7,600	15.8–17.3	1-3/4	1-1/2	2-3/8	2-7/16	2-3/16
3315	4.073	10,000	18.0–18.7	1-25/32	1-15/16	2-1/2	2-13/16	2-1/2
3618	4.500	12,000	22.0–25.4	2-1/4	2-1/16	3-1/8	3	2-9/16
4020	5.000	17,500	33.9-36.0	2-1/2	2-3/4	3-5/8	3-9/16	3-1/16
4824	6.000	23,600	45.0–46.5	3	3	4-1/8	3-7/8	3-1/2
5628	7.000	30,500	66.0–67.5	3-1/2	3-1/4	5-1/4	4-1/2	4

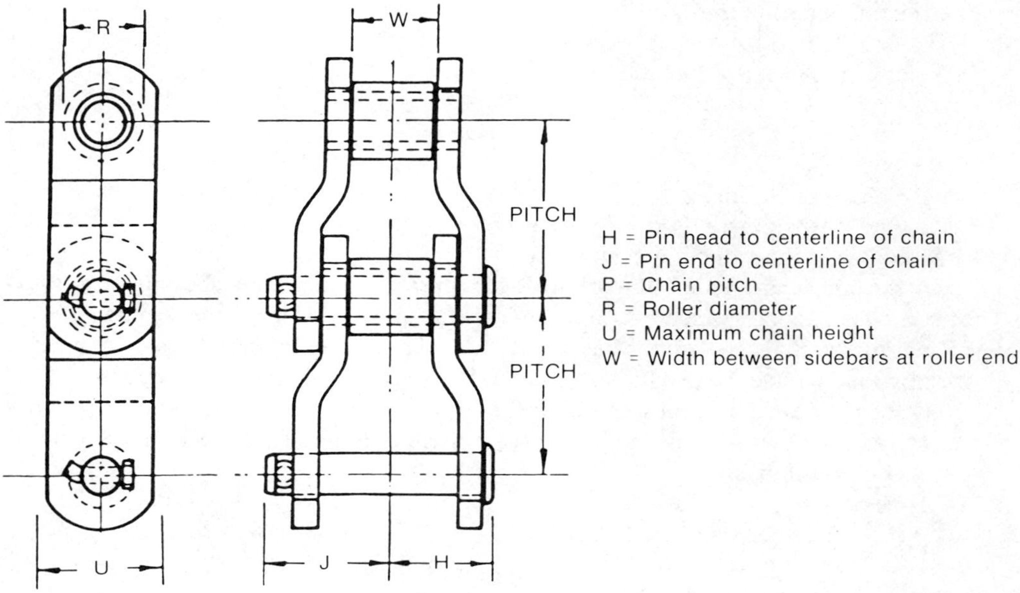

Figure 6-30 Dimensions necessary for selection of standard ANSI offset power transmission chains.

clude horsepower curves in their catalogs, the curves do not usually cover other than the ANSI standard chains. Therefore, working loads will need to be used if chains other than the ANSI series have to be used.

Example 3. Repeat example 2, but calculating using chain maximum recommended working loads rather than horsepower curves and tables. Compare the results.

For the information given and steps 1 and 2 of example 2, the following data may be derived:

Conveyor speed and load: 72 ft/min at 6500 lb

Conveyor head shaft speed: 12 rpm

Reducer output shaft speed: 30 rpm

Speed ratio of drive: 2.5 : 1

Service factors

Moderate shock load:	1.3
Moderately dirty	1.2
8 hr of daily service	1.0

Combined service factor: $1.3 \times 1.2 \times 1.0 = 1.56$

$$HP = \frac{(W_p)(S)}{33{,}000} = \frac{6500 \times 72}{33{,}000} = 14.18 \tag{6-7}$$

Design HP $= 1.56 \times 14.18 = 22.12$

Then, from equation (6-6),

$$WL = \frac{\text{design HP} \times 396{,}000}{P \times n \times RPM_r}$$

where

P = chain pitch, in.
n = number of teeth on the drive sprocket
RPM_r = rpm of the reducer output shaft

In selecting chain by use of working loads, the service factor may be applied as the ratio between actual horsepower and design horsepower, or it may be applied after calculating the chain pull on the basis of the actual horsepower. In that case, required working load will be the product of chain pull times the combined service factor.

Using the design horsepower and the other data known, a function may be set up among the unknowns—working load, chain pitch, and number of teeth in the drive sprocket—as follows:

$$WL = \frac{22.12 \times 396{,}000}{P \times n \times 30} = \frac{292{,}000}{P \times n}$$

Tentatively deciding that the drive sprocket should have 9 to 17 teeth, a tabulation of data may be made, using information from Table 4-45 and 6-29 and Figure 6-30.

It is convenient to calculate sprocket pitch diameters and record clearance height U (see Fig. 6-30) at this time for later selection of drive sprocket hub size and to be sure that the chain selected will clear the sprocket hub.

ANSI standard chain number	Chain pitch, P	Drive sprocket number of teeth, n	Calculated required working load, W_1	Chain rated working load	Csc[a] for number of teeth, n	Required clear height, U	Sprocket pitch diameter
2512	3.067	9	10,578	6,000	2.9238	2-3/8	8.97
2512	3.067	11	8,654	6,000	3.5495	2-3/8	10.88
2512	3.067	13	7,323	6,000	4.1785	2-3/8	12.81
2512	3.067	15	6,347	6,000	4.8097	2-3/8	14.75
2512	3.067	17	5,600	6,000	5.4423	2-3/8	16.69
2814	3.500	9	9,269	7,600	2.9238	2-3/8	10.23
2814	3.500	11	7,584	7,600	3.5495	2-3/8	12.42
2814	3.500	13	6,417	7,600	4.1785	2-3/8	14.62
2814	3.500	15	5,562	7,600	4.8097	2-3/8	16.83
3315	4.073	9	7,965	10,000	2.9238	2½	11.91
3315	4.073	11	6,517	10,000	3.5495	2½	14.45
3315	4.073	13	4,779	10,000	4.1785	2½	17.02

[a]See Table 4-45.

It is nearly always desirable and usually more efficient to select from a tabulation rather than make a single preliminary calculation. The first selection of chain and sprocket teeth often requires recalculation. Even more important, the tabulation is an assurance that the best possible selection of pitch and numbers of teeth with relative working loads have been made with this method of calculation.

From the tabulation, it would appear that the best selection is No. 2814 chain with an 11-tooth drive sprocket, the 7600-lb allowable working load for this chain just exceeding the required 7584 lb required for this combination.

Selection of No. 3315 chain would be an obvious case of overchaining, and according to the calculations, No. 2512 chain would also be satisfactory with a 17-tooth drive sprocket. As the 17-tooth No. 2512 sprocket has a considerably larger diameter than the 11-tooth No. 2814 wheel, and the same ratio of sizes would project to the driven sprockets required, most engineers would probably select the No. 2814 chain, if horsepower curve data were not available.

Comparison with Selection from Horsepower Curves. Returning to example 2, it is found that No. 2512 chain with an 11-tooth sprocket is adequate and would be selected, since manual lubrication, only, is required.

This comparison of chain and sprocket selection brings out quite well the advantages of the use of horsepower curves and tables. As a generality, calculation using horsepower curves will be more appropriate since working loads are based on bearing values and ratios to ultimate or yield strengths rather than on actual test data. The tests used to establish horsepower ratings include, of course, all necessary factors.

Slack-Side Tension and Sag

Satisfactory operation of a drive requires reasonable sag of the slack-side strand in a horizontal drive. This will produce correct slack-side tension, necessary for correct sprocket action. As stated previously, correct tension of the slack side is necessary for all chains, but obviously, sag is not a factor in a vertically arranged drive. To avoid improper tension, the sag from a straight line on the slack-side strand of a drive should be not less than 2 to 3% of the horizontal center distance. High tension will cause rapid wear on both chain and sprockets. Insufficient tension may allow chain to jump teeth on the sprockets or may even cause shock loading or chain breakage resulting from whipping or riding up over the tooth tips.

Figure 6-31 is a diagram defining sag.

$$\text{Sag, in inches } (S) = \sqrt{0.375\,CE} \tag{6-8}$$

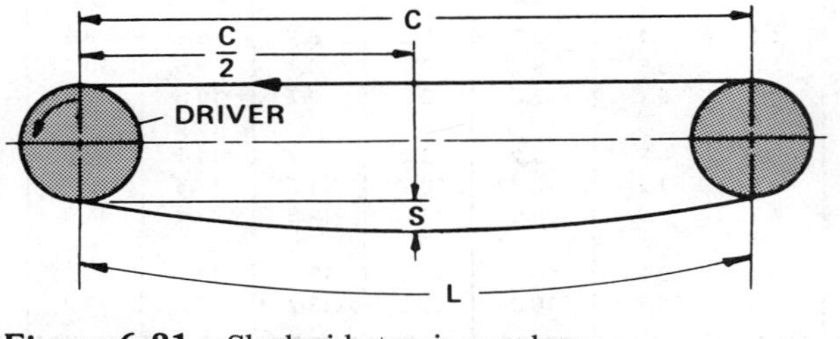

Figure 6-31 Slack-side tension and sag.

where

C = distance between sprocket centers, in.

E = excess chain, L − C

L = return strand length

actual chain used = number of links × chain pitch

Table 6-32 gives maximum chain sag for various centers for given amounts of excess chain. Maximum chain tension or load in the chain on the slack side, due to the weight of the chain, may be found by using Table 6-33. The tension can be calculated using the following formula:

$$\text{Tension} = \frac{C^2 \times W_f}{96 \times \text{sag of chain, in inches}} \qquad (6\text{-}9)$$

where

C = distance between sprockets centers, in.

W_f = weight of chain per foot

Where applications are encountered that do not seem to fit the methods of selection and other information in this chapter, it is suggested that a chain manufacturer be consulted.

Table 6-32 Maximum Chain Sag for Various Centers

WHEEL CENTERS IN INCHES	APPROXIMATE SAG IN INCHES — Excess Chain in Inches											
	1/16	1/8	3/8	1/2	3/4	1	1½	2	2½	3	3½	4
10	0.48	0.68	1.2	1.4	1.7	1.9	2.4	2.7	3.1	3.4	3.6	3.9
20	0.69	0.97	1.7	1.9	2.4	2.7	3.4	3.9	4.3	4.7	5.1	5.5
30	0.84	1.2	2.1	2.4	2.9	3.3	4.1	4.7	5.3	5.8	6.3	6.7
40	0.97	1.4	2.4	2.7	3.3	3.9	4.7	5.5	6.1	6.7	7.3	7.8
50	1.1	1.5	2.6	3.1	3.8	4.3	5.3	6.1	6.9	7.5	8.1	8.7
60	1.2	1.7	2.9	3.4	4.1	4.8	5.8	6.7	7.5	8.2	8.9	9.5
70	1.3	1.8	3.1	3.6	4.4	5.1	6.3	7.3	8.1	8.9	9.6	10.3
80	1.4	2.0	3.3	3.9	4.8	5.5	6.7	7.8	8.7	9.5	10.3	11.0
100	1.5	2.2	3.8	4.3	5.3	6.1	7.5	8.7	9.7	10.6	11.5	12.3
120	1.7	2.4	4.1	4.7	5.8	6.7	8.2	9.5	10.7	11.6	12.5	13.4
140	1.8	2.6	4.4	5.1	6.3	7.2	8.9	10.2	11.5	12.5	13.5	14.5
160	1.9	2.7	4.7	5.5	6.7	7.7	9.5	11.0	12.3	13.4	14.5	15.5
180	2.1	2.9	5.0	5.8	7.1	8.2	10.1	11.6	13.0	14.2	15.4	16.4
200	2.2	3.1	5.3	6.1	7.5	8.7	10.6	12.2	13.7	15.0	16.2	17.3

Table 6-33 Maximum Tension of Chain on Slack Side

WHEEL CENTERS IN INCHES	POUNDS TENSION IN SLACK SIDE OF CHAIN FOR CHAIN WEIGHING ONE POUND PER FOOT — Amount of Sag in Inches													
	.125	.25	.5	1.0	1.5	2.0	3.0	4.0	6.0	8.0	10.0	12.0	14.0	16.0
20	33	16	8	4	3	2	.....	.....	.....	.....	.....	.....	.....	.....
30	76	38	18	9	6	5	3	2	.....	.....	.....	.....	.....	.....
40	133	67	34	17	11	8	4	3	2	.....	.....	.....	.....	.....
50	210	105	53	26	18	13	9	6	5	4	3	.....	.....	.....
60	300	150	75	38	26	18	13	9	7	5	4	.....	.....	.....
70	410	208	102	53	34	25	17	13	9	6	5	4	.....	.....
80	530	266	133	68	44	34	22	17	11	9	7	6	5	.....
100	910	415	208	105	70	53	34	26	17	13	11	9	7	6
120	1200	595	300	150	102	75	51	38	25	19	15	13	11	10
140	1640	750	410	207	137	102	68	51	35	25	21	18	15	13
160	2140	1060	530	265	185	133	90	67	45	33	27	22	19	17
180	2700	1350	675	340	224	168	112	88	56	43	34	28	24	21
200	3330	1650	835	415	275	210	140	105	70	53	42	35	30	26

7
Silent Chain Drives

In previous chapters, the development of Silent Chain (Chapter 1), standards and designations (Chapter 2), components and design (Chapter 3), and sprockets (Chapter 4) have been discussed. This chapter is concerned with the applications of Silent Chain.

Figure 7-1 shows the basic components of a Silent Chain drive: the chain, a pinion or, more properly, a drive sprocket, and a driven sprocket. Figures 7-2 to 7-5 depict four more Silent Chain drives in which power is transmitted to various kinds of equipment or serves as a speed reduction element in a power transmission train.

All these drives demonstrate the same point—Silent Chain is an excellent substitute for gear trains when shafting is too far apart to reach with one set of gears. Generally speaking, Silent Chain is operated at higher speeds and runs more smoothly than other types of chain used for drive service.

The location where a Silent Chain may be used has only a few basic requirements: parallel shafting and space, including enough room for required lubrication, which will include a case around the drive in most applications. Silent Chains, when operated at the higher speeds of which they are capable, frequently require lubrication of types where casings are required, and also, as with precision gearing, require reasonably clean surroundings.

Figures 7-2 to 7-5 show a variety of types and styles of usage.

General Applications Notes

Although each application will have its own individual requirements, Silent Chain, as with all products, has some general limitations and also some elements of selection producing more economic or desirable results than other selections.

Speeds and Horsepower

These are covered in the horsepower tables accompanying this chapter, and are ANSI standard. Speeds range, in terms of the pinion rpm from over 6000 rpm to a maximum of 900 rpm for 2-in.-pitch chain, the largest size covered. Horsepower is rated in terms

Figure 7-1 A Typical Silent Chain drive.

of units *per inch of chain width,* rather than being a figure for individual chains as is necessary for Roller and Engineering Steel drive chains.

On this basis, as 2-in.-pitch chain at 900 rpm is rated at 125 hp, and widths up to 20 in. wide are listed, this represents 2500 hp.

Speed Ratios

Normally, speed ratios up to 12 : 1 can be handled by Silent Chain drives. However, better economics are usually achieved with ratios of 8 : 1 or less; ratios higher than that are often best handled with two drives in sequence.

Generally, large reductions with short centers should use small-pitch, wide chains; with small reductions and long centers, use narrower, larger chains.

Centers

Obviously, the sprockets must clear one another. (To the question: If centers are that close, why not gears?—rotational direction is sometimes a factor.)

At ratios of 2 : 1 or greater, a center distance of the large wheel diameter plus half the small diameter is recommended as a minimum.

Figure 7-2 Silent Chains on an automotive test stand. Two internal combustion engines are involved; one is being tested and the other "run in."

Figure 7-3 Silent Chains are used in automotive engines for timing chains and in transmissions for front-end-drive automobiles and trucks.

Chain Tension Adjustments

Silent Chains require occasional adjustment to attain maximum life.

All chains elongate in service due to wear, and Silent Chain is no exception. To attain maximum service life the drive may be designed to permit movement of one of the shafts to lengthen the center distance and take up accumulated chain slack. If that is not possible, an idler or shoe may be used or one or several pitches of chain can be removed. A chain may have reached the limits of its usefulness when it has elongated 200/N percent, where N is the number of teeth in the large sprocket. As Silent Chain elongates, the effective pitch increases and the chain moves out on the sprocket teeth to a correspondingly larger pitch circle. Eventually, the chain will jump sprocket teeth due to mismatch. Adjustment of the chain slack will not prevent this condition but does prevent chain whipping and backlash, which can greatly reduce chain life. Periodic inspection should be made when the following are present:

Fixed centers
Vertical centers
Shock loadings
Reversing drive

Figure 7-4 Duplex Silent Chain in a serpentine arrangement for multiple rolls in plastic sheet processing.

Numbers of Sprocket Teeth and Chain Links

Even numbers of pitches in the chain length are recommended so that offset links will not be needed. An offset link is an added expense and may be the weakest link in a chain.

Odd numbers of teeth in the drive sprocket will usually result in spreading the wear due to minor inconsistencies over the entire chain length rather than a particular tooth repeatedly contacting a particular pitch of chain. An exception would be where the chain length in pitches is evenly divisible by the number of teeth in the sprocket. Minimum recommended numbers of teeth in the small sprocket are:

Normal operation: 21 teeth; absolute minimum is 12 teeth

Figure 7-5 Silent Chain drives are not limited to conventional positions if the chain type is selected with care.

Speed increasing (or step-up) drive: 23 teeth
High-speed drive (over 1500 ft/min): 25 teeth

Silent Chain Drive Design

The design of a Silent Chain drive consists primarily of the selection of chain size and the sprockets, the determination of center distance and chain length, the lubrication method, and the arrangement of any required casings.

Silent Chain Capacity

The power transmitting capacity of a Silent Chain drive is expressed in units of horsepower per inch of chain width, and varies with the chain pitch and the rotational speed and size of the smallest sprocket in the drive. The ratings take into account the load-carrying ability of the chain over an extended period of time. Chain strength, component wear, metal fatigue, and suitability of the lubrication are examples of the factors contributing to the ratings. Although considerable variation exists in designs of joint components, a common formula for the horsepower ratings has been developed. This formula is based on chain pitch, and is modified by factors that compensate for chordal action, chain velocity, and so on.

Sprocket Size

The minimum number of teeth in a sprocket is limited by the geometry of the tooth form, as discussed above. Also, small sprockets reduce or may even negate the benefits of using Silent Chain.

Chain and Sprocket Selection

The procedure for the selection of Silent Chains and sprockets is similar to that for Roller Chain drives (see Chapter 5), with the modifications indicated in the following paragraphs.

Service Factors and Load Classification

Extensive tables of service factors and load classifications are included in ANSI Standard B29.2. However, acceptable service factor values can be found in Table 7-6, which is a condensation of the standard table.

Horsepower Rating Tables

Figure 7-7 is a Chain Pitch Selection Chart from which a trial pitch may be selected. Horsepower rating tables for Silent Chain pitches from 3/16 through 2 in. are given in Tables 7-8 to 7-16. These show ratings in horsepower per inch of chain width for drives with a service factor of unity.

Steps to Selection of Silent Chain and Sprockets

Using the seven-step process for drive chain selection, as described for other types of chains in Chapters 5 and 6, references as tables and charts must be consulted as indicated in each step.

Step 1. Service factors. Find the load classification from Table 5-2. Then select the service factor from Table 7-6.

Step 2. Design horsepower. Multiply the specified horsepower by the service factor to obtain the design horsepower.

Step 3. Tentative chain pitch selection. Select from the chain pitch selection chart (Fig. 7-7) tentative chain pitches that are indicated as suitable for use at the speed of the faster shaft.

Step 4. Chain and smaller sprocket selection. From Table 7-17 note the smallest sprockets that will accommodate the faster shaft.

Type of Load	Service Factors 10-hr Day	24-hr Day
Uniform	1.0	1.3
Moderate Shock	1.4	1.7
Heavy Shock	1.7	2.0

Table 7-6 Service Factors for Silent Chain Drives (See Table 5-2 for Load Classifications)

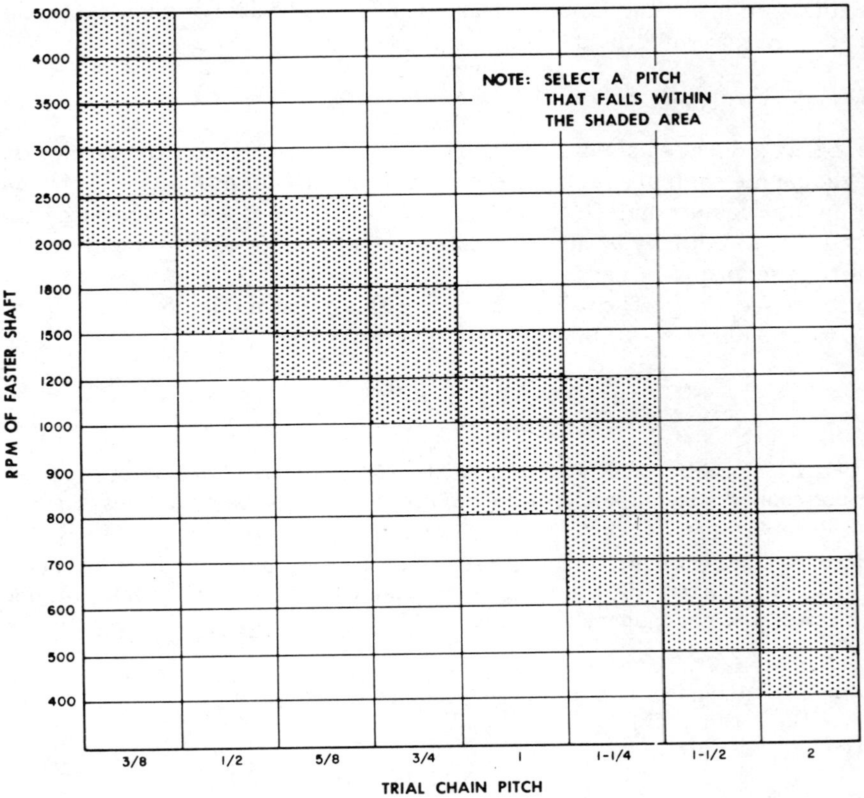

Figure 7-7 Silent Chain pitch selection chart.

From the horsepower rating tables (Tables 7-8 to 7-16), note the horsepower capacities per inch of chain width for the tentative sizes, and divide these into the design horsepower, step 2.

Tabulate these data in the manner indicated in the following example, and, from Table 7-18 note the nearest available widths. Select from these tabulated data the pitch and number of teeth of the most practical smaller sprocket.

Step 5. Selection of large sprocket. Compute the number of teeth for the large sprocket, using the selected small sprocket and specified shaft speed ratio. If the required speed ratio cannot be obtained with an integral number of teeth on the large sprocket, try a larger size for the small sprocket. If space limitations exist, check for clearances.

Step 6. Chain length. Using the specified center distance, compute the chain length of equation (5-11) (p. 134). Although the computed length may be fractional, a chain consisting of an integral number of pitches must be used. The nearest even number of pitches is preferable to an odd number since the latter requires the use of an offset section. The center distance must then be corrected.

Step 7. Correction of center distance. Using the adopted length, recompute the center distance by equation (5-12) (p. 134). If the center distance is fixed and the adopted chain length results in an objectionable slack, an idler may be used or a sprocket combination adopted that will provide a snug fit.

Chain Drive Design Example

Problem: Select a Silent Chain drive to transmit 190 hp at 600 rpm from an electric motor to the countershaft of a rotary screen at 150 rpm. The motor shaft is 4-3/4 in. in diameter and the countershaft is 4-15/16 in. in diameter. Shaft centers must be approximately 60 in., but may be altered a maximum of 2 in. in either direction. The drive will be in service 10 hr per day.

Step 1. Service factor. The drive is listed in Table 5-2 as moderate shock. The corresponding service factor from Table 7-6 is 1:4.

SILENT CHAIN HORSEPOWER RATING TABLES PER INCH OF CHAIN WIDTH

To use the horsepower ratings properly, the following factors must be taken into consideration:

1. Service factors—Table 7-6.
2. Lubrication—The horsepower established from the sprocket and speed combination of the drive under consideration will indicate a method of lubrication. This method or a better one must be used to obtain optimum chain life.

Types of lubrication indicated on tables are:

Type I: Manual or drip
Type II: Bath or disk
Type III: Oil stream

Table 7-8 3/16-In.-Pitch Silent Chain

Number of teeth	Revolutions per minute: 500	600	700	800	900	1200	1800	2000	3500	5000	7000	9000
15	0.28	0.33	0.38	0.43	0.47	0.60	0.80	0.90	1.33	1.66	1.94	1.96
17	0.33	0.39	0.44	0.50	0.55	0.70	0.96	1.05	1.60	2.00	2.40	2.52
19	0.37	0.43	0.50	0.55	0.61	0.80	1.10	1.20	1.80	2.30	2.76	2.92
21	0.41	0.48	0.55	0.62	0.68	0.87	1.22	1.33	2.03	2.58	3.12	3.35
23	0.45	0.53	0.60	0.68	0.75	0.96	1.35	1.47	2.25	2.88	3.50	3.78
25	0.49	0.58	0.66	0.74	0.82	1.05	1.47	1.60	2.45	3.13	3.80	4.10
27	0.53	0.62	0.71	0.80	0.88	1.15	1.58	1.72	2.63	3.35	4.06	4.37
29	0.57	0.67	0.76	0.86	0.95	1.21	1.70	1.85	2.83	3.61	4.40	4.72
31	0.60	0.72	0.81	0.91	1.01	1.30	1.81	1.97	3.02	3.84	4.66	5.00
33	0.64	0.75	0.86	0.97	1.07	1.37	1.90	2.08	3.17	4.02	4.85	
35	0.68	0.80	0.92	1.03	1.14	1.45	2.03	2.21	3.41	4.27	5.16	
37	0.71	0.84	0.96	1.08	1.19	1.52	2.11	2.30	3.48	4.39	5.24	
40	0.77	0.91	1.04	1.16	1.29	1.64	2.28	2.50	3.77	4.76		
45	0.86	1.02	1.15	1.30	1.43	1.83	2.53	2.75	4.15	5.21		
50	0.95	1.12	1.27	1.37	1.58	2.00	2.78	3.02	4.52	5.65		
	Type I						Type II			Type III		

Table 7-9 3/8-In.-Pitch Silent Chain

No. of Teeth Small Sprocket	Revolutions Per Minute – Small Sprocket												
	100	500	1000	1200	1500	1800	2000	2500	3000	3500	4000	5000	6000
* 17	.46	2.1	4.6	4.9	5.3	6.5	6.9	7.9	8.5	8.8	8.8	–	–
* 19	.53	2.5	4.8	5.4	6.5	7.4	7.9	9.1	9.9	10	11	9.8	–
21	.58	2.8	5.1	6.0	7.3	8.3	9.0	10	11	12	12	12	10
23	.63	3.0	5.6	6.6	8.0	9.3	10	12	13	14	14	14	12
25	.69	3.3	6.1	7.3	8.8	10	11	13	14	15	15	15	14
27	.74	3.5	6.8	7.9	9.5	11	12	14	15	16	18	18	16
29	.80	3.8	7.3	8.5	10	12	13	15	16	18	19	19	18
31	.85	4.1	7.8	9.1	11	13	14	16	18	19	20	20	19
33	.90	4.4	8.3	9.8	12	14	15	18	19	21	21	21	20
35	.96	4.6	8.8	10	13	15	16	19	20	23	23	23	21
37	1.0	4.9	9.1	11	14	15	16	20	21	24	24	24	–
40	1.1	5.3	10	12	15	16	18	21	24	25	26	26	–
45	1.3	6.0	11	13	16	19	20	24	26	28	29	–	–
50	1.4	6.6	13	15	18	20	23	25	29	30	–	–	–

Type I Type II Type III

* For best results, smaller sprocket should have at least 21 teeth.

Table 7-10 1/2-In.-Pitch Silent Chain

No. of Teeth Small Sprocket	Revolutions Per Minute – Small Sprocket										
	100	500	700	1000	1200	1800	2000	2500	3000	3500	4000
* 17	.83	3.8	5.0	6.3	7.5	10	11	11	11	11	–
* 19	.93	3.8	5.0	7.5	8.8	11	13	14	14	14	–
21	1.0	5.0	6.3	8.8	10	14	14	15	16	16	–
23	1.1	5.0	7.5	10	11	15	16	18	19	19	18
25	1.2	5.0	7.5	10	13	16	18	20	21	21	20
27	1.3	6.3	8.8	11	13	18	19	21	24	24	23
29	1.4	6.3	8.8	13	14	19	21	24	25	25	25
31	1.5	7.5	10	13	15	21	23	25	28	28	28
33	1.6	7.5	10	14	16	23	24	28	29	30	29
35	1.8	7.5	11	15	18	24	25	29	31	31	30
37	1.9	8.8	11	16	19	25	26	30	33	33	–
40	2.0	8.8	13	18	20	28	29	33	35	35	–
45	2.5	10	14	19	23	30	30	36	39	–	–
50	2.5	11	15	21	25	34	36	40	–	–	–

Type I Type II Type III

* For best results, smaller sprocket should have at least 21 teeth.

Table 7-11 5/8-In.-Pitch Silent Chain

No. of Teeth Small Sprocket	Revolutions Per Minute – Small Sprocket									
	100	500	700	1000	1200	1800	2000	2500	3000	3500
* 17	1.3	6.3	7.5	10	11	14	15	14	–	–
* 19	1.4	6.3	8.8	13	14	16	18	18	–	–
21	1.6	7.5	10	13	15	19	20	20	20	–
23	1.8	7.5	11	15	16	21	23	24	23	–
25	1.9	8.8	11	16	19	24	25	26	26	24
27	2.0	10	13	18	20	26	28	29	29	26
29	2.1	10	14	19	21	28	30	31	31	29
31	2.4	11	15	20	23	30	31	34	34	31
33	2.5	11	16	21	25	33	34	36	36	34
35	2.6	13	16	23	26	34	36	39	39	35
37	2.8	13	18	24	28	36	39	43	41	–
40	3.0	14	19	26	30	39	41	44	–	–
45	3.4	16	21	29	34	44	46	–	–	–
50	3.8	18	24	33	38	48	50	–	–	–

Type I Type II Type III

* For best results, smaller sprocket should have at least 21 teeth.

Table 7-12 3/4-In.-Pitch Silent Chain

No. of Teeth Small Sprocket	Revolutions Per Minute – Small Sprocket								
	100	500	700	1000	1200	1500	1800	2000	2500
* 17	1.9	8.1	11	14	15	16	18	18	–
* 19	2.0	9.3	13	15	18	2C	21	21	–
21	2.3	10	14	18	20	23	24	25	24
23	2.5	11	15	20	23	25	28	28	28
25	2.8	13	16	21	25	29	31	31	30
27	2.9	14	18	24	28	31	34	35	35
29	3.1	15	20	26	30	34	36	38	38
31	3.4	15	21	28	31	36	40	41	41
33	3.6	16	23	30	34	39	43	44	44
35	3.8	18	24	31	36	41	45	46	46
37	4.0	19	25	34	39	44	48	49	49
40	4.4	20	28	36	41	48	51	53	53
45	4.9	23	30	40	46	53	56	58	–
50	5.4	25	34	45	51	58	61	–	–

Type I Type II Type III

* For best results, smaller sprocket should have at least 21 teeth.

Step 2. Design horsepower. The design horsepower is $190 \times 1.4 = 266$ hp.

Step 3. Tentative chain pitch selection. As indicated in Figure 7-7, the chain pitches most adaptable to a driver shaft speed of 600 rpm are 1-1/2 and 2 in. Note that 1-1/4-in. pitch may also be considered. Therefore, investigate the use of 1-1/4- and 1-1/2-in. pitch chains since it is usually wise to use the smallest *practical* pitch of chain.

Consideration should also be given at this point to the fact that smaller-pitch chain may require more frequent adjustments.

Step 4. Chain and smaller sprocket selection. Table 7-17 shows that the smallest 1-1/4-in.-pitch sprocket that will accommodate the 4-3/4-in.-diameter motor shaft has 23 teeth, and that no less than a 19-tooth sprocket may be used on a 1-1/2-in.-pitch sprocket.

Table 7-13 1-In.-Pitch Silent Chain

No. of Teeth Small Sprockets	Revolutions Per Minute – Small Sprocket										
	100	200	300	400	500	700	1000	1200	1500	1800	2000
* 17	3.8	6.3	8.8	11	14	18	21	23	–	–	–
* 19	3.8	7.5	10	13	15	20	25	26	28	–	–
21	3.8	7.5	11	15	18	23	29	31	33	33	–
23	3.8	8.8	13	16	19	25	31	35	38	38	–
25	5.0	8.8	14	18	21	28	35	39	41	41	41
27	5.0	10	15	19	24	30	39	43	46	46	45
29	5.0	11	16	20	25	33	41	46	50	51	50
31	6.3	11	16	23	28	35	45	50	54	55	54
33	6.3	13	18	24	29	38	49	54	59	59	58
35	6.3	13	19	25	30	40	51	56	61	63	61
37	6.8	14	20	26	33	43	54	60	65	66	–
40	7.5	15	23	29	35	45	59	65	70	–	–
45	8.8	16	25	31	39	51	65	71	76	–	–
50	10	19	28	35	43	56	71	78	–	–	–

Type I Type II Type III

* For best results, smaller sprocket should have at least 21 teeth.

Table 7-14. 1-1/4-In.-Pitch Silent Chain[a]

1¼" Pitch

No. of Teeth Small Sprockets	Revolutions Per Minute – Small Sprocket										
	100	200	300	400	500	600	700	800	1000	1200	1500
* 19	5.6	10	15	20	24	26	29	31	34	35	–
21	6.3	11	18	23	26	30	33	36	40	41	–
23	6.9	13	19	24	29	34	36	40	45	46	46
25	7.5	14	20	26	31	36	40	44	50	53	53
27	8.0	15	23	29	35	40	44	49	54	58	58
29	8.6	16	24	31	38	43	48	53	59	63	64
31	9.3	18	26	34	40	46	51	56	64	68	69
33	9.9	19	28	35	43	49	55	60	69	73	74
35	11	20	29	38	45	53	59	64	73	78	78
37	11	21	30	40	48	55	63	68	76	81	–
40	12	24	34	44	53	60	68	74	83	88	–
45	13	26	38	49	59	68	75	81	91	–	–
50	15	29	43	54	65	74	83	90	100	–	–

Type I Type II Type III

* For best results, smaller sprocket should have at least 21 teeth.

Since, in general practice, the small sprocket ranges from 21 to 29 teeth, investigate the horsepower ratings at 600 rpm for the following:

1-1/4-in. pitch: 23-, 25-, 27-, and 29-tooth sprockets

1-1/2-in. pitch: 21-, 23-, 25-, 27-, and 29-tooth sprockets

The amounts given in Tables 7-8 to 7-16 are the horsepower capacities per inch of chain width. Therefore, to determine the width of chain required to transmit the design horsepower, divide 266 by the rating per inch of chain width for each selection. These computations are indicated in the tabulation below, which also gives the chain widths that are indicated as available in Table 7-18.

Table 7-15 1-1/2-In.-Pitch Silent Chain

1½" Pitch

No. of Teeth Small Sprockets	Revolutions Per Minute – Small Sprocket										
	100	200	300	400	500	600	700	800	900	1000	1200
* 19	8.0	15	21	28	31	35	39	40	41	43	–
21	8.8	16	24	30	36	40	44	46	49	49	–
23	10	19	26	34	40	45	49	53	55	56	55
25	10	20	29	38	44	50	55	59	61	65	64
27	11	23	31	40	48	54	60	64	68	70	70
29	13	24	34	44	51	59	65	70	74	75	76
31	14	25	36	46	55	64	70	75	79	81	83
33	14	28	39	50	59	68	75	80	85	88	89
35	15	29	41	53	63	71	79	85	90	93	94
37	16	30	44	59	66	76	84	90	96	99	–
40	18	33	48	66	73	83	90	98	105	–	–
45	19	38	54	68	81	93	101	108	113	–	–
50	21	41	59	75	89	101	111	118	–	–	–

Type I Type II Type III

* For best results, smaller sprocket should have at least 21 teeth.

Table 7-16 2-In.-Pitch Silent Chain

2'' Pitch

No. of Teeth Small Sprocket	Revolutions Per Minute – Small Sprocket								
	100	200	300	400	500	600	700	800	900
* 19	14	26	36	44	50	54	56	–	–
21	16	29	40	50	53	63	65	–	–
23	17	33	45	55	64	70	74	75	–
25	18	35	49	61	70	78	83	85	85
27	20	38	54	66	78	85	91	94	94
29	21	41	58	73	84	93	99	103	103
31	23	44	63	78	90	100	106	110	110
33	25	46	66	83	96	106	114	118	118
35	26	50	71	88	103	114	121	125	125
37	28	53	75	93	110	124	128	131	–
40	30	58	81	101	118	129	138	141	–
45	34	64	90	113	131	144	151	–	–
50	38	71	100	125	144	156	–	–	–

Type I Type II Type III

* For best results, smaller sprocket should have at least 21 teeth.

Table 7-17 Maximum Bores for Silent Chain Sprockets

Number of Teeth	Chain Pitch							
	3/8	1/2	5/8	3/4	1	1¼	1½	2
17	1	1⅜	1¾	2	2¾	3⅜	4⅛	5½
19	1¼	1⅝	2	2⅜	3¼	4	4⅞	6¾
21	1⅜	1⅞	2¼	2¾	3¾	4½	5½	7¾
23	1⅝	2⅛	2⅝	3¼	4⅜	5½	6½	9
25	1¾	2⅜	3	3⅝	4¾	6	7¼	10
27	2	2⅝	3⅜	3⅞	5⅜	6¾	8⅛	11¼
29	2⅛	2 13/16	3⅝	4⅜	5¾	7⅜	9⅛	12½
31	2 5/16	3 1/16	3⅞	4⅝	6⅜	8	9⅞	13½

Table 7-18 Standard Silent Chain Widths

Chain Pitch	Standard Widths																									
3/8	1/2	3/4	1	1¼	1½	1¾	2	2¼	2½		3		4	5	6											
1/2	1/2	3/4	1	1¼	1½	1¾	2	2¼	2½	2¾	3	3½	4	5	6		8									
5/8			1	1¼	1½	1¾	2		2½		3		4	5	6	7	8		10							
3/4			1	1¼	1½		2		2½		3	3½	4	5	6	7	8	9	10	12						
1							2		2½		3		4	5	6	7	8	9	10	12	14	16				
1¼									2½		3		4	5	6	7	8	9	10	12	14	16	18	20		
1½											3		4	5	6	7	8	9	10	12	14	16	18	20	22	24
2													4	5	6	7	8		10	12	14	16	18	20	22	24	30

Selection	Pitch	Teeth	Design hp ÷ by hp/in.		Required chain width	Width available (in.)
A	1¼	23	266 ÷ 34	=	7.82	8
B	1¼	25	266 ÷ 36	=	7.39	8
C	1¼	27	266 ÷ 40	=	6.65	7
D	1¼	29	266 ÷ 43	=	6.19	7
E	1½	21	266 ÷ 40	=	6.65	7
F	1½	23	266 ÷ 45	=	5.91	6
G	1½	25	266 ÷ 50	=	5.32	6
H	1½	27	266 ÷ 54	=	4.93	5
I	1½	29	266 ÷ 59	=	4.51	5

From a review of these selections, it can be seen that selection F is the most practical. The chain pitch is the mean recommended in Figure 7-7. The number of teeth in the small sprocket is quite close to the minimum recommended and within the range of normal application. The small sprocket will adequately accept the bore requirement, and the chain width to be used is within 3% of width calculated.

Step 5. Selection of larger sprocket. Since the driver shaft operates at 600 rpm and the driven shaft at 150 rpm, the speed ratio = 600/150 = 4.0. Therefore, the large sprocket will have 23 × 4 = 92 teeth.

Step 6. Chain length. Using 23- and 92-tooth sprockets on 60-in. centers (60-in. ÷ 1-1/2 in. = 40 pitches), calculations to determine chain length per equation (5-11) (p. 000) are as follows:

$$L = 2C + \frac{N+n}{2} + \frac{(N-n)^2}{4\pi 2C}$$

where

L = chain length, pitches
C = shaft centers, pitches
N = number of teeth in large sprocket
n = number of teeth in small sprocket

Substituting values for C, N and n, we have

$$L = 2 \times 40 + \frac{92+23}{2} + \frac{(92-23)^2}{39.48 \times 40}$$

$$= 140.515 \text{ pitches}$$

Use 140 pitches.

Step 7. Correction of center distance. Since the chain is to couple at an even number of pitches, use 140 pitches for L and recompute the centers in pitches from equation (5-12).

$$C = \frac{L - \frac{N+n}{2} + \sqrt{\left(L - \frac{N+n}{2}\right)^2 - 8\frac{(N-n)^2}{4\pi^4}}}{4}$$

$$= 140 - \frac{92+23}{2} + \sqrt{\left(140 - \frac{92+23}{2}\right)^2 - 8\frac{(92-23)^2}{39.48}}$$

$$= 39.73 \text{ pitches}$$

or

$$39.73 \times 1\tfrac{1}{2} = 59.60 \text{ in.}$$

Summary

Since none of the other possible selections satisfy the drive requirements quite as closely, our final selection is 23- and 92-tooth sprockets mounted on 59.60-in. centers using 1-1/2-in. pitch 6-in.-wide chain.

In the horsepower rating table for 1-1/2-in.-pitch chain (Table 7-15) the recommended method of lubrication can be provided by either a disk mounted adjacent to one of the sprockets, or by a circulating pump. Assuming the relative shaft elevations would permit the design of an oil sump in the drive housing, the disk method will be satisfactory.

Installation, Maintenance, and Lubrication

Service of Silent Chain drives, including connecting and disconnecting the chain, is covered in Chapter 12.

The Silent Chain of the S.S. *Great Britain*

In Chapter 1 reference was made to an important early use of a Silent Chain drive. In the historic S.S. *Great Britain,* it was used as a speed-increasing drive between the steam engines and the propeller shaft.

A great deal of data are available on the details of that drive, which for obvious reasons was quite well documented at the time. The *Great Britain* had a great impact in its field, being the first iron screw steamer intended for major transatlantic commercial service, and it was the first screw steamer in such service, also. Her chain drive was just one of many innovative designs first placed in service on the *Great Britain.*

The service in which the *Great Britain* chains were placed was very large, even by modern standards. As compared to modern chain installations, chain speed was low and pull high. But the reason for use of Silent Chains was as good then as today. Noise aboard a new passenger vessel, particularly in an era when passengers were accustomed to sail and not the sounds of machinery, was a major consideration for the *Great Britain* design. Reliability was also a factor. The engines of the *Great Britain* were expected to run on a 24-hr basis, just as with a modern ship, sail being only a standby. Smoothness of operation was also necessary, and it is worth noting that the *Great Britain* was expected to set records for fast passage, as she had a hull of excellent design.

Setting up the data of the *Great Britain's* power train, the following data are available:

Horsepower: 1800 hp total output of two inverted "vee" (60°) "triangle" steam engines as (invented and patented by Sir Marc Brunel, father of the builder of the *Great Britain*) of two cylinders each, with a drive wheel on a common shaft between (see Fig. 7-19).

Shaft speeds: engine shaft output—18 rpm at engine design speed; desired maximum propeller shaft speed—53/54 rpm. (Note. This is derived partially from a reported speed ratio of the drive of 2.95 : 1.)

Load and power source: Variable, but steady load, 24 hr per day for from 14 to 20 days (or until out of coal), continuously. (Note: From Table 7-6, a service factor of 1.3 would have been required.)

Center distance: Approximately 21 ft. The depth of hold was 32 ft. The upper drive wheel protruded about 3 ft above the deck, it being a source of attention through glass windows of an enclosing cabin. The lower wheel, down near the keel, was 6 ft in diameter.

Of the design, which was apparently an effective solution by Sir Isambard Kingdom Brunel, the designer, many facts concerning the wheels are available. The upper

Figure 7-19 The power train of the S.S. *Great Britain* of the 1840s: steam engines, crankshaft and drive wheel, silent chain, driven wheel and propeller shaft with propeller.

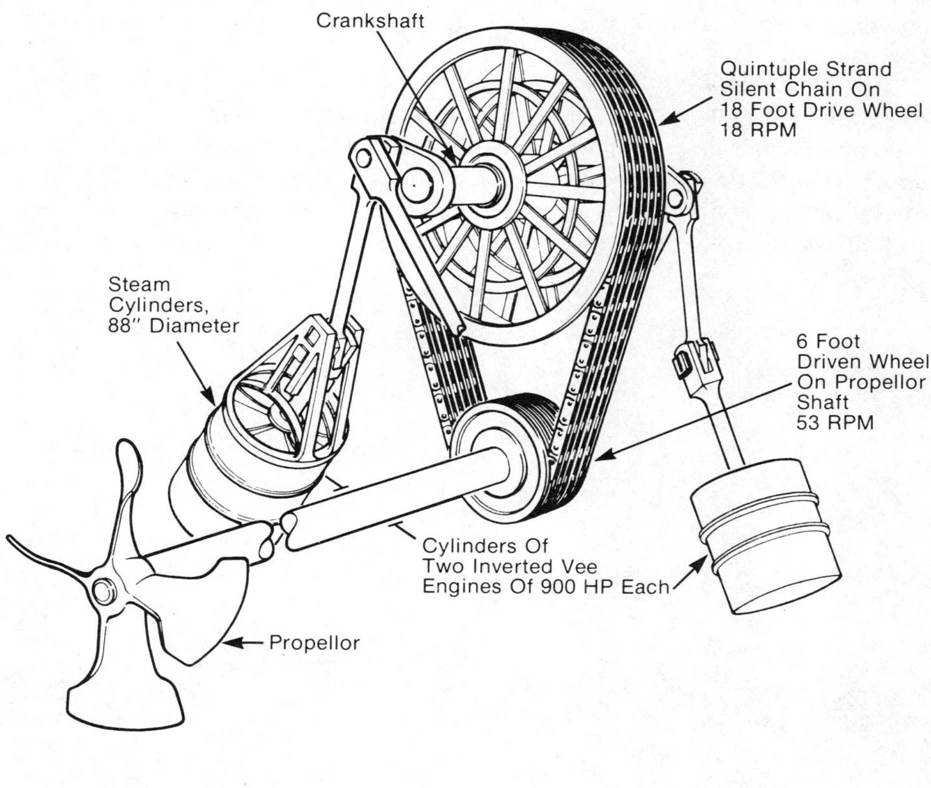

or drive wheel, some 18 ft, 3 in. in outside diameter, had a pitch circle of 18 ft. The lower wheel was approximately 6 ft, but neither the pitch of the chains nor the numbers of teeth in the sprockets were reported. The drive wheel apparently had iron spokes and a teakwood rim, with flanges. Teeth were wooden, inserted and pinned in holes in the rim. The lower wheel was lignum vitae.

The common engine shaft, a huge forging with crank pins on either end, was 28 in. in diameter with 24-in. bearings and was hollow, so that it could be water cooled. The propeller shaft was 18 in.

The chains were made in three separate strands run close together in a mass 38 in. wide on the wheels. Each strand consisted of alternate links having two and three link plates each. The link plates for the two-plate links were probably thicker than those in threes, as the cross-sectional area was reported to have been an apparently uniform 24 in.[2] The chains weighed 7 tons.

The links were apparently of forged steel, and according to a contemporary account, were brought to pitch by bringing to red heat and then carefully inspecting and stretching the links by 1/8 in. They were then bored and planed, and finished (probably by hand) "against a gauging tool." The links were case hardened.

The Silent Chain drive of the *Great Britain* operated quietly and smoothly until she finally went aground after eight trips across the Atlantic. Operations had been somewhat marred by occasional loss of propeller blades, which was a problem the designer never really solved. No reports of any trouble with the chains stemming from losing propeller blades at maximum speed were reported.

After going ashore, which event bankrupted her owners, the *Great Britain,* badly damaged in her drive train, was salvaged after several months in the sands. After a few years, she was sold and new machinery installed, as auxiliary steam power, for the Australian trade.

The *Great Britain* still survives, having been restored for exhibition in England, but her innovative Silent Chain drive exists now only as a model.

Comparing this drive with modern Silent Chain drives, shaft torque is very high and rotative speed slow, but the drive is a very large one even by modern standards. If there is a lesson in reviewing this drive, it is that the designer can use chain for applications well away in concept from the conventional "catalog"-style applications.

8
Tension Linkage Chains

Tension linkage describes a spectrum of chain applications in which the chain's main function is to move a load slowly or intermittently through a short distance, or to hold the load in place when it is not moving. In applications of this kind, the chain generally moves back and forth, rather than continuously in one direction.

The function of this type of chain is thus to transmit a moving force as chain tension; hence the term "tension linkage." Applications are probably more frequently a form of power transmission than conveying, but either, or both in combination, are not unusual in tension linkage applications.

Tension linkage chains are used for hoisting and positioning various loads, supporting counterweights, and drawing metal. Tensile loads on such chains can range from very light to the heaviest loads found on any chain products. As may be seen in Figure 8-1, size of the chains used may also vary accordingly. Frequency of chain movement or travel, even with continuous loading, may be fairly frequent, as with a lift truck, or may remain nearly immobile for months, as with the chains supporting a gate on a river dam.

Special classes of Roller Chain and leaf and laminated steel block chains, made by manufacturers of Roller Chains, are offered for tension linkages. ANSI standards exist on some of these types, but many more are listed in manufacturers' catalogs. A brief discussion of leaf chain appeared in Chapter 2.

Engineering Steel Chains, mainly in the various styles of bar-link or block-and-bar chain, are extensively used for tension linkages.

The applications of the various kinds will be taken up by types, starting with Roller Chains.

Roller Chains Used for Tension Linkages

Standard Roller Chain is frequently used for static or slow-moving tensile applications where compactness and flexibility or engagement with sprockets is needed. Supporting floating machine tool heads (as shown in Fig. 8-5), lifting private residence

Figure 8-1 Comparison of chains to show the size of some tension linkage chains.

elevators, and linking remote steering mechanisms on lift trucks are three examples of the many ways that Roller Chain may be used in tension linkage applications.

Rollerless Roller Chain, often called *lift* or *hoist* chains, has identical basic outer dimensions as Roller Chains. Whereas most of these chains are made with curled "rolls" and to rated strength, for some, bushings and rollers are provided as one solid piece. Such chains frequently are furnished with higher strength than Roller Chains of equivalent pitches.

Tension linkages in general, as stated, have high loads at very low operating speeds. These data do not normally fall within the ranges of the horsepower tables given for standard Roller Chain in Chapter 5, and extrapolation or projection of data from those tables for tension linkage applications should not be attempted. Rather, manufacturers of such chains should be consulted. Figure 8-2 shows an installation having several examples of Roller Chain in tension linkage service.

Rollerless Roller Chains and extended pitch Roller Chains of the power transmission and conveyor types are also used, as are many special variations of standard Roller Chain such as a specially made high-capacity chain commonly used on portable hoists. Because of the large number of products and the various ways these may be used, the reader is encouraged to consult Roller Chain manufacturers for use on specific products and application recommendations for them.

Tension Linkages Using Leaf Chain

The most widely used chain for tension linkage applications is Leaf Chain, which is made specifically for such usages.

Leaf Chain, described in Chapter 2, together with illustrations indicating those dimensions standardized by ANSI Standard B29.8, was shown in Figure 2-13. Figure 8-3 shows what is probably the most common application of Leaf Chain; use on a hydraulic lift of an industrial lift truck. Fixed at the ends with clevises, the chain passes over sheaves on the hydraulically activated platen to multiply the travel of the "fork." Figure 8-4 shows a photo of a typical Leaf Chain of this type, and Figures 8-5 and 8-6 show additional applications.

Dimensions and Arrangements of Leaf Chain

Figure 8-7 illustrates Leaf Chain lacings covered by Standard B29.8 and in general usage. The dimensions are shown in terms of the proportioning rules, and as may be read from the chain number, the first digit (after the series designation letters) represents the chain pitch expressed in eighths of inches, and the remaining digits indicate the numbers of plates in alternate links.

Leaf Chain Strengths. The minimum ultimate tensile strengths specified in ANSI Standard B29.8 are tabulated in Table 8-8.

Any chain test performed to verify the minimum tensile strengths in this table should be considered destructive. Even though the chain survives the test, its load-carrying capability will have been compromised and the chain left in unfit condition for use.

Leaf Chain Dimensions. The general dimensions for Leaf Chains as specified in the ANSI standard are shown in Table 8-9.

Figure 8-2 Actuation accomplished by means of roller chain. Any chain slack eliminated by introducing turnbuckles.

Figure 8-3 Leaf Chain in use on a hydraulic lift truck. (Courtesy Clark Equipment Co.)

Tolerance for chain length is plus or minus 1/32 in. per foot of chain under a measuring load of 1% of the minimum ultimate tensile strength, but not exceeding 1000 lb.

Leaf Chain Clevises. The general proportions of terminal clevises for BL series Leaf Chains are shown in Figure 8-10.

Care must be exercised in the manufacture and attachment of clevises to assure equal load distribution across the chain. Failure to do so will compromise the load-carrying capability of the chain.

The clevis and pins used to anchor the chain should accept the breaking strength of the chain.

Sheaves. The general proportions for pulleys or sheaves to be used with Leaf Chains are shown in Figure 8-11. Sheave diameters smaller than the recommended minimum 5P may be used to suit special design needs, but the design should be verified by testing.

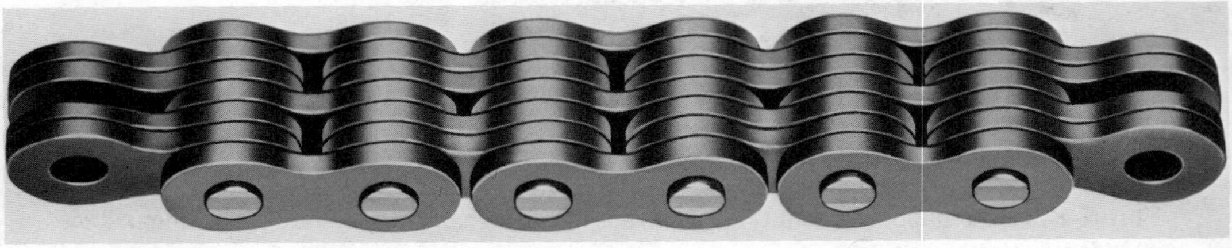

Figure 8-4 Leaf Chain as used in tension linkage service.

Figure 8-5 Leaf Chain is sometimes used as a tension linkage connection between machine tool heads and counterweights.

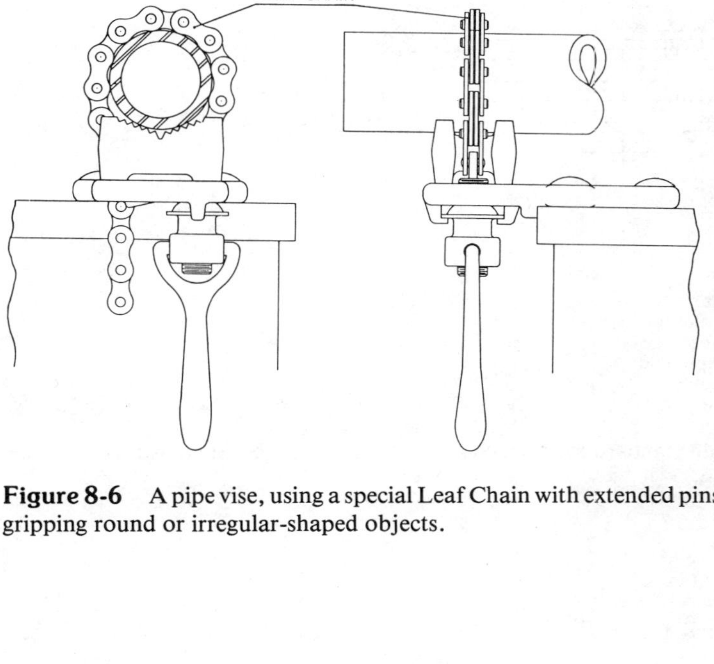

Figure 8-6 A pipe vise, using a special Leaf Chain with extended pins. Such vises are used for gripping round or irregular-shaped objects.

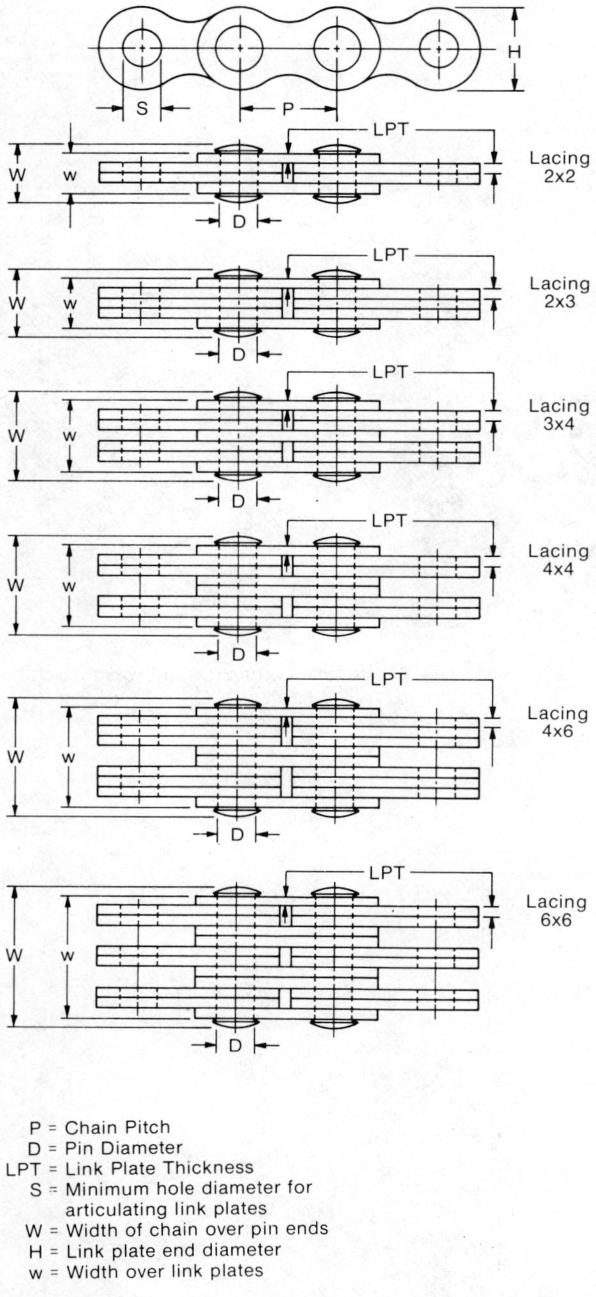

Figure 8-7 Leaf Chain standard lacings and dimensions, per ANSI Standard B29.8.

Table 8-8 Leaf Chains—Minimum Ultimate Tensile Strengths

Chain number	Minimum ultimate tensile strength (lb)	Chain number	Minimum ultimate tensile strength (lb)	Chain number	Minimum ultimate tensile strength (lb)
BL 422	5,000	BL 822	19,000	BL 1422	43,000
BL 423	5,000	BL 823	19,000	BL 1423	43,000
BL 434	7,500	BL 834	29,000	BL 1434	71,000
BL 444	10,000	BL 844	38,000	BL 1444	86,000
BL 446	10,000	BL 846	38,000	BL 1446	86,000
BL 466	15,000	BL 866	57,000	BL 1466	130,000
BL 522	7,500	BL 1022	26,000	BL 1622	65,000
BL 523	7,500	BL 1023	26,000	BL 1623	65,000
BL 534	11,000	BL 1034	41,000	BL 1634	99,000
BL 544	15,000	BL 1044	52,000	BL 1644	130,000
BL 546	15,000	BL 1046	52,000	BL 1646	130,000
BL 566	22,500	BL 1066	78,000	BL 1666	195,000
BL 622	11,000	BL 1222	34,000	BL 2022	97,500
BL 623	11,000	BL 1223	34,000	BL 2023	97,500
BL 634	17,000	BL 1234	55,000	BL 2034	146,000
BL 644	22,000	BL 1244	68,000	BL 2044	195,000
BL 646	22,000	BL 1246	68,000	BL 2046	195,000
BL 666	33,000	BL 1266	102,000	BL 2066	292,500

Sheaves are turning points; therefore, the following factors should be considered when designing a sheave for a particular application.

1. The sheave should be fitted with a bushing or bearings, or be keyed to a freely rotating bearing-mounted shaft.
2. Bearings or bushings should be provided with a means of lubrication.
3. The sheave face should be parallel to the centerline of rotation to assure full support of the chain width.

Other Styles of Leaf Chains. There are other types of Leaf Chains designed for tension linkage applications. Among these are laminated block Leaf Chain, as shown at the left in Figure 8-12. This type of chain is used when loads must be applied to the chain at central points as well as at the ends. The chains are arranged with outside plates only in alternate links, so that sprockets or other "toothed" devices may be used with the chain.

Wrench chain, shown at the right in Figure 8-12, is another style of Leaf Chain used in the same way, except that intermediate loads are applied through the pins, which are extended at each joint from both sides of the chain.

These and other specialized styles may be found listed in the catalogs of most manufacturers.

Table 8-9 Leaf Chains—General Dimensions, BL Series

Chain number	Pitch	Lacing	w Max.	W Max.	D Max.	S Min.	H Max.	LPT Max.
BL 422	0.500	2×2	0.334	0.434	0.2005	0.2015	0.475	0.082
BL 423		2×3	0.418	0.518				
BL 434		3×4	0.585	0.685				
BL 444		4×4	0.668	0.768				
BL 446		4×6	0.835	0.935				
BL 466		6×6	1.002	1.102				
BL 522	0.625	2×2	0.390	0.507	0.2345	0.2355	0.594	0.096
BL 523		2×3	0.488	0.605				
BL 534		3×4	0.683	0.800				
BL 544		4×4	0.780	0.897				
BL 546		4×6	0.975	1.092				
BL 566		6×6	1.170	1.287				
BL 622	0.750	2×2	0.528	0.684	0.3125	0.3135	0.713	0.130
BL 623		2×3	0.660	0.816				
BL 634		3×4	0.924	1.080				
BL 644		4×4	1.056	1.212				
BL 646		4×6	1.320	1.476				
BL 666		6×6	1.584	1.740				
BL 822	1.000	2×2	0.652	0.840	0.3755	0.3765	0.950	0.161
BL 823		2×3	0.815	1.003				
BL 834		3×4	1.141	1.329				
BL 844		4×4	1.304	1.492				
BL 846		4×6	1.630	1.818				
BL 866		6×6	1.956	2.144				
BL 1022	1.250	2×2	0.780	0.999	0.4375	0.4385	1.188	0.193
BL 1023		2×3	0.975	1.194				
BL 1034		3×4	1.365	1.584				
BL 1044		4×4	1.560	1.779				
BL 1046		4×6	1.950	2.169				
BL 1066		6×6	2.340	2.559				
BL 1222	1.500	2×2	0.916	1.166	0.5005	0.5015	1.425	0.227
BL 1223		2×3	1.145	1.395				
BL 1234		3×4	1.603	1.853				
BL 1244		4×4	1.832	2.082				
BL 1246		4×6	2.290	2.540				
BL 1266		6×6	2.748	2.998				
BL 1422		2×2	1.010	1.321				
BL 1423		2×3	1.300	1.581				

Table 8-9 *(continued)*

Chain number	Pitch	Lacing	w Max	W Max.	D Max.	S Min.	H Max.	LPT Max.
BL 1434	1.750	3×4	1.820	2.101	0.5625	0.5635	1.663	0.258
BL 1444		4×4	2.080	2.361				
BL 1446		4×6	2.600	2.881				
BL 1466		6×6	3.120	3.401				
BL 1622		2×2	1.192	1.536				
BL 1623		2×3	1.490	1.834				
BL 1634	2.000	3×4	2.086	2.430	0.6875	0.6885	1.900	0.296
BL 1644		4×4	2.384	2.728				
BL 1646		4×6	2.980	3.324				
BL 1666		6×6	3.576	3.920				
BL 2022		2×2	1.568	2.037				
BL 2023		2×3	1.960	2.429				
BL 2034	2.500	3×4	2.744	3.213	0.9375	0.9385	2.375	0.390
BL 2044		4×4	3.136	3.605				
BL 2046		4×6	3.920	4.389				
BL 2066		6×6	4.704	5.173				

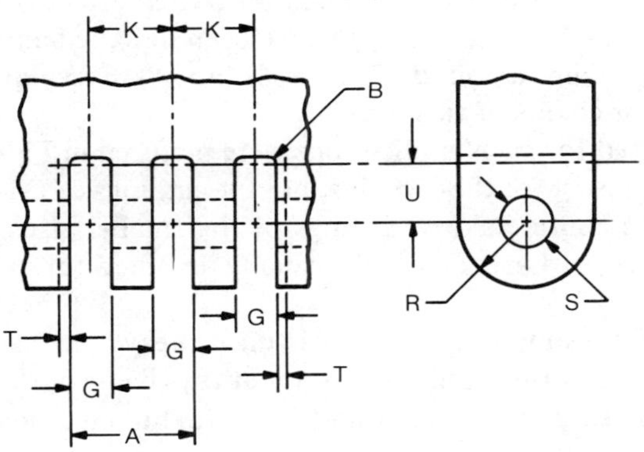

P = Chain pitch
LPT max. = Maximum link plate thickness
S = Minimum hole diameter
R = End radius = 0.5 P
B = Fillet radius
T = Minimum thickness of outside flange = LPT max
U = Minimum depth of slot for clearance = 0.5 P
CL = Clearance = 0.0015 inch for ⅝-inch pitch and smaller
= 0.002 inch for ¾-inch pitch and larger

Lacing	
2 x 2	G = 2 LPT max + 4 CL + 0.008 P
2 x 3	G = 3 LPT max + 5 CL + 0.008 P
3 x 4	A = K + G K = 3 (LPT max + CL) G = K - (LPT max - 0.004 P)
4 x 4 6 x 6	A = K + G K = 4 (LPT max + CL) G = K - 2 (LPT max - 0.004 P)
4 x 6	A = K + G K = 5 (LPT max + CL) G = K - 2 (LPT max - 0.001 P)

Figure 8-10 General proportions of clevises for Leaf Chains.

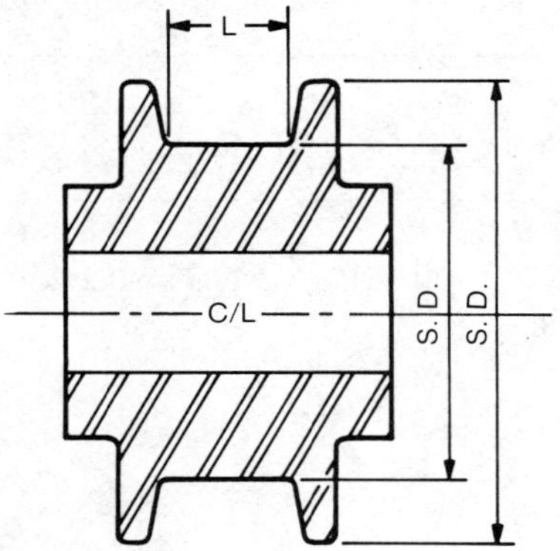

S D = Minimum recommended sheave diameter = 5P*
L = Minimum distance between flanges = 1.05 W max (see Table 8-9)
F D = Flange diameter = SD + H max
F D max = SD + 1 inch
H max = 0.95 P (see Table 8-9)

*Other sheave sizes may be used at the discretion of the end user to suit particular designs needs, but the validity of the usage should be supported by testing.

Figure 8-11 General proportions of Leaf Chain sheaves.

Figure 8-12 Two special-purpose types of Leaf Chain.

Tension Linkage Applications Using Engineering Steel Chains

Tension linkage applications using Engineering Steel Chains usually involve relatively heavy chains. The chains used, usually of the bar-link (or block-and-bar) style, are much larger in size and much stronger than the Leaf and other chains used for lighter tension linkage service.

Some of the heavy tension linkage chains of the Engineering Steel bar-link type are among the heaviest, most massive, and durable chains ever built. Their effectiveness depends on strength and not speed. Chains of this type have been designed and built with an ultimate strength of over 3 million pounds. Figures 8-1 and 8-16 display the massiveness sometimes required for chains of this type.

Some Engineering Steel Chains used for such applications were pictured in Figures 1-29, 2-19, and 2-20. No standards exist for these chains, but the catalogs of Engineering Steel Chains usually contain comprehensive listings of the smaller sizes, which still include some very heavy chains. Larger sizes are usually designed for an application and built to order.

Most heavy tension linkage chains are simple designs consisting of heavy blocks, bars, and leaves with heat-treated pins, the other components not always being hardened. Bar-link and Leaf Chains are very similar in concept and function, but methods of manufacturing differ greatly.

Characteristics of Engineering Steel Chain Tension Linkages

In many cases there are no rigid rules to govern the engineer's choice of the type of tension linkage to be used for a given application; however, chain has the following advantages:

1. Where positive traction is necessary, a chain meshed with a sprocket is preferable.
2. With chain, articulation (or flexing) is confined to specially treated bearing surfaces.
3. Chains are often the better choice for very heavy tension linkages because they can be designed to handle very high loads.
4. Chains used in tension linkage applications are easy to lubricate and their life is extended through relubrication at regular intervals.
5. Chains are easy to install. End terminals, or connectors, simplify the attachment of the chain to the load.

Draw Bench Applications

Draw benches (see Fig. 8-13) are machines used in metal processing industries to manufacture wire and tubular shapes from such materials as steel, copper, and aluminum. On continuous benches a heavy block, with a hook that engages the chain, slowly draws the metal through a series of dies to produce the desired size of wire or tubing. At the completion of the stroke, the hook disengages from the chain and the draw block returns to its starting position. In this application, the chain operates continuously, picking up the load instantly with each pass of the draw block.

On reversing benches (Fig. 8-14), the block is connected to the chain and is pulled with a high-capacity chain (lower chain) and retracted with a smaller, more economical chain (above).

Most draw bench chains are heavy-duty types of the bar-link chain previously described. In chains intended for draw bench service, all parts are hardened; bearing surfaces are machined, and pins securely riveted to endure the stresses of draw bench oper-

Figure 8-13 A draw bench chain in a steel mill used for drawing tubing through a series of dies.

Figure 8-14 Reversing draw benches have high-capacity bar-link chains for the heavy pull, and smaller chains, above, to return the block.

ation. Figure 8-15 shows typical draw bench chains. As shown in Figure 2-19, designs with a single block alternating with two sidebars are used for the least rigorous kinds of draw bench work. Their ultimate strength range is from 30,000 to 200,000 lb. Designs in which two and three sidebars alternate may have an ultimate strength well in excess of 3 million pounds and are used for very heavy duty work.

The pitches of bar-link chains used in draw bench service may be as short as 2 in. or as long as 12 in. The weights of such chains may run from 4 to 400 lb/ft.

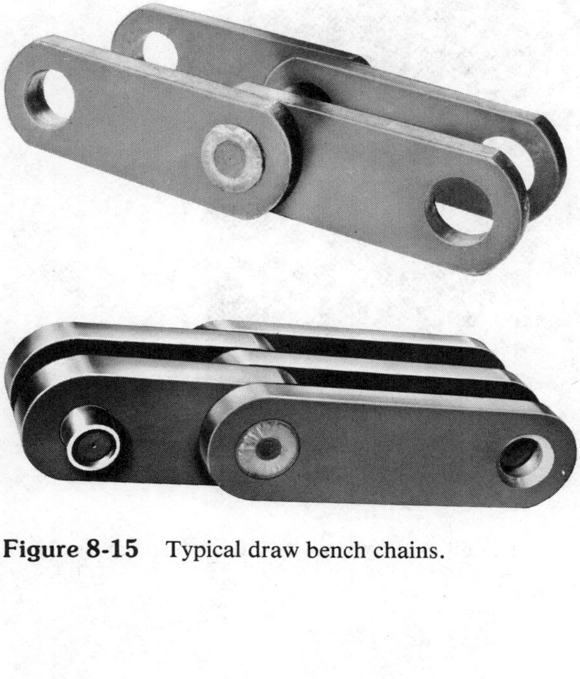

Figure 8-15 Typical draw bench chains.

Giant bushed rollerless chains (Fig. 8-1) are being used with increasing frequency on higher-speed draw benches. This construction provides more generously proportioned wear areas hardened to high levels for maximum wear resistance. The construction also allows use of a full complement of sprocket teeth for smoother operation or more compact drives.

Tension Linkage Chains for Dam Gates

Tension linkage chains are widely used to raise and lower dam gates and draw bridges, and to open and close canal lock gates. Bar-link chain has been popular for these services because of its simple design and relatively low cost, but other varieties of tension linkage chain are being used with increasing frequency. The giant bushed rollerless dam gate chain shown in Figure 8-16 incorporates relubrication features, allowing greasing of the inner chain joint for protection against corrosion and wear. Giant roller chain also incorporating relubrication features is shown in Figure 8-17 as installed to lift and hold a frequently used navigation lock gate that weighs 1 million pounds.

Other Applications

Other tension linkage applications for Engineering Steel Chain include the hold-down apparatus on rock drills, ladle-tilting devices in steel mills and foundries, hydraulic lifts, counterweight balances on the arms of radial drills, and transmission systems between double-acting hydraulic cylinders and their shafts. A group of such applications is shown in Figures 8-18 to 8-21.

The kind of chains chosen for these applications depends on the circumstances. When chain movement is infrequent, a bar-link chain should be considered. When the chain rotates frequently, relative to sprockets, rollers, and sheaves, a bushed roller chain, with its smoother sprocket action, may be the better choice. The accessibility and environment, considering lubrication, is a factor here also.

Because of the many variables to be considered when selecting chains for tension linkages requiring Engineering Steel Chains, it is suggested that manufacturers' catalogs, at the least, and in most cases the manufacturers themselves, be consulted.

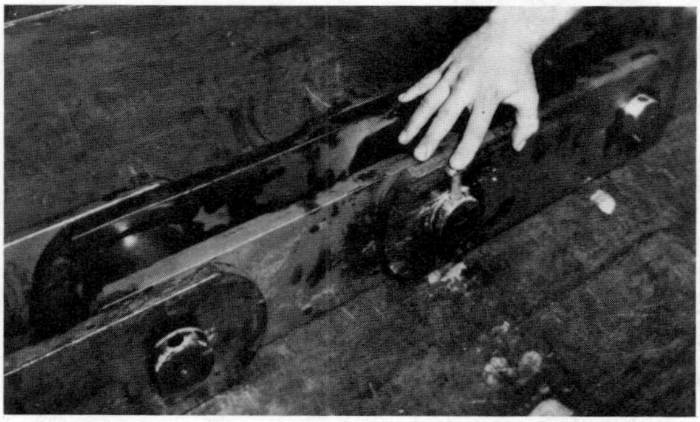

Figure 8-16 Huge tension linkage chain constructed for operation of the gates on a Columbia River dam project.

Figure 8-17 Tension linkage chain constructed for and in service on locks of a Mississippi River dam. The chain has an ultimate strength of more than 1,100,000 lb.

Figure 8-18 Reciprocating motion of these Engineering Steel Chains raise steel billets for black light inspection, then lower the billets for transfer to other operations. Roller Chain drives in the foreground power the conveyor shafts.

Figure 8-19 Two huge strands of bar-link chain operate a pipe expander which cold works the inside diameter of oil country pipe. Each strand of chain has an ultimate strength rating of 1,700,000 lb.

Catenary Tension and Chain Sag

The return strands should be considered with care on installations using heavy bar-link chains running over sprockets and horizontally arranged. Most steel mill installations are in this category. Generally, numbers of teeth in drive sprockets are minimal, loads are often very high, and the chains frequently operate in an environment saturated with good high-pressure lubricants. In such a situation it is essential that adequate slack-side chain tension at the drive be maintained so that the chain cannot jump teeth at peak loads.

Slack-side tension is usually maintained with a deep catenary at the head shaft end

Figure 8-20 Engineering Steel Chain used on a draw bench in a tube mill.

Figure 8-21 Bar-link chain operates the clamshell digging mechanism on an ore boat.

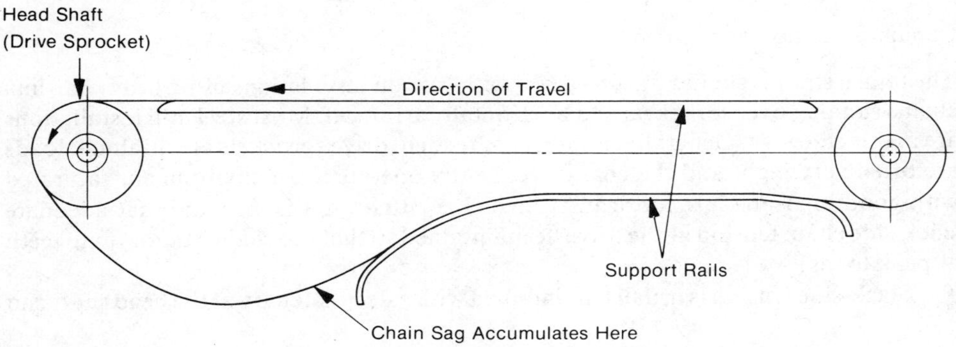

Figure 8-22 Diagram of method for assuring adequate slack-side tension for horizontally arranged bar-link chain conveyors or draw benches by providing deep catenary-type returns.

of the return strand, as shown in Figure 8-22. In most steel mills, so doing is relatively easy, as ample space usually exists in the basement under the mill main floor.

The amount of sag necessary to maintain adequate tension on the slack side of the drive sprocket should be determined from the manufacturer of the chain. A method of calculating such tension versus sag is given in Chapter 10, using data included in Tables 10-22 and 10-23.

Sprockets

Sprockets for heavy bar-link chains should either be purchased from the manufacturer of the chain or their recommendations secured as to the desirable number of teeth for a particular application, together with recommended tooth contours.

Sprockets for tension linkage chains in general are covered in Chapter 4.

9

Engineering Steel Chains for Conveying

Types of Chain Conveyors

Engineering Steel conveyor chains are the workhorse in the field. The outstanding characteristics of applications where they are used are heavy duty needs, including shock loadings, and exposure to environments conducive to corrosion and heavy abrasive wear with minimal protection from any of these factors.

The need for heavy duty ranges from carrying heavy weights such as automobiles on final assembly lines to the ability to absorb a blow from a log, perhaps 18 in. in diameter and 10 ft long, being dropped endwise on the conveyor from more than the length of the log above the conveyor.

Conveyors using Engineering Steel Chains operate with exposure to all kinds of weather, sand and saltwater, and the chemical and abrasive action of everything from coal, taconite, and granite to sugar cane bagasse and hot lime.

Engineering Steel Chains are used on conveyors in an enormous variety of applications, although most installations can be resolved into one of the types of conveyors described in this chapter. Conveying applications may be classified in many ways—whether materials handled are "bulk" or "unit" as sand and TV sets—or whether the conveyor carries the material or pushes or scrapes it along, for example.

The material conveyed and how it is to be taken off and placed on the conveyor really dictates the type of conveyor to be selected. For example, in selecting a conveyor to handle individual large objects, the size and weight of the objects, and whether they could be slid on and off the conveyor, would usually dictate a floor or a carrier or a slat conveyor. If the material to be handled is a bulk material, such as coal or sawdust, or even woodyard logs, the material will be carried by the conveyor or pushed along in a trough by a drag or scraper conveyor chain.

Many of the styles of conveyors are furnished with any of several types of Engineering Steel Chains. However, the capacity required, expressed in terms of conveyor speed as well as load, and other detailed requirements for the application will usually narrow the selection of the chain type down to a decision involving projected life and economics.

The following section of this chapter describes with drawings, and with photographs of applications, the more common basic types of conveyors normally furnished with standard Engineering Steel Chains and attachments. The key to understanding these types of conveyors is to understand how they function relative to what is conveyed, and also to recognize the basic chain attachments required to interface with the material in supporting and moving and discharging it from the conveyor.

For this reason, the chain conveyors discussed in this chapter are arranged according to the types of attachments added to Engineering Steel Chains to permit them to perform their required function, or by the type of service the conveyors perform.

Plain Chain Conveyors

Many objects are adapted to movement on a system of plain Engineering Steel Chains without requiring any attachments, slats, or appurtenances other than the chains themselves. Figure 9-1 is a sketch of a typical application. Here, the objects being conveyed, pulpyard logs, could be rolled onto the conveyor at one end, and will be discharged over the head shaft of the conveyor. Although no protruding attachments are present, as the logs rest on the edges of the chain sidebars, chains with high sidebars are sometimes used.

The Engineering Steel Chains used on such conveyors may be such as to roll or to slide, dependent on the details of the operation. Engineering Steel roller chains, steel bushed rollerless chains, welded steel chains, and bar-link steel chains are all frequently furnished for this style of conveyor.

Figures 9-2 to 9-9 illustrate the wide variety of applications on which plain chain conveyors may be used. In all these applications, two and often more parallel strands of chains are used. The chains carry the weight of the material, and the frictional drag of the chains on the material carries it along. Such conveyors are usually level or nearly so, slopes being relatively mild and usually downhill toward the head shaft. Any lubrication of the chain must be provided with care for the obvious reasons.

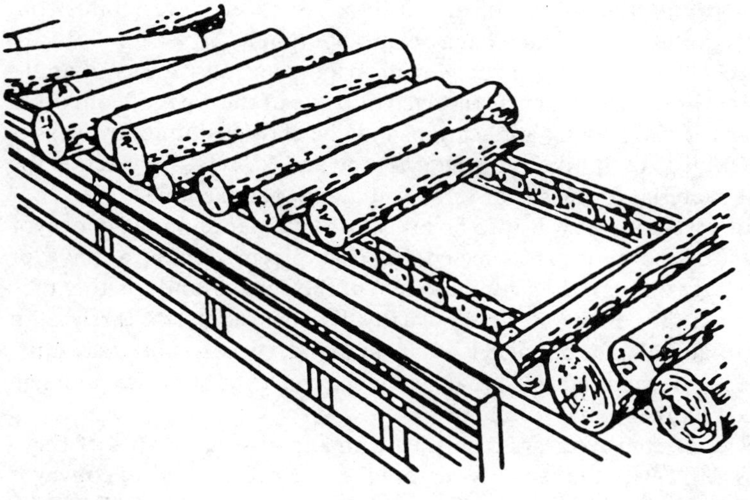

Figure 9-1 Plain chain conveyor. The objects resting on the strands of chain simply move with the chain. They may be discharged over the end, or be pushed off the side of the conveyor.

Figure 9-2 Chain conveyors in a large cannery transfer baskets of cans from cooling tanks to floor level.

Figure 9-3 Engineering Steel bar-link conveyor chain going around a curve in a lumber mill.

Figure 9-4 Four strands of welded steel conveyor chain move cants from sawmill to kraft mill on this inclined conveyor.

Figure 9-5 Four strands of bushed rollerless chain carry sheets of stainless steel to an annealing furnace, where they are transferred to another chain conveyor within the furnace.

Figure 9-6 Engineering Steel bar link chain conveys stacked cases to storage in a dairy plant.

Figure 9-7 Feeding export boxes on hydraulic lift tables in an automotive plant, this four-strand conveyor uses bushed rollerless chain. Six-strand delivery conveyor using bushed roller-less chain receives boxes from hydraulic lift.

Figure 9-8 Engineering Steel roller chain storage conveyors used to feed truck bodies to automatic transfer conveyor.

Only one of the applications shown has chain with rollers. Although roller conveyor chains have the advantage of low friction, means such as high sidebars must be provided to prevent the rollers from making contact with the materials conveyed.

As shown in Figure 9-3, some Engineering Steel conveyor chains are designed so that they can traverse large-radius bends in a plane parallel to their pins. However, not all such chains will accept this type of application without rapid wear, and manufacturers' recommendations should be secured for these applications.

As illustrated in Figures 9-7 and 9-8, plain chain conveyors are often used in conjunction with mechanized transfer devices. The lack of attachments on plain chain conveyors greatly reduces the possibility of hang-ups if a misaligned object is transferred to the conveyor.

Figure 9-9 Pulpwood logs are conveyed from woodyard to barking drum on a six-strand conveyor using steel bushed rollerless chain.

Pusher Chain Conveyors

When an article or part can slide or roll or be moved on rollers or wheels, the most economical method of conveying it in a definite path is frequently a pusher chain conveyor. These conveyors have one or more endless strands of chain with suitable attachments to push the loads.

By definition, pusher chain conveyors do not usually support the materials conveyed. Rather, the load slides or rolls on rails with the top of the chain below the level of the rails and the chain pusher attachments rising between them to push the material along. A typical such arrangement is shown in Figure 9-10.

Considerable care should be given to laying out conveyors of this type with respect to entry and discharge of material, so that the latter, particularly, will not be restricted. Some conveyors of this type, used in highly mechanized manufacturing systems, are at times furnished with movable "pusher dog" attachments which can be retracted below the tops of the rails to disengage the material. Chain manufacturers' recommendations should be secured for the installation of conveyors of this type.

Photographs of typical pusher chain conveyors are given in Figures 9-11 to 9-15. As may be noted, the applications shown, all from either the steel or the forest products industries, mostly involve long, narrow objects. Figure 9-12 shows what superficially seems an exception to the concept that pusher chain conveyors do not support the material conveyed. However, in this type of V-trough installation, the chain is protected by the vee configuration from being encountered to any large degree by the larger logs, the sides of the trough actually supporting the load. In effect, the chain slides in a groove at the bottom of the trough.

Carrier Chain Conveyors

Carrier chain conveyors are built with attachments, either connected to or integrally a part of the conveyor chain, which form a carrying medium or continuous surface for

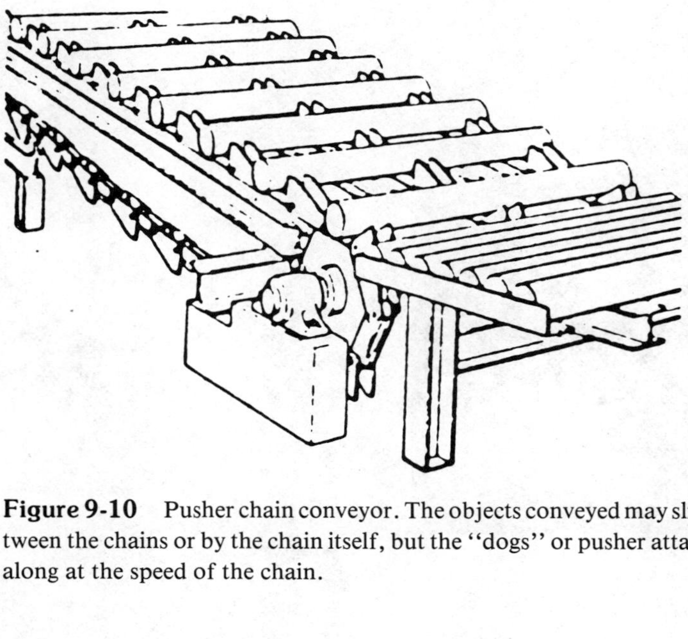

Figure 9-10 Pusher chain conveyor. The objects conveyed may slide or be carried by rails between the chains or by the chain itself, but the "dogs" or pusher attachments force the material along at the speed of the chain.

Figure 9-11 Engineering Steel roller chain conveyor with pusher attachments used to transfer pipe in mill from inspection department to cutoff.

Figure 9-12 Inclined conveyor using steel chain with pusher attachments at intervals to deliver hardwood to debarker and chipper.

Figure 9-13 Cooling hot pipe on rack conveyor in large pipemill.

an individual part. Figure 9-16 is a sketch of a carrier chain conveyor using roller conveyor chain with flat-top attachments every link. It is carrying large coils of steel. Carrier chain conveyors have an extremely broad application in conveying and transporting parts and assemblies. Many of the attachments used in carrier chains are of special construction, being designed to conform to the shape of the objects being handled. Chains are usually standard roller conveyor or bar-link chains. They can be arranged to travel paths in horizontal, inclined, or combination patterns and contours, and are

Figure 9-14 Straight sidebar steel bushed rollerless chain with pusher attachments as used in the paper industry.

Figure 9-15 A jack ladder conveyor consisting of two strands of steel bushed rollerless chain carrying logs to a revolving table feeder.

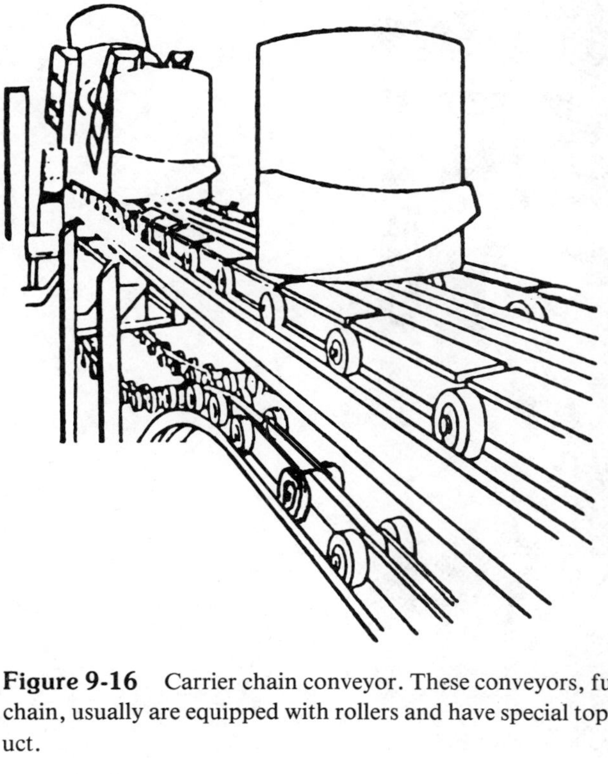

Figure 9-16 Carrier chain conveyor. These conveyors, furnished in many different types of chain, usually are equipped with rollers and have special tops or attachments to carry the product.

built in a wide range of types and sizes. Special designs can be furnished for handling objects from medicinal pills to 100,000 lb coils of strip steel.

Figures 9-17 and 9-18 show two carrier conveyors handling coils of strip steel, but in totally different modes. The conveyor shown in Figure 9-17 handles the coils nested in shallow vee-block attachments into which the cylindrical steel coil can be lowered. The attachments are spaced every sixth link in a bar-link Engineering Steel Chain equipped with outboard (probably antifriction bearings) rollers. The coils could be discharged from the conveyor by taking the chain below the floor level and rolling the coils away.

The application illustrated in Figure 9-18 transports the coils setting on their ends on a roller conveyor chain with flat-top attachments every link. Coils are picked up from the conveyor by a machine made for the purpose, the whole arrangement being planned to keep scratches or marring of the finished steel surfaces to a minimum.

Slat Conveyors

Figure 9-19 is a drawing of a basic slat conveyor. This type of conveyor consists of two or more endless strands of chain with attached noninterlocking wooden slats or metal flights for carrying material loads.

Unit handling applications are the major use for conveyors of this style. The most common uses are to move products for shipping and for processing applications, such as appliance assembly lines.

Figure 9-17 Engineering Steel roller chain handling steel coils in mill.

Figure 9-18 Hot strip coil is handled on-end on this double-strand flat-top engineering roller chain conveyor with attachments to side-discharge down-tilter equipped with gravity rollers.

Roller conveyor chains, with closely spaced attachments of styles A or K (see Figs. 2-25 and 2-16) to which the slats or flights may be attached, are generally used for slat conveyors.

Most slat conveyors have a chain at each side with flights between. If loads are fairly heavy or the conveyor wide, additional strands of chain in the center are sometimes necessary to prevent excessive flight deflection and chain wear. Slats or flights

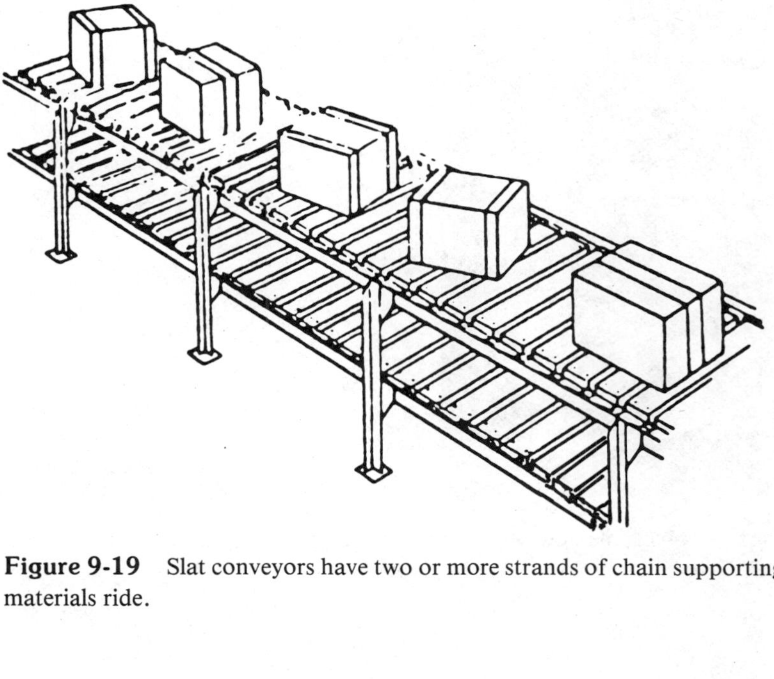

Figure 9-19 Slat conveyors have two or more strands of chain supporting slats upon which materials ride.

are usually extended to cover the chains at the side for safety reasons and to prevent material from dropping through the conveyor.

Where conditions are such that lubrication cannot be provided, such as on a meat packing line where frequent washdown is necessary, chains having rollers with plastic bushed rollers are sometimes used. Several of the chain manufacturers can provide such chain as a part of their Engineering Steel standard product line.

Figures 9-20 to 9-23 show several applications of slat conveyors. The conveyor shown in Figure 9-20 is quite large and is also unusual in that the material handled, pineapples, is a bulk material rather than the unit product handled by most slat conveyors. The problem of retention on the conveyor was solved in this instance by providing skirt boards above the conveyor edges. The slat conveyor illustrated in Figure 9-22 is operated as an assembly line, fixtures on which the product is manufactured being spaced attached to the slats of the conveyor. Figure 9-23 shows a large, slow-moving conveyor on which furniture is assembled.

Figure 9-20 Unloading pineapples into bulk storage at a cannery, this slat conveyor utilizing Engineering Steel roller chain is 128 ft long and can store up to 42 tons of fruit.

Figure 9-21 Maple slat conveyor with Engineering Steel roller chains handle packaged meats in cold storage room of large meat packing plant.

Apron or Pan Conveyors

Apron conveyors or feeders consist of die-formed steel aprons or pans mounted on two or three strands of chain. The effect in operation is that of a moving trough, with discharge over the head shaft as shown in Figure 9-24. This type of conveyor is used extensively for handling all types of loose and bulk materials, ranging from fines to large irregular lumps. The flexibility of the concept permits apron conveyors to convey materials such as coal, coke, lime, sand, stone, sugar cane, and ingot molds. Apron conveyors are commonly used where impact, abrasion, and high temperatures must be considered.

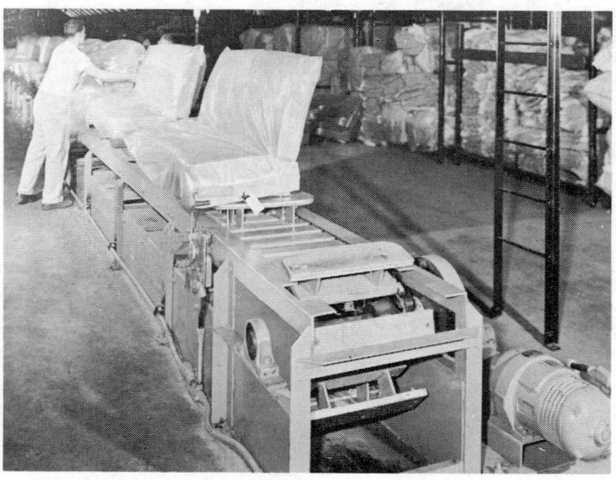

Figure 9-22 Seat cushions are assembled on this long slat conveyor, acting as a traveling worktable in a vehicle assembly plant.

Figure 9-23 Slats mounted on an Engineering Steel roller chain move furniture through an assembly line in a modern factory.

These conveyors can be furnished in extremely heavy duty versions with sufficient strength and rigidity to accept severe operating conditions that other types of conveyors will not withstand.

Usually, ends are welded to the die-formed sections of apron conveyor pans. Chain attachments are of types specifically made to interface efficiently with the apron flights. Engineering Steel conveyor roller chains are used for apron conveyors

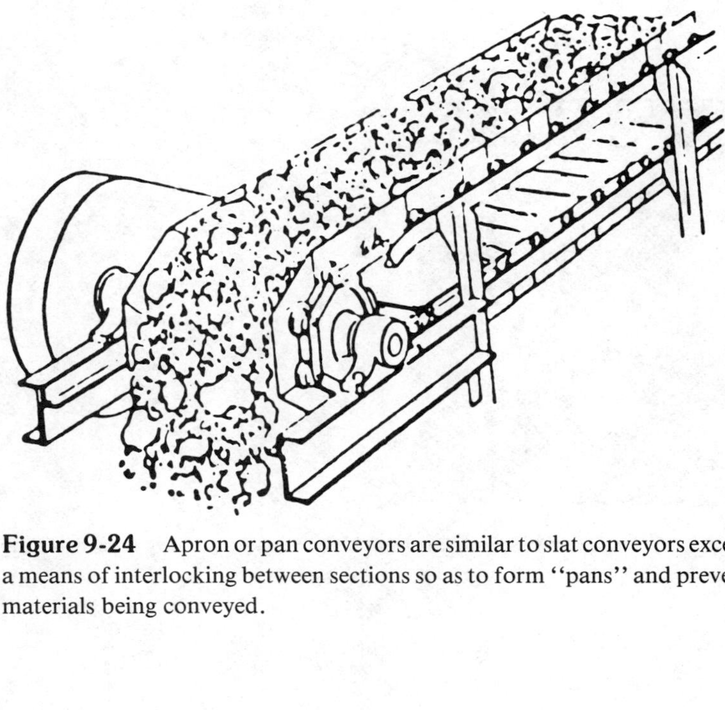

Figure 9-24 Apron or pan conveyors are similar to slat conveyors except for having sides and a means of interlocking between sections so as to form ''pans'' and prevent leakage of fine bulk materials being conveyed.

with the exceptions of some heavy-duty conveyors using steel bushed rollerless chain with outboard rollers.

"Leakproof" apron conveyors, so called because the flights and sides are formed with interlocking "beads" and offsets such that minimal clearances allow the conveyor to retain very fine material, have been used in the mining and foundry industries for years.

Apron conveyors may be used as inclined conveyors, with extra "lifter" blades available for conveyors operated at an angle steeper than is compatible with the angle of repose of the material handled. Aprons are particularly adaptable as feeders for rock crushers, for elevating systems, picking systems, and in power plants using coal conveyors, such as that shown in Figure 9-25.

Scraper Flight Conveyors

A sketch of a scraper flight conveyor is shown in Figure 9-26. This type of conveyor is used to move bulk materials, the flights being suspended into the material from above to force the material conveyed to slide along a trough or deck. Thus the trough and not the chain of the conveyor supports the material. The usual arrangement of the strands is as in Figure 9-26, the conveying strand being below and the return on top. Conveyors so arranged are usually loaded from the sides; the alternative is to drop the material through the return strand. Discharge is through the conveyor trough just in front of or below the head shaft.

In this type of conveyor, the chains are usually above the material and, if a two-strand conveyor, at the sides. For single-strand conveyors, the chain is, of course, centered, but in both cases, the chain does not usually operate in the material con-

Figure 9-25 Engineering Steel roller chain with aprons being used to convey coal in a power plant.

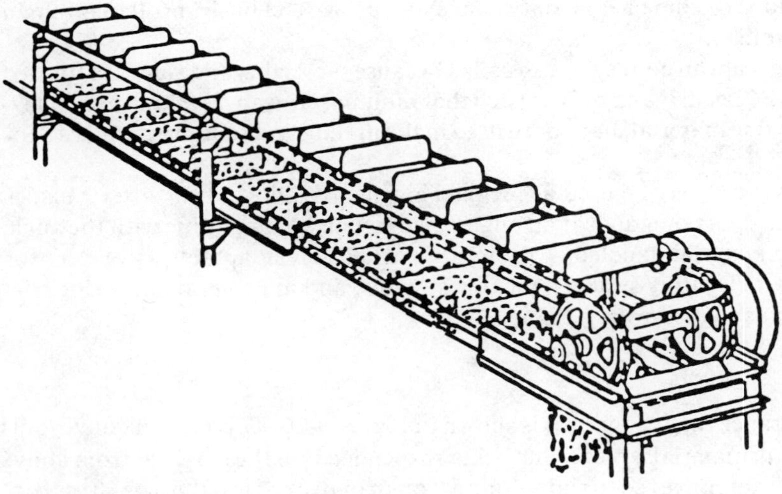

Figure 9-26 A scraper flight conveyor has one or, usually, two strands of chain with spaced flights arranged to scrape bulk material through a trough.

veyed. Engineering Steel roller conveyor chain, steel bushed rollerless chain, welded steel, and bar-link chain are all used in varying scraper conveyor applications. Nearly an equally wide variety of attachments are used; style G (see Fig. 2-25) is popular and type A used with flight wings on the flights is also an arrangement encountered frequently. In the latter design, the chain and attachment connections are not rigid. Flights are made from a wide range of materials—steel fabricated, cast steel or cast iron, plastic injection molded or extruded, plastic sawn, wooden—all these materials have been used successfully, depending on the characteristics of the material conveyed and the other details of the application.

Scraper conveyors are widely used to handle bulk materials such as coal, ashes, potash, lime, grains, and other fine materials. As shown in Figures 9-27 and 9-28, short heavy duty scraper conveyors are used for efficient loading of elevator scrapers used on construction jobs, where large amounts of dirt must be moved.

Drag Conveyors

Drag conveyors are used to handle bulk materials, and are very similar to scraper conveyors both as to function and the types of materials handled. The drag conveyor sketched in Figure 9-29 is quite typical, displaying the major difference between scraper and drag conveyors in that the chains of the latter run in the material on the bottom of the conveyor trough. Operation of several drag chain strands side by side in a trough is not unusual in this type of conveyor design.

Chains for drag conveyors may have separate flights connected to attachments on the chain. Such a conveyor is shown in Figure 9-30, the flights being wooden and placed flat on the conveyor deck. The chain, an Engineering Steel bushed rollerless type, travels on top of the flights, being connected to them with A attachments, but both chain and flights are in the material.

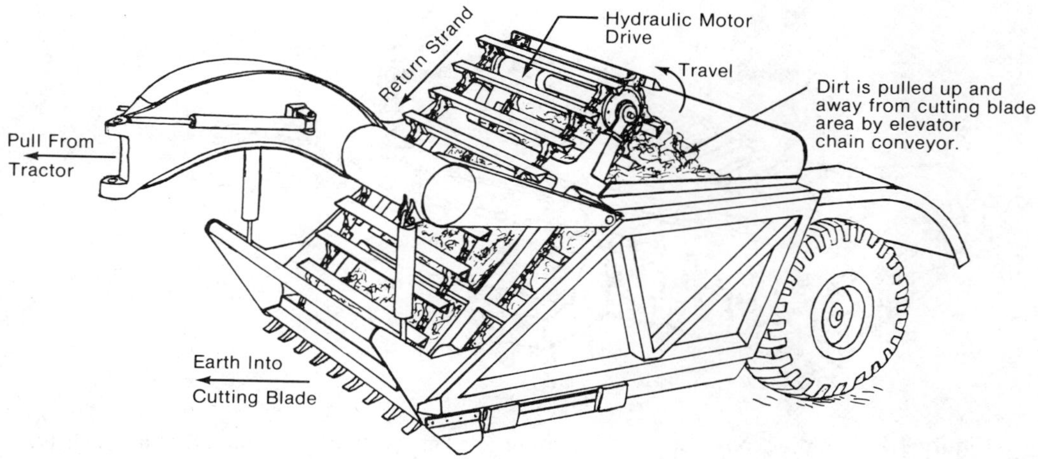

Figure 9-27 Sketch of an elevator scraper-type earthmover, showing how the scraper conveyor pulls the dirt away from the cutting blade so that the unit does not requrie a pusher tractor.

Probably, the most common type of drag conveyor chain are those with the drag feature built into the construction of the links. Engineering welded steel drag chain (see Fig. 2-22), depicted in Figure 9-29, is such a chain. Conveyors using such chains normally have a low initial cost and provide good service life under severe conditions.

Drag conveyors may be arranged as shown in Figure 9-29, with the return on top and discharge through the conveyor trough at or just before the head shaft. Some drag conveyors have the loaded strand on top and discharge over the head shaft. In such cases, the sprockets serve as a continuation of the conveyor deck. For conveyors with wide troughs, sprockets with extended flanges or "wings" (see Fig. 4-21) are sometimes used to assure discharge over the head shaft.

Drag conveyors are used to handle abrasive bulk material such as sand, coal, ashes, coke, and cement, and other materials, such as sawdust, wood chips, and, in fact, just about any bulk material that can be discharged by gravity.

Figure 9-28 Elevator scraper-equipped earthmover in its normal function as road construction equipment.

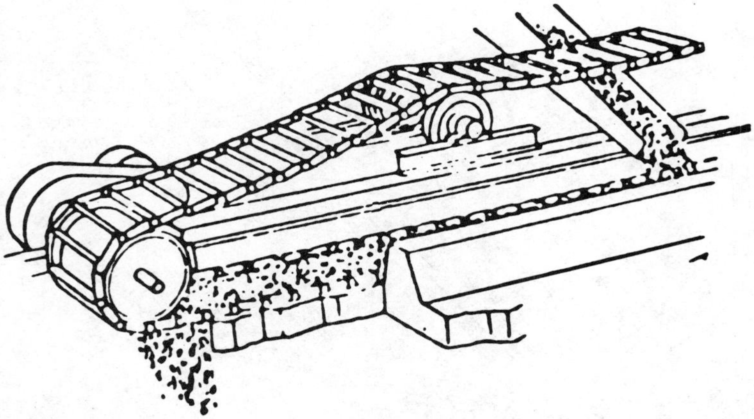

Figure 9-29 A drag conveyor has flights, often as an integral part of the link, forming pockets to pull or drag bulk materials along the bottom of a trough.

Figure 9-30 Straight sidebar conveying chain equipped with A attachments for moving wood chips in the forest products industry.

Cross-Bar Conveyors

Cross-bar conveyors have many applications and are found in practically every type of manufacturing facility. Their chief use is in surface treating and finishing lines, where parts suspended from the conveyor are dipped and dried.

Cross-bar conveyors, typically arranged as shown in Figure 9-31, have two strands of chain interconnected with cross bars. Parts hung from the cross bars can thus be dipped in a series of treating and preparation tanks, with drying between, and then passed through a paint dip and drying oven, all without removing the parts from the conveyor. The double-strand design with cross bars permits better control and positioning of parts in process than on similar conveyors with but a single strand of chain.

The chains used on such conveyors are Engineering Steel roller conveyor or steel bushed rollerless chains. The latter are usually used when the chain is subjected to elevated temperatures in drying ovens.

Cross-bar conveyors are used in the appliance and automobile industries extensively, and also in foundries and the agricultural equipment industry. The conveyors can be arranged in various paths, from inclined to serpentine. Figure 9-32 pictures a cross-bar conveyor used in a continuous paint dipping and drying line.

In-Floor Conveyors

In-floor conveyors have a wide variety of industrial applications. One popular type, sketched in Figure 9-33, is the tow cart type frequently used to transfer boxes of materials between warehousing and shipping areas. This style of conveyor, which has to traverse horizontal curves, often uses bar-link type chain lying on its side, or, alter-

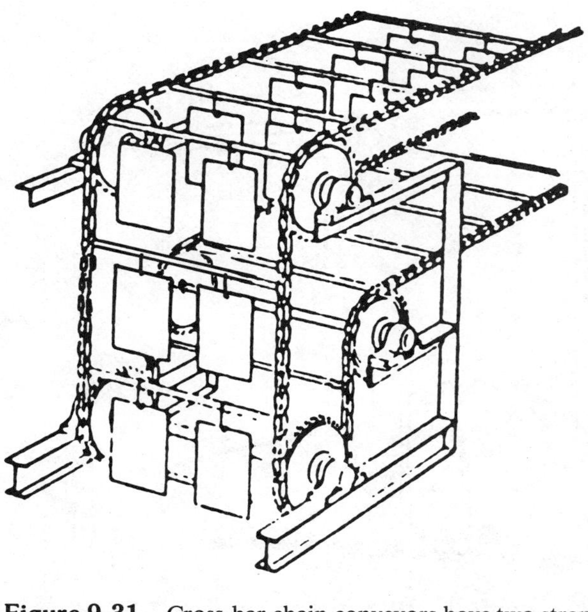

Figure 9-31 Cross-bar chain conveyors have two strands of chain with cross bars between, often at every joint, from which objects to be treated or painted may be hung.

Figure 9-32 Painted metal parts are passed through an industrial paint drying oven in this conveyor, which uses two strands of bushed rollerless chain.

Figure 9-33 In-floor chain conveyors, the standard for vehicle final assembly lines, have many other uses due to their low profile.

Figure 9-34 Engineering bar-link steel chain makes up the in-floor conveyors in the storage area of a modern dairy plant.

natively, bar-link chain with special features enabling it to be flexed both vertically and horizontally. Some conveyors of this type are part of sophisticated systems with remote computer-controlled features to disengage or pick up the carts at will.

Some types of in-floor conveyors handle objects directly on the chains, unloading with diverter bars, as shown in Figure 9-34.

Another type of in-floor conveyor, using heavy steel bar-link type chain with closely spaced flat tops, is widely used to move automobiles and other vehicles down final assembly lines. A typical such line is shown in Figure 9-35. The chains in these conveyors, with the weight of the vehicle included, is supported by antifriction bearing equipped outboard rollers on the chain, running on tracks under the floor.

In-floor conveyors are indispensible to highly mechanized production facilities

Figure 9-35 In-floor conveyors such as those illustrated are indispensable in the production line of an automotive factory.

producing aircraft, automobiles, trucks, and other vehicles. Not only do the final assembly lines use these conveyors, but also rely on them for an orderly flow and storage of materials, components, and subassemblies.

Other Types of Conveyors

Most conveyors using standard chains and attachments of the Engineering Steel types will fall into one of the categories described in this chapter. Some do not; this may be by reason of specialized needs of a complex nature or simply because an attachment of not quite the right form is listed as a standard item. In either case, manufacturers of Engineering Steel Chains for conveyors should be consulted. Probably less than half the conveyors ever installed are from totally standard conponents, and the manufacturers are very familiar with the need for many kinds of conveyors never pictured in any catalog.

Engineering Steel Chain conveyors play a major role in materials handling applications in all the major industries.

When properly selected and specified, chain conveyors can provide one of the most economical and trouble-free conveying systems available.

Considerations for Conveyor Selection

Conveyor Width and Height

The chain conveyor must be of sufficient width and height to accommodate the dimensions of the objects or materials being handled. Careful consideration must be given to any physical restrictions when determining the size of the chain conveyor required.

Although this requirement would seem to apply primarily to conveyors handling unit products, selection of the width and height is also important for those handling bulk materials. Consideration should be given to the dimensions of the chutes or other equipment from which the material is delivered to the conveyor, as well as those into which it discharges.

Conveyor Speeds

Obviously, the capacity of a conveyor is a product of the speed at which it travels and the material carried per unit of length of the conveyor. The nature of the material on the conveyor, how it is picked up or loaded or unloaded on the conveyor, what happens while it is on the conveyor, and how it is discharged will often dictate the conveyor speed.

Table 9-36 lists basic conveyor types and the ranges of speed typical of the operation of each type. From this table, an estimate can be made of the speed of the conveyor to be selected.

Conveyor Capacity

Chain Conveyors Handling Parts and Packages. The capacity of these conveyors can be determined by multiplying the items per lineal foot of conveyor by the feet per hour of the conveyor speed.

Table 9-36 Recommended Operating Speed Ranges for Basic Conveyor Types.

Conveyor type	Speed (ft/min)
Slat or flat-top conveyor	50–150
Carrier conveyor[a]	50–150
Assembly line conveyor	5–15
Drag and scraper conveyors	50–100
Apron conveyor	10–60

[a]Material conveyed directly on chain.

Chain Conveyors Handling Bulk Material. When the capacity required to be conveyed is known in tons per hour, the weight of the conveyed load per foot of conveyor length equals

$$W = \frac{(T)(2000\ \text{lb/ton})}{(60\ \text{min/hr})(S)} = \frac{33.3T}{S} \tag{9-1}$$

where

W = weight in pounds of conveyed load per foot of conveyor
S = speed of conveyor, ft/min
T = weight, tons/hr

In the previous calculation, the conveyor speed has been estimated, and some juggling of conveyor speed and capacity per foot may now be necessary.

Conveyor Modes and Basic Conveyor Types

To obtain the most usable and economical chain conveyor for a specific application, careful consideration must be given to selecting which of the two basic conveying modes should be used. These are broken down as follows:

1. The material conveyed is supported and carried entirely on the chain and attachments of the following conveyor types: carrier, apron, pan, slat, cross-bar or infloor conveyors.
2. The material conveyed is pushed, scraped, dragged, or otherwise slid along a conveyor deck, without the chain supporting the material in the following types of conveyors: pusher, drag, and scraper conveyors.

The final decision as to which basic conveyor type to select should be made only after a thorough investigation of the following:

1. Advantages and disadvantages of each type
2. Operating and initial installation costs
3. Service life required—maintenance costs

4. Physical and chemical properties of material or items to be conveyed
5. Conveying speed required
6. Distance to be conveyed
7. Loading methods

One of the best guidelines to use is a similar chain conveyor installation that has given good or exceptional service life. A chain manufacturer should be consulted for assistance in recommending and selecting the proper type of chain conveyor for a specific application.

Chain Conveyor Design Practices

The design, including chain selection together with the necessary calculations for selection of the components, requires careful consideration of many factors. The essential components of a typical conveyor installation are shown in Figure 9-37. This drawing shows and identifies most of the important features of a conveyor. Not all of these will be used on all conveyors, but most will at least be considered.

Two important factors of the design of a conveyor that should be considered very early in the design process are the means of transmitting power to the conveyor and where this power is to be applied.

Wherever possible, power should be applied to the conveyor at the head shaft, usually the discharge end of the conveyor. This arrangement is recommended so that only the carrying run of the conveyor chains will be under maximum tension in operation. Chains of conveyors have wear life as a function of magnitude of load and time under such loads, so that conveyor chains driven from points where the entire continuous loop is always under load have shorter chain life.

In normal practice, chain conveyors are driven by another chain-sprocket arrangement. For best operating conditions, this supplementary drive should have its tensioned chain section parallel to the tensioned run of the chain conveyor (see Fig. 9-38).

It is sometimes useful to use a compensating sprocket to drive the head shaft of long-pitch chain conveyors, as shown in Figure 9-38. This arrangement minimizes conveyor surging due to chordal action and permits the use of smaller-diameter sprockets.

Chain Types and Selection

After deciding which of the two basic conveying modes are to be used—whether the conveyor is to carry the material as well as cause it to move—a chain selection must be made to suit the design. The first step in the process is to make an educated guess as to the probable weight per foot of the entire conveyor, including any attachments, the chain itself, the flights or slats, and any other moving component traveling with the chain. Table 9-39 lists ranges of average conveyor weights in pounds per foot of several of the conveyor types listed in this chapter with specific kinds of chains. This estimated weight of the composite conveyor will be reexamined after a specific chain selection has been made in step 5 of the following section.

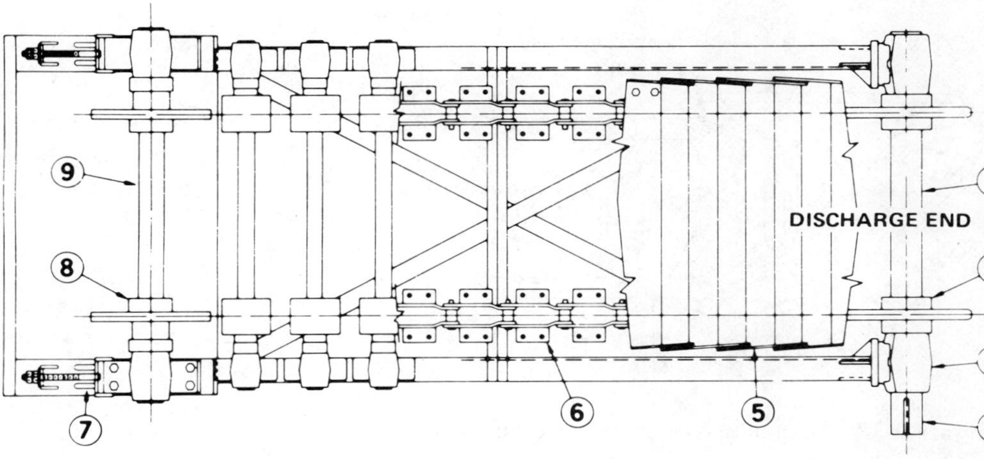

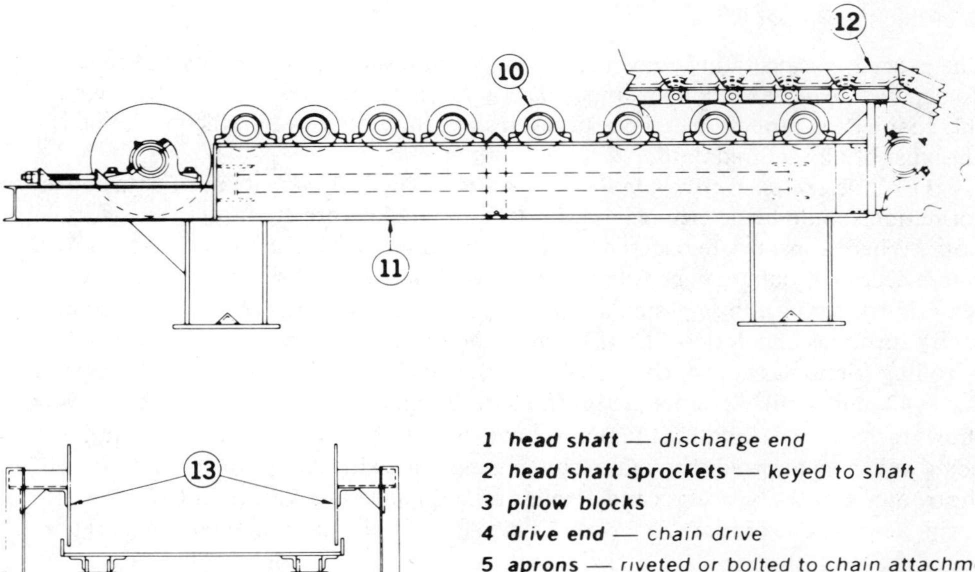

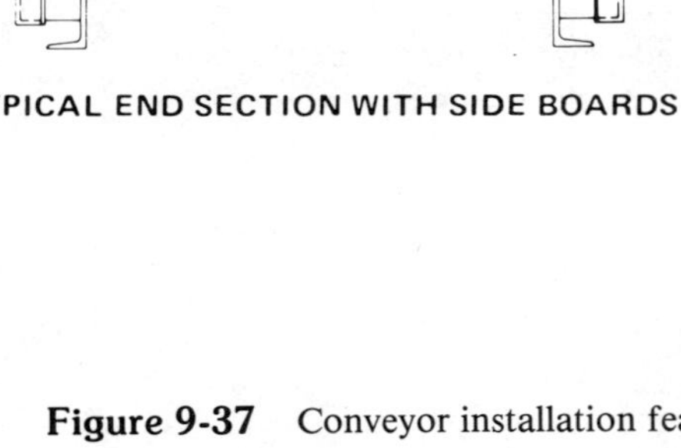

TYPICAL END SECTION WITH SIDE BOARDS

1 **head shaft** — *discharge end*

2 **head shaft sprockets** — *keyed to shaft*

3 **pillow blocks**

4 **drive end** — *chain drive*

5 **aprons** — *riveted or bolted to chain attachments*

6 **conveyor chain** — *offset sidebar or straight sidebar (two or three strands normally used)*

7 **take up** — *normally located on tail shaft end*

8 **tail shaft sprockets** — *only one keyed to shaft. (other sprockets locked in place with collars)*

9 **tail shaft**

10 **intermediate rollers** — *help support chain sections when conveyor is heavily loaded*

11 **conveyor structure**

12 **pan ends** — *welded to aprons*

13 **stationary sideboards** — *offer additional capacity*

Figure 9-37 Conveyor installation features.

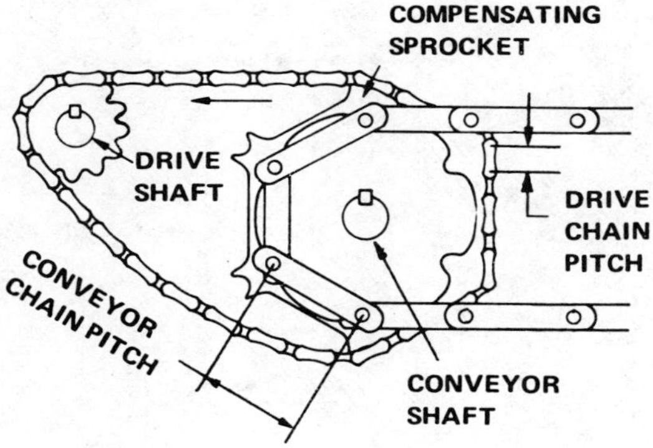

Figure 9-38 Diagram showing arrangement of compensating sprocket and head shaft.

Selecting the Proper Chain

The process of calculating conveyor loadings and resulting chain loads and required horsepower to drive the conveyor may be organized into a series of sequential steps. In this chapter the process is laid out in seven steps, not the least of which is the first, a checklist of all required data.

First, Determine Conveying Requirements. The following items of data and information should be determined or at least estimated before attempting any calculations. Where appropriate, reference is made to tables in this chapter or other sources where necessary data may be found. Figure 9-40 shows the various conveyor layouts requiring differing although similar methods of calculation. Included also are the necessary formulas and definitions of terms in the formulas. Friction factors for sliding or rolling friction as appropriate to conveyor chain operation are given in Tables 9-41 and 9-42, and similar factors are given for bulk materials in Table 9-43. Table 9-44 shows factors for calculating the chain conveyor pull caused by the action of materials being pulled or dragged through a trough in contact with the stationary sidewalls of the trough. For the basic data and details of the Engineering Steel Chains used in conveying service, one or more American Chain Association manufacturers' catalogs are

Table 9-39 Ranges of Average Weights per Foot of Types of Chain Conveyors

Conveyor type and conveyor chain type	Average weight range (lb/ft) of chain conveyor
Carrier, apron, pan, or slat conveyors: roller-type engineering chain	10–20
Pusher, scraper, or drag conveyors	
Rollerless-type engineering chain	5–20
Drag chain	8–20
Bar-link-type chain	10–50

HORIZONTAL

(9-2)

$$P = (Mfm + Wfw)\ C + 1.1\ Mfm\ C + J^*$$

$$= C\ (2.1\ Mfm + Wfw) + J^*$$

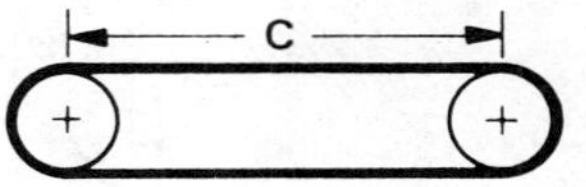

*where applicable from Table 9-44

NOTE: The 1.1 factor is to allow for tailshaft friction

INCLINED

when the angle of incline is known:

$$P = (Mfm\ Cos\ \alpha + Wfw\ Cos\ \alpha + MSin\ \alpha + WSin\ \alpha)\ C + (Mfm\ Cos\ \alpha - MSin\ \alpha)C + J^* \quad (9\text{-}3)$$

when distances a and b are known:

$$Cos\ \alpha = \frac{b}{C} \qquad Sin\ \alpha = \frac{a}{C} \qquad (9\text{-}4)$$

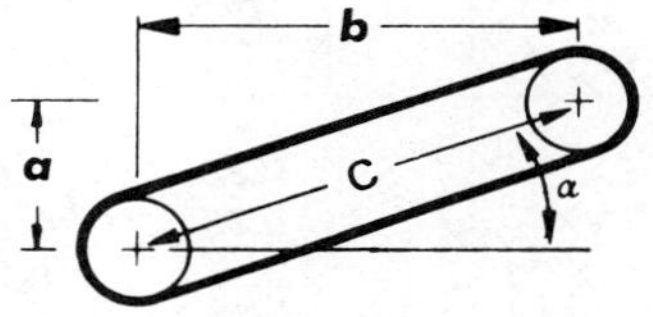

*where applicable from Table 9-44

NOTE: When ($Mf_m Cos\ \alpha$ − $MSin\ \alpha$) is a positive quantity, multiply the difference by 1.1 to allow for tailshaft friction

VERTICAL

$$P = (M + W)C + \frac{P_1}{2} \qquad (9\text{-}5)$$

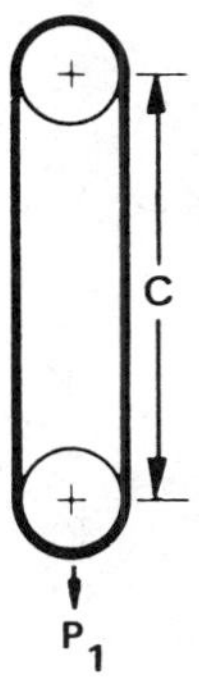

Does not apply to Bucket Elevators See Chapter II

Where:

- M = weight, in pounds per foot, of conveying moving elements
- W = weight of material in pounds per foot of conveyor
- E = speed correction factor Table 9-46
- J = additional pull on chain Table 9-44
- P = conveyor pull
- S = conveyor speed in feet per minute
- V = service factor Table 9-45
- P_1 = take-up tension
- fm = coefficient of friction of chain sliding or rolling Table 9-41 (fs) or Table 9-42 (fr)
- fw = coefficient of friction of material sliding in trough Table 9-43 (fw) (NOTE: When material is carried on chain fw = fm)
- C = conveyor centers

Figure 9-40 Chain conveyor arrangements and formulas for calculation.

Table 9-41 Coefficients of Sliding Friction, Chain, f_s.

	COEFFICIENT fs	
	DRY	LUBRICATED
Steel on Steel	0.33	0.20
Cast Iron or Cast Steel on same surface	0.50	0.40
Steel on Bronze	—	0.15
Steel on Hardwood	0.35	0.25
As Cast: Iron or Steel on Wood	0.44	—

Table 9-42 Coefficients of Rolling Friction, Chain, f_r*.

ROLLER OUTSIDE DIAMETER	COEFFICIENT, fr*	
	DRY	LUBRICATED
1-1/2"	0.22	0.16
2"	0.20	0.15
2-1/2"	0.16	0.12
3"	0.14	0.09
4"	0.12	0.08
5"	0.11	0.07
6"	0.10	0.06
Roller Bearing	—	0.015
Ball Bearings in Hardened Races	—	0.01

* Approximate values developed from standard chains. To calculate rolling friction (fr) for other size rollers use the following formula:

$$fr = fs \times \frac{d}{D} + \frac{0.1}{D} \qquad (9\text{-}6)$$

where:
fr = Coefficient of rolling friction
D = OD of roller
d = ID of roller
fs = sliding coefficient (0.33 for steel chain, dry)

Table 9-43 Coefficients of Sliding Friction, Conveyed Materials, f_w.

	COEFFICIENT, fw	
	DRY	LUBRICATED
Coal on Steel	0.33	—
Crushed Stone or Sand on Steel	0.33	—
Cement on Steel	0.80	—
Wood on Wood	0.55	—

Table 9-44 Calculation of Conveyor Pull J, Resulting from Drag of Material Conveyed on Stationary Skirtboards

$$J = \frac{Ch}{R} \tag{9-7}$$

Where:
h = height of material in inches
R = variable factor for different materials
C = length of conveyor in feet

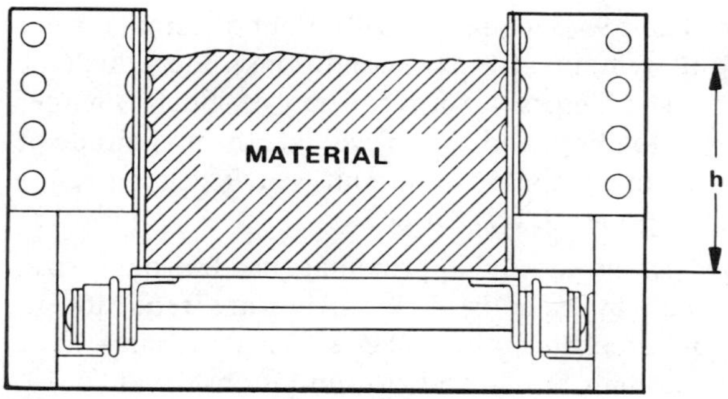

MATERIAL	R
Coal	14.0
Coke	35.0
Limestone	7.5
Gravel	7.0
Sand	5.5
Ashes	14.0

necessary. These catalogs give the necessary information on rated working loads, pitches, other dimensions and weights, and other details concerning the chains needed.

Most of the catalogs include engineering data or applications sections showing processes of selection and calculation similar to those shown here. Terminology and details may differ, but all have been proven over many years. A reader of this book following the steps laid out in this chapter will have no difficulty understanding the similar sequences suggested in any manufacturer's catalog.

Step 1. Summing up the data required:

a. Type of conveyor, including mode of conveying.

b. Class of chain, with pitches and attachments available. The source here is other parts of this book, mainly Chapter 2, and also at least one American Chain Association manufacturer's catalog listing and describing the types of standard conveyor chains discussed in this book.

c. Estimated weight of chain and all attachments and other accompanying moving parts of the conveyor (M in pounds per foot).

d. Conveyor length in feet (C).

e. Height of lift (a in feet) or angle of incline (α in degrees) of the conveyor.

f. Weight of material carried on the conveyor (W in pounds per foot of conveyor length). When the conveyor does not carry the material, as in conveying mode 2, as defined previously, W will be zero. As a practical matter, only in scraper flight conveyors will the weight of the conveyor be wholly independent of the weight of the material. In most mode 2 applications, this factor may be disregarded as not being significant.

g. Coefficients of friction f from Tables 9-41 to 9-43. In these tables, the various coefficients of friction are identified as to whether they are sliding (f_s) or rolling (f_r) friction of the chain on the conveyor ways, or the sliding friction (f_w) of the material sliding on the conveyor deck. In the formulas of Figure 9-40, f_s and f_r are used interchangeably with the common designation f_m as the friction factor for the chain.

h. Conveyor speed (S) in feet per minute. This figure should be a final determination at this point, having been checked against the desired capacity, the conveyor cross section, and the recommended speeds of the type of chain selected, from manufacturers' catalogs.

i. Service factor (overall, V) from Table 9-45, equation (9-8). Four different service factors are given in Table 9-45, all stemming from the conditions under which the conveyor is expected to operate. The factors V_s for frequency of shock, whether in terms of chain pull or conveyor loading, and V_L, to evaluate the magnitude of shock resulting from conveyor loading, should not be confused. The latter refers to the severity of the shocks; the former to how often.

j. Head shaft sprocket revolutions per minute and approximate head sprocket size. Correlate with class of chain and pitch, under (b) above. Drives are frequently already available and the space for sprockets may also be restricted in many ways. Both of these factors are related to chain speed, and any limitations should be established. Sprockets' dimensions should be secured from manufacturers' catalogs.

k. Conveyor pull J resulting from contact of material conveyed with skirtboards. Calculate from Table 9-44, equation (9-7). Although friction related, this factor cannot be evaluated in terms of conventional friction factors, qualities other than weight also being factors in the frictional drag of the material conveyed on the stationary skirtboards. For example, a material of low flowability might not impinge upon the skirtboards to any degree, even though heavy; and bagasse, being resilient, can cause a great deal of drag even though relatively light.

Step 2. Calculate conveyor pull P. Use the applicable formula from Figure 9-40, and solve for the conveyor pull P of a horizontal, inclined, or vertically arranged conveyor. The formula for a vertical conveyor should not be used for bucket elevators; appropriate formulas for such are given in Chapter 11.

V_S FREQUENCY OF SHOCK	V_L CHARACTER OF CONVEYOR LOADING	V_O CONDITIONS OF OPERATION	V_T DAILY OPERATING PERIOD
INFREQUENT (1)	Uniform or steady (1)	Relatively clean and moderate room temperature (1)	8–10 Hours (1)
FREQUENT (1.2)	Moderate shock (1.2)	Moderately dusty (1.2)	24 Hours (1.2)
	Heavy shock (1.5)	Unprotected from weather, dirty corrosive conditions or unusual temperatures within permissible operating range (1.4)	

(9-8)

*Service Factor $V = V_s \times V_L \times V_0 \times V_t$

As the formulas are arranged, they work with either mode 1 (material carried on the conveyor) or mode 2 (material rides on conveyor deck) type conveying, as the material weight on the conveyor is zero and the term drops out for mode 2 designs.

Step 3. Select sprocket size. In Table 9-46, under the conveyor operating speed, read down to the number nearest 1.00. This will be in the vicinity of the heavy dividing line. Read across to the left to find the number of sprocket teeth. This is the ideal sprocket size.

If space limitations, as determined in item j of step 1 dictate a smaller sprocket, read across from the smaller number of teeth to the proper conveyor speed column to find the speed correction factor E. This factor will serve as a multiplier to the calculated conveyor pull P to compensate for the smaller sprocket.

Step 4. Apply service factors, speed correction factor, and additional conveyor pull factor J to calculate the design working load:

$$\text{Design working load} = P \times V \times E \tag{9-9}$$

Multiply the conveyor pull P from step 2, by the combined service factor V from Table 9-45 and also by the speed correction factor E from Table 9-46 and step 3.

If the conveyor is arranged so that the material drags on skirtboards or other stationary components, add the calculated resulting conveyor pull J as calculated from Table 9-44. This is the conveyor design working load.

Step 5. Select a chain size. The conveyor working load applies to all the strands of chain in the conveyor. If two or more strands of chain are used, which is usually the case, multiply the conveyor design working load by 1.2 and divide by the number of strands of chain in the conveyor. This will be the required working load per strand of chain.

Catalogs of American Chain Association chain manufacturers supplying Engineering Steel Chains for conveyors provide rated working loads for such chains. These rated working loads should be interpreted as *maximum allowable working loads,* from the standpoint that consistent operation at loads above the rating will usually limit the service life.

A chain should be selected having a rated working load at least equal to and preferably greater than the required working load per strand of chain calculated above.

Chain pitch should be as long as is found to be compatible with sprocket size, as selected in step 3. Generally speaking, manufacturers will list standard chains with two or three popular pitches with similar or identical working loads. The best economics are obtained by use of standard chains of the longer pitches, but sprockets of the same number of teeth will be proportionally larger for the longer pitch chains. If sprockets with fewer teeth are then chosen, the speed correction factor will increase the conveyor pull accordingly.

Step 6. Calculate actual working load. Repeat steps 2 and 4 using actual chain and attachment weights, as taken from the manufacturer's catalog from which the chain was selected. Having determined a specific chain to use, the calculations of steps 2 and 4 will produce a somewhat different conveyor design working load and required chain working load.

After rechecking the calculations, the listed chain working load, as taken from the catalog, should at least equal and preferably be greater than the calculated required chain working load for the application. If it is not, a chain of proportionally

Table 9-46 Speed Correction Factors E

NUMBER OF TEETH	CONVEYOR SPEED/FEET PER MINUTE														
	10	25	50	75	100	125	150	175	200	225	250	275	300	400	500
6	.917	1.09	1.37	1.66	2.00	2.40	2.91	3.57	4.41	5.65	7.35	10.6	16.7		
7	.855	.971	1.13	1.27	1.44	1.61	1.31	2.04	2.29	2.60	2.96	3.42	3.95	8.62	
8	.813	.909	1.04	1.16	1.26	1.37	1.49	1.63	1.76	1.93	2.10	2.29	2.48	3.62	6.21
9	.794	.870	.980	1.07	1.17	1.26	1.36	1.45	1.55	1.65	1.76	1.88	2.00	2.56	2.94
10	.775	.840	.943	1.02	1.09	1.16	1.24	1.31	1.37	1.45	1.53	1.61	1.68	2.03	2.41
11	.758	.820	.901	.971	1.03	1.09	1.15	1.22	1.28	1.34	1.40	1.46	1.52	1.78	2.05
12	.741	.787	.862	.926	.990	1.05	1.10	1.16	1.21	1.26	1.32	1.37	1.42	1.63	1.84
14	.735	.769	.833	.885	.935	.980	1.02	1.07	1.11	1.15	1.19	1.24	1.28	1.47	1.61
16	.725	.763	.813	.855	.893	.935	.971	1.01	1.05	1.08	1.12	1.16	1.19	1.34	1.48
18	.719	.752	.800	.833	.877	.909	.943	.980	1.01	1.04	1.08	1.11	1.14	1.27	1.40
20	.719	.746	.787	.826	.855	.893	.917	.952	.980	1.01	1.04	1.07	1.10	1.22	1.34
24	.714	.735	.769	.800	.820	.847	.877	.901	.935	.962	.980	1.01	1.04	1.15	1.26

higher working load (and, probably, weight) should be selected and the calculations again be rechecked until a chain of satisfactory working load is selected.

Step 7. Calculate horsepower

Horizontal and inclined conveyors:

$$HP = \frac{(P)(S)(1.1)}{33,000} \tag{9-10}$$

Vertical conveyors:

$$HP = \frac{(WC + 0.1MC)(1.1)(S)}{33,000} \tag{9-11}$$

This is the horsepower required at the input shaft of the conveyor.

Specification of Conveyor Chains and Components

Generation of a suitable set of specifications from which the components for a conveyor may be purchased is about like the work of a baseball umpire. When done well, it isn't noticed. Many a conveyor problem can be traced to an omission in the specifications for the chain or other components of a feature that could have been secured at little or no cost, being readily available from the manufacturer—and which would have prevented the problem.

The features covered in this section are available from all American Chain Association manufacturers of Engineering Steel conveyor chains. Most of these are not only mentioned in their catalogs, but are emphasized therein.

While specification of all the features noted in this section is not always required in specifying the chain and other components for a conveyor, every one of them should always be considered when a conveyor is specified.

Matched and Tagged Strands. Use of matched strands helps assure correct alignment across the strands and is particularly important where through rods, scraper flights, or other carrying attachments are used. Strands operating out of alignment will throw a greater load on one chain, causing excessive wear on the chain and/or attachments, and the sprockets. Periodic checks of multiple strand chain conveyors should be made to make sure that the chains are operating in proper alignment with each other.

When chains are required for multiple strand operation, it is important to specify to the manufacturer "matched and tagged chain" together with the number of strands required. The chains will then be measured at the factory, and a number tag will be attached to each strand (see Fig. 9-47). Each matching group of strands will be tagged with the same number and, whenever possible, will be wired and shipped together. The tags should not be removed until the chain is assembled. The strands must be coupled so that those with the same number are installed side by side.

All chain, just as with any other manufactured product, is made to established tolerance specifications for assembled pitch. The matching of strands assures that sections of chain, made with length at opposite ends of the tolerance range, do not get placed opposite one another in the conveyor installation.

Figure 9-48 shows the effect of two strands that are not equal in pitch. One strand becomes overloaded or the chain becomes twisted when engaging the sprockets.

Right- and Left-Hand Conveyor Strands. Right- and left-hand strands are required in all multiple strand installations where the chain attachments are not symmetrical. In this situation right- and left-hand strands must be specified when ordering (see Fig. 9-49).

Specifying of right- and left-hand conveyor strands should not be confused with the concept of left- and right-hand attachments, discussed in Chapter 3 and also under the section "Direction of Travel" later in this chapter.

Matched Sprockets. On multiple strand conveyors, correct head sprocket alignment is extremely important (see Fig. 9-50). Sprockets should be ordered in sets, keyed

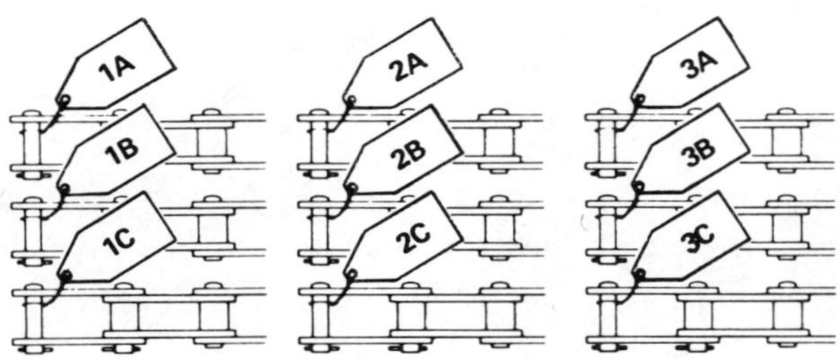

Figure 9-47 "Matched and tagged chain" strands. This specification assures that all strands are the same length and will not throw all the load on one strand.

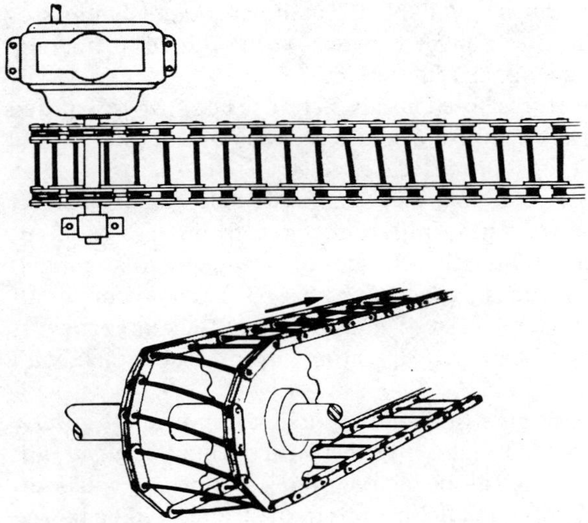

Figure 9-48 Diagram showing the effect of using two strands of chain unequal in pitch.

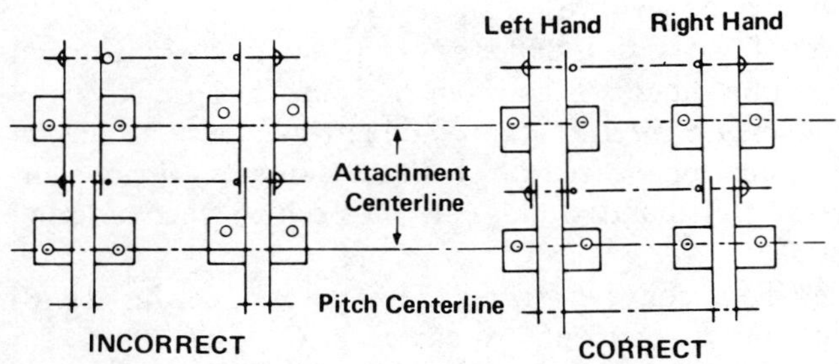

Figure 9-49 Right- and left-hand strands are required in all multiple strand installations where the chain attachments are not symmetrical.

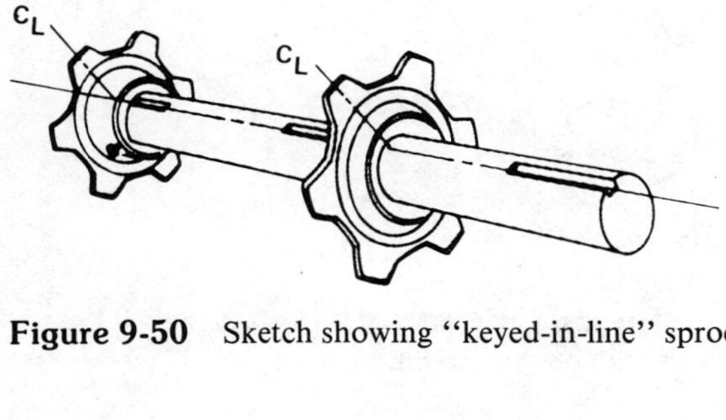

Figure 9-50 Sketch showing "keyed-in-line" sprockets on shafts.

in line, matched, and tagged. One foot shaft sprocket should be keyed to the shaft to turn the shaft in its bearings. The other(s) should be allowed to turn freely between set collars to position automatically should uneven wear occur between the strands.

Shipping and Storing of Chain; Rust Preventative Measures. Manufacturers of Engineering Steel conveyor chains normally coil sections, usually 10-ft lengths, and wire the ends, connecting pins and tags attached. The coils are then stacked and wired together on pallets. Often, since chains awaiting installation may set outside at construction sites for some time, the chains will be dipped in rust preventatives. This is because internal corrosion of the chain joints, if water gets entrapped, can be extensive before the chain is operated and will have the net effect of heavy early wear and elongation.

Consideration of these facts should be recognized in writing specifications. Alternatives may be desired. Some chains, particularly those used in handling food materials, chemicals, or finishing lines cannot be treated with the normal rust preventatives. Others cannot be lubricated in service, but may require special lubrication of parts prior to or during assembly. The type of handling desired, even if standard, should be a part of the specifications.

Design Practices for Use of Conveyor Chains

Although this chapter cannot attempt even to outline all the details of good practice necessary in designing a conveyor, there are certain design considerations important to effective performance of the chains selected.

All of the following items should be considered in the design stages of a new installation, and most of them should at least be looked at in a replacement project. In a new installation, compliance is best assured by drawings, drawing notes, or specifications placed in the hands of contractors.

The items discussed are intended to supply guidance, but should not be considered as a checklist.

Shaft Center Distances. Long shaft centers can result in chain conveyor surging due to the elastic qualities of chain. In this situation, two conveyors with a transfer point may be required.

Chain Supports. It is usually considered desirable to support the return strand of chain conveyors, although support is not always necessary on conveyors under 15 ft in length. However, an unsupported return strand, hanging as a catenary, can substantially increase the chain pull, in contrast to the pull when the return strand is supported. In addition, a sagging return strand, elongated from wear, sometimes will "hang up" on obstructions it cleared when the conveyor was new. Supporting the return strand also minimizes pulsation and whip.

Load-carrying strands are normally supported by tracks or guides, although spaced rollers are used occasionally. Return strands are often supported by guides or spaced rollers (see Fig. 9-51).

Figure 9-51 illustrates in cross section the carrying and return strands of four different styles of conveyors.

In designing the support members for a conveyor, considerable attention should be given to match-up of the span of centerline to centerline of sprockets on shafts and pairs of matching chains and tracks or ways.

When flanged rollers are used together with crane rails to support the conveyor, it

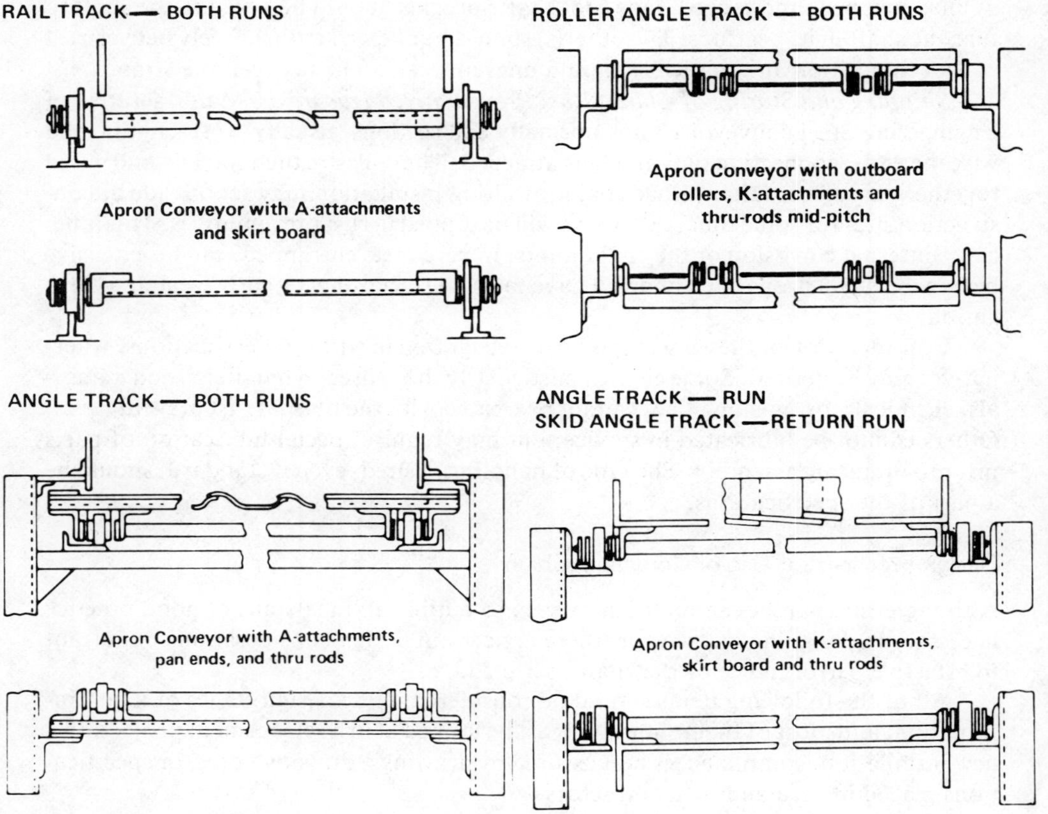

Figure 9-51 Typical methods of supporting conveyor on run and return run.

should be recognized the gauge of the track will be longer than the centerline to centerline of the chains. The difference will often be found as a listed dimension of such types of chain in manufacturers' catalogs under the heading "track gauge." The reason for the difference is that the track is gauged to the center of the faces of the rollers, excluding the space occupied by the flanges. The flanges are normally on the inside of the rails.

Certified Prints. Certified prints showing important dimensions related to the fit of Engineering Steel Chains with the other components of the conveyor are normally available from the manufacturers.

Direction of Chain Travel and Left- and Right-Hand Strands. As has been noted elsewhere, not all chains and attachments used in conveyor service are symmetrical. It should be noted that many such chains have a recommended direction of travel, relative to the head sprockets. Where such recommendations exist, they should be clearly indicated on the conveyor drawings or be a part of the contractor's specifications.

Considering that left- and right-hand strands are invariably involved in such cases, with repair parts usually referring to left- and right-hand attachments (see Chapter 3), lack of adequate specifications can often prevent a conveyor chain from realizing its full usefulness.

Tension Takeup Devices. Most conveyors require control of chain tension through the use of a takeup device. Takeups, which are normally installed at the loading end of a conveyor, are a mechanical device for adjusting shaft center distances. Takeups should be provided on all conveyor and elevator installations to compensate for elongation as joint wear occurs and to provide temporary slack for installation or maintenance. Adjustment of takeups should be made while the conveyor is in operation to ensure the best possible conditions. There are four basic types: screw takeup, spring takeup, gravity takeup, and catenary takeup. These are illustrated in Figure 9-52.

Loading Conveyors. Loading any conveyor should be accomplished as gently as possible to reduce the impact. The load should, if possible, be placed or slid onto the conveyor to reduce the pulsations and surging caused by rough loading.

In laying out chutes or hoppers to direct discharge onto a conveyor, it is desirable to place the load as near the center of the conveyor span between the chains as is possible. Otherwise, the chain on one side of the conveyor may wear more rapidly than that on the opposite side.

Access. When required, walkways or other means of access should be provided for maintenance and inspection of the conveyor. Particular attention should be given, in the design stage, to access to chain pin ends or rollers at some point on the conveyor where pressure-equipped or other lubricating devices may be used conveniently.

Further information on lubrication of Engineering Steel conveyor chains is included in Chapter 12. That chapter also includes information on the installation of chains and conveyors.

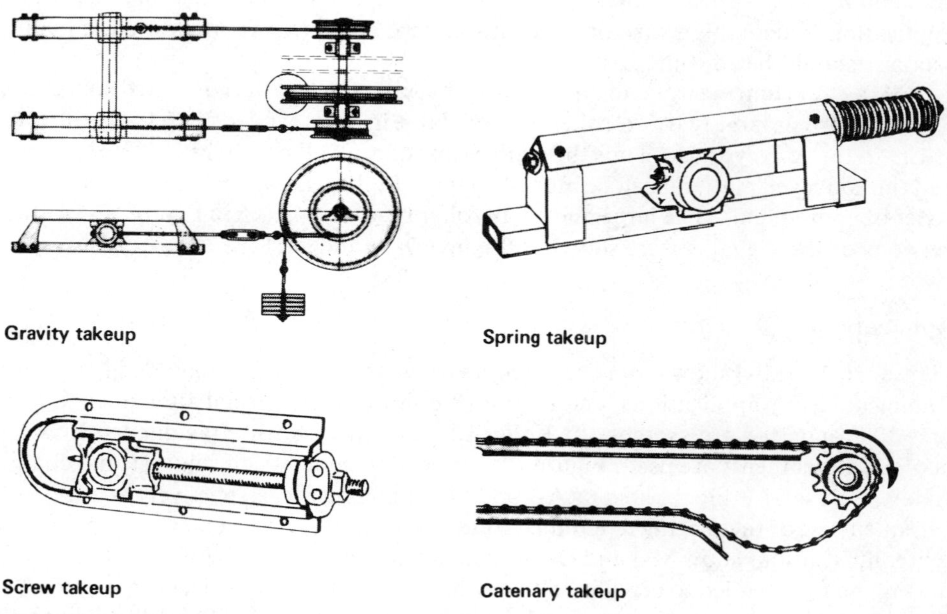

Figure 9-52 Four basic types of chain tension takeups.

10

Roller Chain Conveyor Applications

Precision Roller Chains, both of the base pitch design (ANSI Standard B29.1) and double pitch conveyor chain design (ANSI B29.4), are widely used as conveyor components. A variety of standard and made-to-order attachments are available for these chains. Some of the standard attachments are shown in Figures 10-1 to 10-5. Modified attachments can also be provided for special conditions such as variations in hole size and locations, or the attachment design altered to provide special geometry of bend, contour, or other forms. Figure 10-6 shows a group of special attachments. In applications where these variations from standard attachments are desired, a manufacturer should be consulted.

Many attachment sizes and types are available in stainless steel or plated condition.

Standard or large (carrier) rollers are available in the extended pitch series of Roller Chain for conveyor use. Thus the chains can slide or roll on tracks or support rails, and the conveyed material can be moved by bars, flights, slats, pusher lugs, and other styles of components. The large and small roller types are shown in Figure 10-5; other versions of these chains were shown in Figures 2-10 and 2-11.

Applications

Figures 10-1 to 10-4 shows some of the more popular attachments intended for Roller Chain conveying applications, and Figure 10-6 shows a few special attachments. Figures 10-7 and 10-8 are sketches of Roller Chain sections with attachments showing how the attachments are used. Figure 10-7 shows attachments on single strand chain. At left, the use of V-blocks held by A-1 and M-1 attachments, respectively, are shown, a frequent use of these being to handle cylindrical objects. In the center, the upper and center illustrations show M-1 and D-1 attachments used to mount rollers on the chain, the M-1 being intended to provide rolling contact with either the objects conveyed or a "hold-down" and the D-1 setup usually being to support the chain in rolling contact. The lower cut shows a D-3 attachment with a rubber block used to provide a high-fric-

ONE SIDE OF CHAIN

BOTH SIDES OF CHAIN

Figure 10-1 Small roller series, straight link plate extension with one attachment hole.

EXTENSION ONE SIDE OF CHAIN

EXTENSIONS BOTH SIDES OF CHAIN

Figure 10-2 Small roller series, straight link plate extension with two attachment holes.

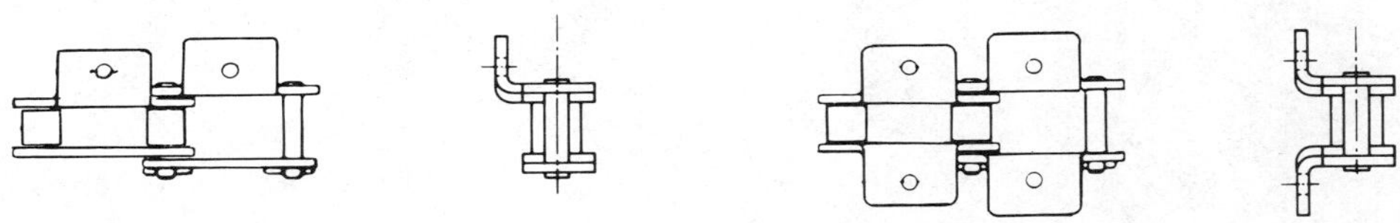

EXTENSION ONE SIDE OF CHAIN

EXTENSIONS BOTH SIDES OF CHAIN

Figure 10-3 Small roller series, bent link plate extension with one attachment hole.

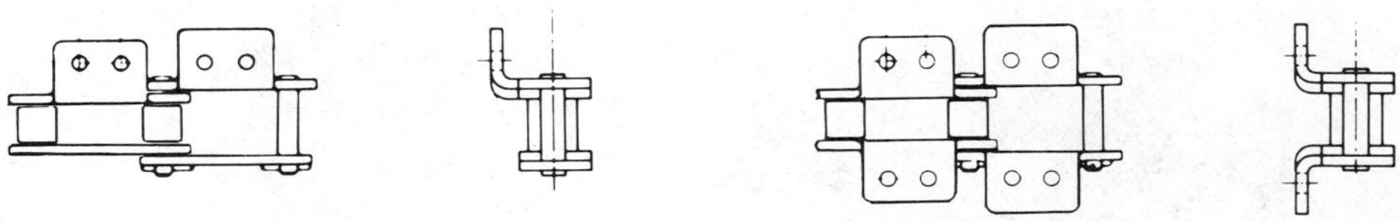

EXTENSION ONE SIDE OF CHAIN

EXTENSIONS BOTH SIDES OF CHAIN

Figure 10-4 Small roller series, bent link plate extension with two attachment holes.

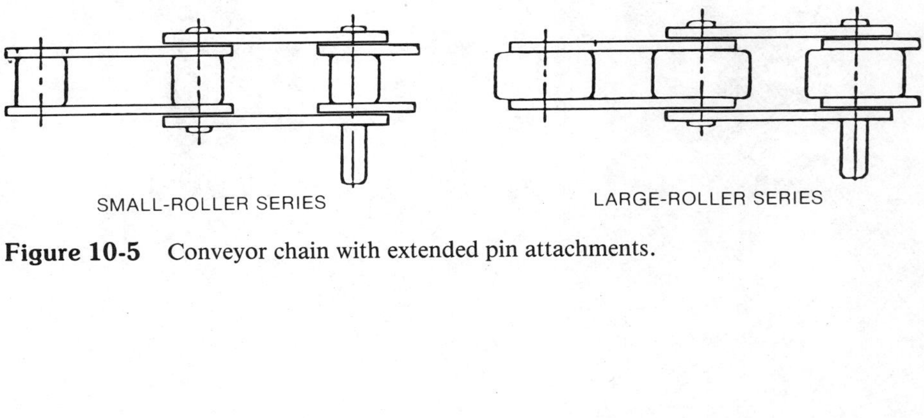

SMALL-ROLLER SERIES

LARGE-ROLLER SERIES

Figure 10-5 Conveyor chain with extended pin attachments.

Figure 10-6 Typical special conveyor attachments for Roller Chain.

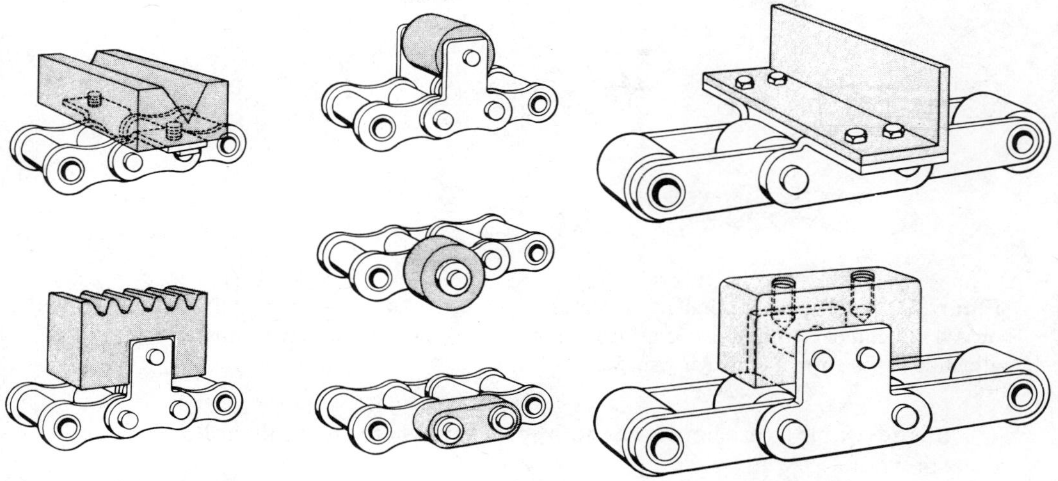

Figure 10-7 Use of attachments on single-strand Roller Chains. At left, V-blocks on A-1 and M-1 attachments; center, M-1 and D-1 attachments used to mount rollers, with a D-3 attachment mounting a rubber block; at right, an angle iron on a K-2 and a fixture mounted on an M-1 attachment.

tion conveying surface. The two illustrations at the right in Figure 10-7 show an angle iron attached to a K-2 attachment to serve as a pusher and a tapped block mounted on a chain with M-2 attachments.

Figure 10-8 shows double strand setups, left to right, slats mounted with A-1 attachments, M-35 attachments used as spacers to convey and position long objects, and tubing assembled on D-3 attachments to form a conveying surface.

Figures 10-9 and 10-10 show two typical applications of the type where Roller Chain is most commonly used in conveying service, special unit handling machines. Figure 10-9 shows a packaging machine used in the food industry, and Figure 10-10 illustrates a top and bottom case sealing unit.

Parallel Strand Operation

Two or more chains operating in parallel in a conveyor should be matched for length for smooth operation and maximum life. In this case, the sprockets, especially those on the head shaft, must be keyed to a common shaft so that the teeth of each sprocket are aligned to assure equal load distribution on all chains.

Sprockets

Conveyor Roller Chain sprockets are described in Chapter 4. They are generally selected so as to be the same size on both head and tail shafts of a conveyor. With double-pitch chain, as few as six effective teeth can be used on very slow speed conveyors where smooth operation is not essential; however, 15 effective teeth minimum is desirable, and no fewer than 15 is recommended for single-pitch Roller Chain in conveyor service. Where smoothness of operation is of prime consideration, 21 or more effective teeth are recommended to reduce the effect of chordal action (see Chapter 4).

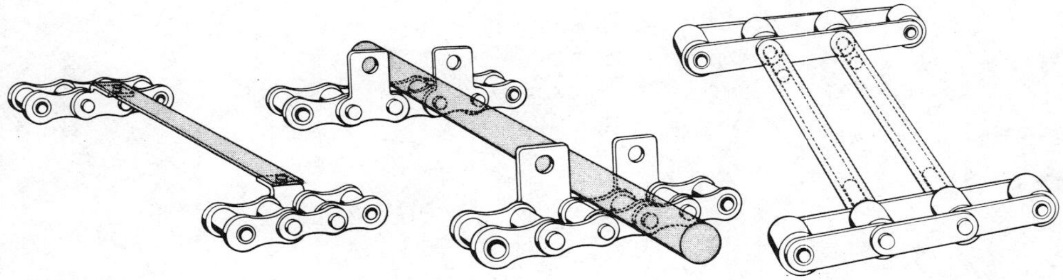

Figure 10-8 Use of attachments on parallel strands of Roller Chain. At left, slats mounted on A-1 attachments; center, M-35 attachments used as spacers; at left, tubing assembled on D-3 attachments to form a conveying surface.

It is advisable to use hardened tooth sprockets where any of the following conditions prevail:

Abrasive environment

Small numbers of teeth

High speeds

High loads

Takeups

For maximum service life, chain takeup tension should only be sufficient for proper conveyor functioning. To maintain proper tension on long conveyors, takeup devices should be provided, preferably at the tail shaft end. Screw takeups are most common

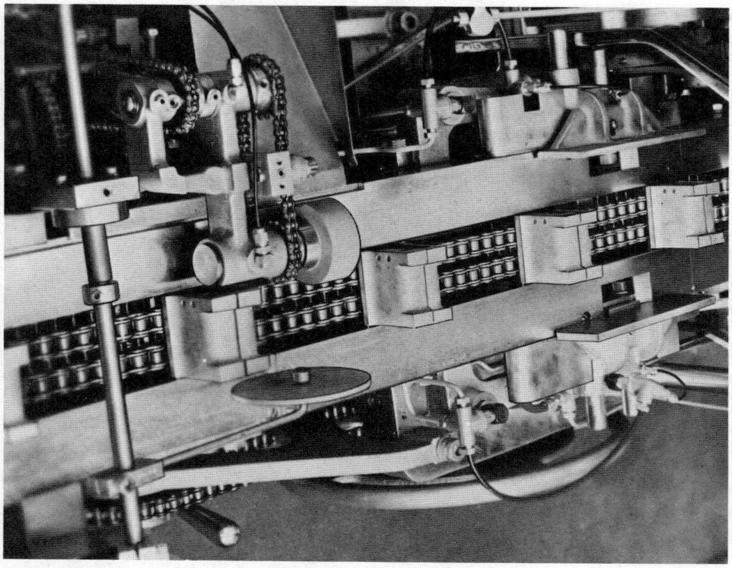

Figure 10-9 Roller Chain used on a packaging machine.

Figure 10-10 Roller Chains in use on a top and bottom case-sealing machine. The chains in a horizontal plane guide the packed cases through the gluer section; others are used for sealing; still others for the main drive of the machine.

(see Fig. 9-52) and used where periodic manual adjustments are sufficient. Spring or gravity takeups are useful where fixed tension is required on conveyors subject to temperature or other variations in operating conditions. Spring takeups are also especially useful when shock loading is inherent. Where chains are operating in parallel on common head and tail shafts, it is essential that these shafts be parallel after adjustment. In long conveyors, automatic tensioning can be provided by a catenary arrangement at the drive sprocket on the return run (Fig. 10-21).

Conveyor Drive

To enhance chain and sprocket life, the power input should be at the discharge end (head shaft) so that only the working span will be under maximum tension (see Figure 10-21).

Temperature Limitations

Steel Roller Chains can be used up to 350 °F for continuous operations and up to 400 °F for intermittant operation. For temperatures up to 500 °F, extra clearance is required: however, some tempering of the steels in the chain will occur at this temperature, the pins and bushings softening and becoming less resistant to wear. Chain manufacturers should be consulted on applications in this temperature range in view of the various types of stainless steel available for chains.

Corrosive Environments

Carbon and alloy steel and certain stainless steel chains are susceptible to stress corrosion cracking when exposed to corrosive environments. Failure can result when operating in or near acidic or caustic mediums or atmospheres. In cases of carbon steel chain, even rusting can promote this type of failure. Failure generally initiates as a crack at a maximum stress point such as the chain link plate aperture, and propagates to the outside of the plate. Depending on the hostility of the environment, the failure could be rather sudden or can be eventual. In view of the many chemicals used in industrial processing, the chain manufacturer should be consulted for material recommendations when a hostile environment is known or suspected. Depending on the extent of the environment, several remedies are available, such as:

1. Standard chains but with reduced hardness (reduces susceptibility to stress corrosion cracking)
2. Plated chain (nickel, zinc, cadmium, etc.)
3. Various types of stainless steel

Design Information for Chain and Sprocket Selection

For proper chain and sprocket selection, the following information should be determined, where applicable, for the particular conveyor application under consideration.

1. Type of conveyor design (see Figs. 10-15 to 10-17)
   a. Horizontal, inclined, vertical, or a combination of these
   b. Whether the chain is to slide or is to roll on ways
   c. Whether the material is to be conveyed by sliding or is to be carried on the conveyor
2. Conveyor speed, in feet per minute (S)
3. Approximate center distance between shafts (C)
4. Approximate diameter and bore of sprockets
5. Weight of material to be handled per foot of conveyor length or per cubic foot (W)
6. Kind of material to be handled
7. Weight of carrier (slats, flights, pans, etc.) and approximate chain weight (including attachments) per foot of conveyor (M)
8. Size and spacing of attachments and/or carriers
9. Additional loading imposed by material drag or containment walls, if applicable (J) (see Table 10-11)
10. Running coefficient of friction for chain sliding or rolling (f) (see Tables 10-12 and 10-14 for f_s or f_r)
11. Sliding coefficient of friction of material in trough (f_w), if the material conveyed is slid and not carried (see Table 10-13)
12. Operating conditions:
   a. Temperatures (see p. 263) for limitations)
   b. Hours per day, days per week, of operation (consult manufacturer for conditions that exceed 8 hr per day, 5 days per week)
   c. Lubrication details, including restrictions, if any (refer to the section "Lubrication of Conveyors" in Chapter 12)

Chain Conveyor Layouts and Calculation of Conveyor Pull (P)

Selection for specific operating conditions is usually on the basis of the maximum anticipated chain pull, in pounds. Chain pull will be the conveyor pull divided among the number of strands of chain used.

The desired spacing of attachments considered for use can be the most consideration and therefore the controlling factor in dictating a specific chain pitch.

Referring to the foregoing design information, select the appropriate formula (see Figs. 10-15 to 10-17 and select the designated conveyor arrangement) for calculation of the appropriate conveyor and chain pulls in pounds. For these calculations, it is first necessary to determine the additional conveyor pull (J) and the friction factors (f_r, f_s, and f_w) from the appropriate tables.

Additional Pull (J)

Bulk materials sliding against containment walls of conveyor troughs cause an additional pull on the conveyor. This factor, J, can be calculated from the formula

$$J = \frac{Ch^2}{R} \qquad (10\text{-}1)$$

where h is the height of the material in the trough and C the distance the material is carried on the conveyor, both in inches. The factor R is taken from Table 10-11, which also illustrates and describes the calculation.

Coefficients of Friction

Chain Sliding Friction (f_s). Where the chain slides on the edges of its plates on ways of steel or other materials, the coefficient of friction f_s, as taken from Table 10-12, should be used. It applies to the total conveyor weight.

Table 10-11 Trough Drag Friction Factor J for Materials

$$J = \frac{Ch^2}{R} \qquad (10\text{-}1)$$

Where:
- h = height of material in inches
- R = variable factor for different materials
- C = length of conveyor in feet

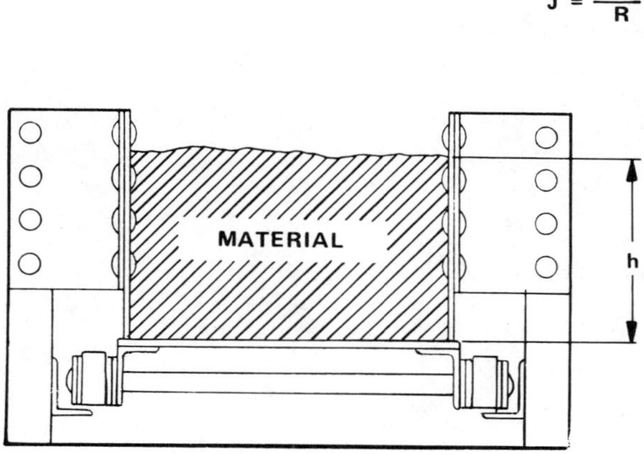

MATERIAL	R
Coal	14.0
Coke	35.0
Limestone	7.5
Gravel	7.0
Sand	5.5
Ashes	14.0

Table 10-12 Friction Factors f_s for Sliding Roller Conveyor Chains

Condition	Dry	Lubricated
Static[a]	0.33	0.24
Sliding	0.27	0.21

[a]Use static coefficient of friction for speeds of 3 ft/min or less.

The factor only applies to Roller Chains of the extended pitch conveyor series (which have straight-sided link plates) when the conveyor carries the material.

Material Sliding Friction (f_w). Table 10-13 shows a few typical coefficients of friction of well-known bulk materials sliding on steel or wooden troughs. For further data on materials not covered, standard handbooks, manufacturers' catalogs, or the manufacturers should be consulted.

Chain Rolling Friction (f_s). When the chain rolls on rails, the coefficients of friction f_r shown in Table 10-14 should be used. If the material is pushed and is not carried by the weight of the chain, these factors are applied only to the weight of the conveyor. If the material is carried by the rollers of the conveyor, this factor applies to the weight of both conveyor and material.

Note that the coefficients of Table 10-14 do not apply to conveyors using Roller Chains with standard-diameter rollers. Supporting of the conveyed load by standard-diameter rollers is not normally recommended since some of the rollers may not turn, due to the small ratio of roller outside diameter to roller inside diameter, often resulting in the wearing of flat spots on the rollers.

Calculation of Conveyor Pull (P)

The weight of the conveyor or carrier and the material (the latter, although previously recorded, should be reduced to weight per foot of conveying distance) and other needed factors secured, calculation of conveyor pull may be made as follows.

1. *Horizontal conveyors (see Fig. 10-15)*
   a. Conveyed material moved but not supported by the chain:

Table 10-13 Friction Factors f_w for Sliding of Materials

	Coefficient
Coal on steel	0.33
Crushed stone or sand on steel	0.33
Cement on steel	0.80
Wood on wood	0.55

Table 10-14 Friction Factors f_r for Double-Pitch Roller Chains Equipped with Carrier Rollers

	Static[a]		Rolling	
Chain number	Dry	Lubricated	Dry	Lubricated
C-2042	0.17	0.12	0.14	0.10
C-2052	0.16	0.11	0.13	0.09
C-2062H	0.16	0.11	0.13	0.09
C-2082H	0.15	0.10	0.12	0.08
C-2102H	0.14	0.09	0.11	0.07
C-2122H	0.14	0.09	0.11	0.07
C-2162H	0.13	0.08	0.10	0.07

[a]Use static coefficient of friction for speeds of 3 ft/min or less.

$$P = C(2.1Mf + Wf_w) + J \tag{10-2}$$

b. Conveyed material supported by chain. In this case, $f_w = f$ (of the chain—that is, f_s or f_r) and the formula becomes

$$P = Cf(2.1M + W) + J \tag{10-3}$$

(J applies only when sidewalls are stationary).

2. *Inclined conveyors (see Fig. 10-16)*

a. Conveyed materials moved but not fully supported by the chain conveyor:

$$P = C(2Mf \cos\alpha + Wf_w \cos\alpha + W \sin\alpha) + J \tag{10-4}$$

b. Conveyed material fully supported by the conveyor. In this case, $f_w = f$ and the formula becomes

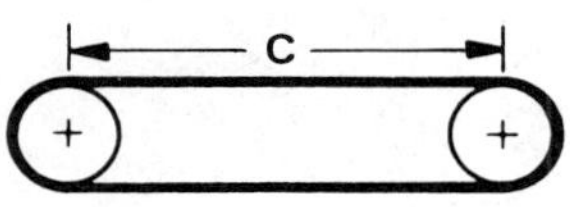

Figure 10-15 Horizontally arranged conveyor.

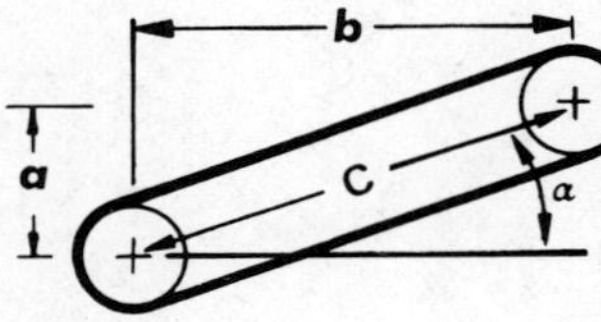

Figure 10-16 Inclined conveyor.

$$P = Cf\cos\alpha(2M + W) + (CW\sin\alpha) + J \tag{10-5}$$

3. *Vertical conveyors (see Fig. 10-17)*

$$P = C(M + W) \tag{10-6}$$

Notes (applicable to preceding conveyor pull formulas):

1. When a and b, referring to Figure 10-16, are known:

$$\cos\alpha = \frac{b}{c} \quad \text{and} \quad \sin\alpha = \frac{a}{c} \tag{10-7}$$

2. When ($Mf\cos\alpha - Mf\sin\alpha$) is a positive quantity, multiply the difference by 1.1 for tail shaft friction.
3. Summing up quantities used in calculating P:

   M, weight, in pounds per foot, of the moving elements of the conveyor as carried by the chains.

   W, weight of material carried in pounds per foot of conveyor. (For unit materials, sum up the average weight of units expected to be on the conveyor at maximum capacity, and divide by the conveyor length in feet.)

   f, coefficient of conveyor, sliding or rolling on chains (f_s or f_r), as given in Tables 10-12 and 10-14.

   f_w, coefficient of material sliding in trough, as given in Table 10-13 or from other references. (Note: When material is entirely carried by conveyor, $f_w = f$.)

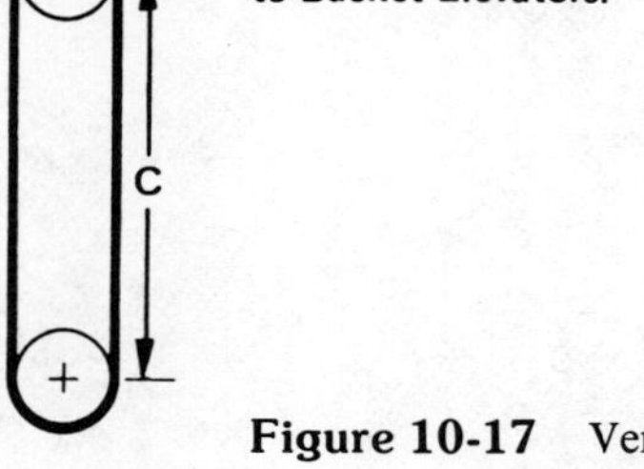

Figure 10-17 Vertical conveyor.

J, additional pull from drag of material on stationary sides of trough, given in Table 10-11.

S, conveyor speed, in feet per minute.

P, conveyor pull, in pounds.

Working Load Required of Chain

Few conveyors are centrally loaded on a consistent basis, and sprocket alignment is not always perfect nor is wear on opposing strands of chain in a multistrand conveyor always equal. For this reason, a multistrand factor should be applied to the conveyor pull before computing the required working load for the chain selected. For conveyors having cut-tooth sprockets, a factor of 1.1 should be used; for all others, use 1.2. The required working load for the chain selected is calculated as follows:

$$\text{Required chain working load} = \frac{\text{multistrand factor} \times \text{P}}{\text{number of strands of chain used}} \tag{10-8}$$

Recommended Working Loads for Roller Conveyor Chains

Once the required chain working load has been calculated for a specific application, the chain size (pitch) can be selected from Table 10-18. Note that Table 10-18 gives working loads (with a built-in speed factor) for double-pitch conveyor chains of ANSI Standard B29.4, as listed in Table 10-19. The pitch of these chains is twice that of Standard B29.1 chains having a comparable pin diameter. Standard B29.1 chains do not have recommended working loads for conveyor service, since they are rated in horsepower for power transmission. However, B29.1 chains are quite capable of accepting loads recommended for B29.4 chains of comparable pin diameter and are used extensively in conveyor service when provided with suitable attachments. Comparable data of these chains are shown in Table 10-20.

It should be noted that all the tensile strengths of B29.4 and B29.1 chains, given in Tables 10-19 and 10-20, respectively, refer to Roller Chains of carbon/alloy steel construction. The chain sizes specified are also available in stainless steel and plated (zinc,

Table 10-18 Working Loads for Double-Pitch Roller Conveyor Chains (lb)

Pitch (in.)	Chain speed (ft/min) 5	25	50	75	100	200	300	400	500
1.00	530	525	510	490	465	335	230	160	115
1.25	870	865	840	805	765	555	380	265	190
1.50	1215	1205	1170	1125	1065	775	530	370	265
2.00	2070	2055	2000	1915	1815	1320	905	630	455
2.50	3425	3400	3310	3175	3000	2180	1500	1040	750
3.00	4855	4815	4690	4495	4250	3090	2125	1480	1065
4.00	8585	8210	8000	7670	7250	5275	3625	2520	1815

Table 10-19 Technical Data; Double-Pitch Roller Conveyor Chains

Chain number	Chain pitch (in.)	Average ultimate tensile strength (lb)	Weight per foot (lb) without attachments
Small roller series			
C-2040	1	3700	0.34
C-2050	1¼	6100	0.58
C-2060H	1½	8500	1.05
C-2080H	2	14500	1.40
C-2100H	2½	24000	2.48
C-2120H	3	34000	3.60
C-2160H	4	58000	6.42
Large roller series			
C-2042	1	3700	0.50
C-2052	1¼	6100	0.81
C-2062H	1½	8500	1.42
C-2082H	2	14500	2.13
C-2102H	2½	24000	3.51
C-2122H	3	34000	5.48
C-2162	4	58000	9.34

Table 10-20 Technical Data; Standard Pitch Precision Roller Chains

Chain number	Chain pitch (in.)	Average ultimate tensile strength (lb)	Weight per foot (lb) without attachments
40	1/2	3700	0.41
50	5/8	6100	0.68
60	3/4	8500	0.99
80	1	14500	1.73
100	1-1/4	24000	2.51
120	1-1/2	34000	3.69
140	1-3/4	46000	5.00
160	2	58000	6.53

nickel, etc). Chain manufacturers should be consulted for working loads in view of the differences in tensile strength for various types of stainless steels.

The loads in Table 10-18 apply to single strand chain conveyors that are evenly loaded, have at least adequate lubrication, and are operating in nonabrasive environments at temperatures not exceeding 350 °F. These loads should be decreased for unfavorable environments or elevated temperatures, as per consultations with chain manufacturers.

After a specific chain has been selected for the conveyor, and the pitch has been found to be compatible with desired attachment spacing, the conveyor pull P should be recomputed, considering the actual rather than estimated data as chain weight (from Table 10-19), weight of attachments, weight of transported load, and appropriate coefficients of friction as determined from Table 10-14 for sliding or rolling conveyors as appropriate.

The working load of the chain finally selected must equal or exceed the required chain pull as finally computed.

Calculation of Chain Length

In most conveyor systems, both driving and driven sprockets have the same number of teeth. If so, the chain length, most conveniently expressed in number of pitches required, can be calculated from the formula

$$\left(\begin{array}{c}\text{Chain length,}\\ \text{in pitches}\end{array}\right) = \left(\begin{array}{c}\text{effective num-}\\ \text{ber of teeth,}\\ \text{one sprocket}\end{array}\right) + 2 \times \text{center distance, in pitches} \qquad (10\text{-}9)$$

If the driver and driven sprockets are of unequal size, calculate the chain length from the formula and tables used for Roller Chain drives in Chapter 5.

Power Required

Calculate the input horsepower needed to drive the conveyor from the formula

$$HP = \frac{P \times S \times 1.1}{33{,}000} \qquad (10\text{-}10)$$

In vertical conveyors, conveyor pull P will be equal to the unbalanced weight, which is to say, the material lifted, and thus

$$P = CW \qquad (10\text{-}11)$$

Catenary Tension and Chain Sag

In most long conveyors the return span of chain is supported over most of its length. The weight of the unsupported span or portion of a span would otherwise add to the chain pull, and must be considered, as well as the bearing and shaft loads, when calculating the total working load on the chain. Catenary pull is not a factor when calculating the horsepower required to operate the conveyor.

Some length of the return span should remain unsupported, creating a catenary span which provides a place for any accumulation of excess chain as wear takes place. This catenary also assures adequate slack-side tension so that the likelihood of the conveyor jumping teeth under a heavy load is minimal. A diagram of the catenary arrangement is shown in Figure 10-21.

Ample space should be provided for the catenary. Attempting to operate with too little depth of sag in a long catenary will result in high bearing loads as well as increasing the loading on the chain.

Depth of sag is often important in determining necessary clearances. The calculation of catenary sag and tension is quite complex. However, referring to the quantities given in Figure 10-21, these values may be determined approximately by the following simplified formulas:

$$D = \frac{\sqrt{3S^2 - 3L^2}}{4} \tag{10-12}$$

$$T = \frac{M}{12}\left(\frac{S^2}{8D} + \frac{D}{2}\right) \tag{10-13}$$

$$T = \frac{M}{12}\left(\frac{28D^2 + 3L^2}{24D}\right) \tag{10-14}$$

where

T = catenary tension, lb
M = weight of conveyor catenary, lb/ft
S = length of catenary in slack span, in. or
S = L + excess chain
D = depth of sag of conveyor, in.
L = straight-line distance between points of support, in.

The catenary tension and depth of sag may also be determined from Tables 10-22 and 10-23.

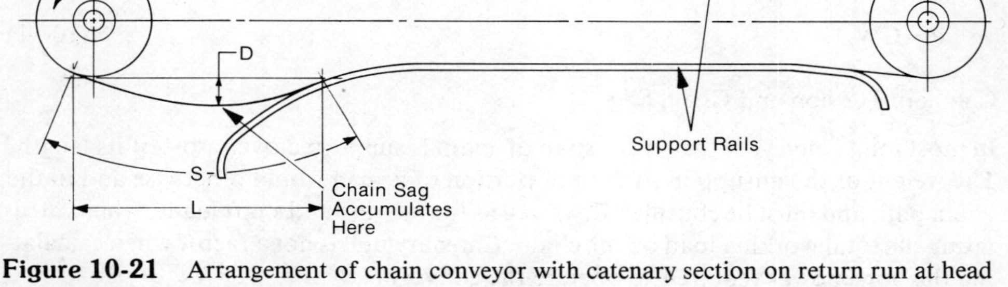

Figure 10-21 Arrangement of chain conveyor with catenary section on return run at head shaft end.

Table 10-22 Pounds Tension for Slack-Side Span for Roller Chain Weighing 1 lb/ft

Center distance (in.)	Amount of excess chain in slack span (in.)											
	0.01	0.05	0.10	0.20	0.30	0.40	0.60	0.80	1.00	2.00	3.00	4.00
20	16	7	5	3	3	2	2	2	2	1	1	—
40	43	19	14	10	8	7	6	5	5	3	3	2
60	79	35	25	18	14	13	10	9	8	6	5	4
80	122	54	39	27	22	20	16	14	13	9	8	6
100	171	76	54	33	31	27	23	20	17	13	11	9
110	197	88	62	44	36	31	26	23	20	14	12	10
120	224	100	71	50	41	36	29	26	23	16	14	11
130	252	113	80	57	46	40	33	29	26	19	16	13
140	282	126	89	64	52	45	37	32	29	21	17	15
150	312	139	99	71	53	50	41	36	32	23	19	16
160	344	153	109	78	64	55	45	39	35	25	21	13
170	377	168	120	85	70	60	49	43	38	28	23	20
180	411	183	130	93	76	65	54	46	42	30	25	21
190	446	199	141	100	82	71	58	50	45	32	27	22
200	432	215	152	108	88	77	63	54	49	35	29	24
210	513	231	164	116	95	83	68	58	53	38	31	25
220	555	243	176	124	102	89	72	63	57	40	34	28
230	593	265	188	133	109	95	77	67	60	43	36	30
240	632	283	200	142	116	101	83	72	64	46	38	32
250	671	301	213	151	123	107	88	76	68	49	40	34

[a]To find catenary tension for any other chain weighing M pounds per foot, multiply figures shown by M.

Table 10-23 Depth of Sag of Chain in Slack-Side Catenary Strands (in.)

Center distance (in.)	Amount of excess chain in slack span (in.)											
	0.01	0.05	0.10	0.20	0.30	0.40	0.60	0.80	1.00	1.50	2.00	3.00
10	0.19	0.43	0.61	0.87	1.07	1.24	1.52	1.76	1.99	2.46	2.87	3.60
20	0.27	0.61	0.87	1.23	1.51	1.74	2.14	2.47	2.77	3.42	3.97	4.92
30	0.34	0.75	1.06	1.50	1.84	2.13	2.61	3.02	3.38	4.16	4.83	5.95
40	0.39	0.87	1.23	1.73	2.13	2.45	3.01	3.48	3.90	4.79	5.55	6.83
50	0.43	0.97	1.37	1.94	2.37	2.74	3.36	3.89	4.36	5.34	6.19	7.61
60	0.47	1.06	1.50	2.12	2.60	3.00	3.63	4.25	4.77	5.84	6.77	8.32
70	0.51	1.15	1.62	2.29	2.81	3.24	3.98	4.59	5.15	6.30	7.30	8.97
80	0.55	1.23	1.73	2.45	3.00	3.47	4.25	4.91	5.50	6.74	7.80	9.58
90	0.58	1.30	1.84	2.60	3.19	3.63	4.50	5.21	5.83	7.14	8.27	10.15
100	0.61	1.37	1.94	2.74	3.36	3.83	4.74	5.49	6.14	7.53	8.71	10.69
110	0.64	1.44	2.03	2.87	3.52	4.06	4.98	5.75	6.44	7.89	9.13	11.20
120	0.67	1.50	2.12	3.00	3.68	4.24	5.20	6.01	6.73	8.24	9.53	11.69
130	0.70	1.56	2.21	3.12	3.83	4.42	5.41	6.25	7.00	8.57	9.92	12.16
140	0.72	1.62	2.29	3.24	3.97	4.59	5.61	6.48	7.26	8.89	10.29	12.61
150	0.75	1.68	2.37	3.35	4.11	4.75	5.81	6.71	7.52	9.21	10.65	13.05
160	0.77	1.74	2.45	3.46	4.24	4.90	6.00	6.93	7.77	9.51	11.00	13.48
170	0.80	1.79	2.52	3.57	4.37	5.05	6.19	7.14	8.01	9.80	11.33	13.89
180	0.82	1.84	2.60	3.67	4.50	5.20	6.37	7.35	8.24	10.08	11.65	14.28
190	0.84	1.89	2.67	3.77	4.62	5.34	6.54	7.55	8.46	10.36	11.98	14.67
200	0.86	1.94	2.74	3.87	4.74	5.43	6.71	7.75	8.68	10.63	12.29	15.05
210	0.89	1.98	2.81	3.97	4.86	5.61	6.88	7.94	8.89	10.89	12.59	15.42
220	0.91	2.03	2.87	4.06	4.97	5.74	7.04	8.13	9.10	11.15	12.88	15.78
230	0.93	2.08	2.94	4.15	5.09	5.87	7.20	8.31	9.30	11.40	13.17	16.14
240	0.95	2.12	3.00	4.24	5.20	6.00	7.35	8.49	9.50	11.64	13.45	16.48
250	0.97	2.16	3.06	4.33	5.30	6.12	7.50	8.67	9.69	11.88	13.71	16.82

Roller Chain Conveyor Design Examples

Example 1

A conveyor consisting of parallel chain strands interconnected by carrying slats, desired on 3-in. centers. Slats are 1 in. wide and weigh 1.91 lb each. The chain is to utilize carrier-type rollers supported and returned on guide rails. The conveyor will transport cartons of canned foodstuff 78 ft between the carton-filling and tape-sealing stations. Filled cartons weigh 48 lb and measure 1 ft × 2 ft × 10 in. high and will be spaced approximately 4 in. apart on the conveyor. Cartons will be conveyed with the 2-ft dimension perpendicular to direction of travel. The desired conveyor speed is 125 ft/min.

Step 1. Establish conveyor requirements.

a. Tentatively select smaller 1-in.-pitch conveyor series in view of the light loading involved. Modify selection to C2042 large roller type in view of the required support by rollers.

b. Calculate weight of the conveyor construction:

Chain weight—2 strands × 0.50 lb/ft	= 1.00 lb	(Table 10-19)
Slat weight—4 pieces/ft × 1.91 lb each	= 7.64 lb	
Weight per foot of conveyor (M)	= 8.64 lb	

c. Determine conveyed product weight per foot of conveyor:
48 lb ÷ 1.33 ft (spacing) = 36 lb/ft (W)

d. Conveyor length = 78 ft (C)

e. Coefficient of rolling friction, from Table 10-14, for
C2042 chain = 0.10 (f_r)

f. Speed of conveyor = 125 ft/min (S)

Step 2. Calculate conveyor pull P.

$$P = Cf_r(2.1M + W) \quad \text{[equation 10-3)]}$$
$$= 78(0.10)[2.1(8.64) + 36]$$
$$= 422 \text{ lb}$$

Then, applying a multistrand factor of 1.1, as cut-tooth sprockets will be used, and the conveyor will have two strands of chain:

$$422 \text{ lb} \times 1.1 \div 2 \text{ strands} = 232 \text{ lb per strand} \quad \text{[equation (10-8)]}$$

Refer to Table 10-18 and note that 232 lb is well within the recommended maximum working load for C2042 chain in the specified speed range.

Step 3. Calculate the required horsepower.

$$HP = \frac{P \times S \times 1.1}{33{,}000} \quad \text{[equation (10-10)]}$$
$$= \frac{422 \times 125 \times 1.1}{33{,}000}$$
$$= 1.76 \text{ hp}$$

Example 2

A conveyor consisting of two parallel strands of chain, 3 ft apart, is to carry bales weighing 1200 lb each up an incline of 10°. It is intended to use a standard attachment type such as is shown in Figure 10-3 with extension on both sides of each chain and spaced on every pitch of both chains to provide a continuous carrying surface on each. It is also intended to utilize the larger roller series as carrying rollers supported and returned on guide rails. Bales will be loaded rapidly enough to allow an average of 6 in. between them. Bales measure 4 ft × 4 ft × 6 ft and should be conveyed at 35 ft/min. Horizontal travel distance is 48 ft.

Step 1. Establish conveyor requirements.

a. Tentatively select a strong chain commensurate with the loading imposed. Evaluate a C2122H chain at 5.48 lb/ft:

Chain weight = 2 strands × 5.48 lb/ft = 10.96 lb
Cross members = none = 0.0 lb
Weight per foot of conveyor = 10.96 lb (M)

b. Determine product weight:

1200 lb ÷ 4.5 ft = 267 lb/ft (W)

c. Sprocket center distance:

$\frac{48\text{ ft}}{\cos 10} = \frac{48}{0.985}$ = 48.73 ft (C)

d. Coefficient of rolling friction: = 0.07 (f_r)

e. Conveyor speed: = 35 ft/min (S)

Step 2. Calculate conveyor pull P.

$P = Cf_r(\cos 10°)(2M + W) + CW(\sin 10°)$ [equation (10-5)]
$= (48.73)(0.07)(0.985)[2(10.96) + 267] + 48(267)(0.174)$
$= 3234$ lb

Assuming that flame-cut sprockets are used, use a multistrand factor of 1.2; the requried working load WL will then be, for chain selection:

$P = 3234 \times 1.2 \div 2$ strands $= 1940$ lb [equation (10-8)]

This is the required working load for chain selection.

Referring to Table 10-18, between 25 and 50 ft/min. it is obvious that the tentative selection of C2122H, rated at 4690 lb chain working load at 50 ft/min, is a far heavier chain than is needed. It also appears from inspection in this speed range that a C2082H chain appears more in line with the tentative working load calculation.

Step 3. Revise the changed requirements for C2082H.

a. Modify chain weight M to 2.13 lb/ft

2 strands × 2.13 lb/ft = 4.26 lb/ft (M)

b. Modify coefficient of rolling friction = 0.08 (f_r)

Step 4. Recalculate conveyor pull P.

$P = (48.73)(0.08)(0.985)[2(4.26) + 267] + 48.73(267)(0.174)$ [equation (10-5)]
$= 3321$ lb

If flame-cut teeth sprockets are used, the required chain working load WL will be $3221 \times 1.2 \div 2$ strands $= 1993$ lb [equation (10-8)]

If cut-tooth sprockets are used, the required chain working load WL will be $3221 \times 1.1 \div 2$ strands $= 1826$ lb [equation (10-8)]

C2082H Roller Chain, rated in Table 10-18 at 2055 lb working load at 25 ft/min and at 2000 lb at 50 ft/min, is just adequate with flame-cut sprockets and more than adequate with cut-tooth sprockets. In view of the smaller area of the C2082H attachments compared with those used with C2122H chain, bale-carrying cross members could be used and should be considered for their more positive carrying ability, and without any great addition of weight. Wider spacing of attachments could also be considered if cross members are to be used. If cross members are to be used, the cut-tooth sprockets probably also should be used both because of the slight added weight of the cross members in view of the borderline working load, and for better alignment.

Step 5. Calculate required horsepower.

$$HP = \frac{P \times S \times 1.1}{33,000} \quad \text{[equation (10-10)]}$$

$$= \frac{3321 \times 35 \times 1.1}{33,000}$$

$$= 3.87 \text{ hp}$$

11

Chains for Bucket Elevators

The bucket elevator, defined broadly, is a specialized type of conveyor used to lift bulk materials in a vertical or nearly vertical plane through the use of carrying buckets. It will handle nearly any loose material that does not contain lumps too big for the buckets and is not too sticky to be discharged. Typical bucket elevators are shown in Figures 11-1 to 11-3.

Typically, a bucket elevator consists of three basic components: a head end, a foot end, and intermediate sections (see Figs. 11-2 and 11-3). The head end assembly includes the head shaft and sprocket, bearings, drive assembly, supporting structure, buckets, chain, and casing. The foot end includes the foot shaft and sprocket, bearings with a takeup, supporting structure, casing, buckets, and chain. Intermediate sections are normally considered to consist of the elevator supporting structure and casing, plus buckets and chain. In most instances, elevator casings are used to support the machinery and control dust. However, when conditions permit, open elevators may be used and include supporting structure only (Fig. 11-2).

Elevator buckets are bolted to one or two strands of endless carrying chain. Material is loaded into the buckets via a hopper and loading leg at the ''boot'' in a continuous elevator, or is scooped up by the buckets as they round the lower sprocket in a centrifugal discharge elevator. The material is discharged by centrifugal force or gravity as the buckets overturn at the head sprocket.

Elevators Using Engineering Steel Chains

Typical examples of the types of chain used in elevators are shown in Figure 2-24. In most applications, straight sidebar, hardened bushing, rollerless chain is used; however, most continuous high-capacity elevators use straight sidebar, hardened bushing roller chain whose rollers ride on vertical guides integral with the elevator casing. In some applications, instead of straight sidebar chain, the offset sidebar type can be used. Table 11-4 lists chains of various pitches and designs with typical working loads.

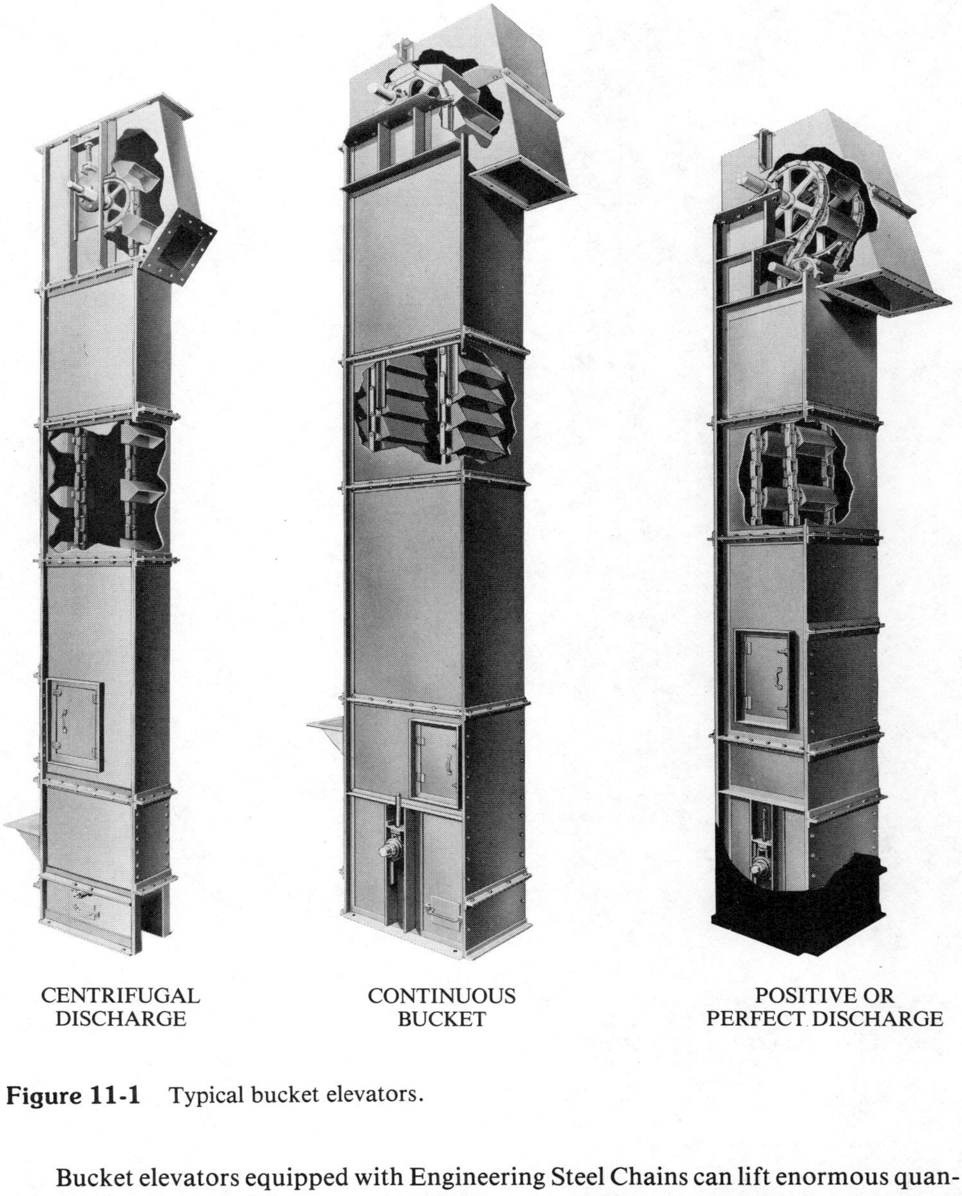

CENTRIFUGAL DISCHARGE — CONTINUOUS BUCKET — POSITIVE OR PERFECT DISCHARGE

Figure 11-1 Typical bucket elevators.

Bucket elevators equipped with Engineering Steel Chains can lift enormous quantities of heavy materials to considerable heights. "Super-capacity" elevators have been designed to lift up to 750 tons/hr of material weighing 125 lb/ft^3 as high as 120 ft. Special chains can be engineered for an even higher-rated performance.

Since there are many different elevator designs, the following discussion will be limited to medium or heavy-duty types, with emphasis on chain, attachments, buckets, and sprockets.

There are four basic types of bucket elevators:

1. Centrifugal discharge

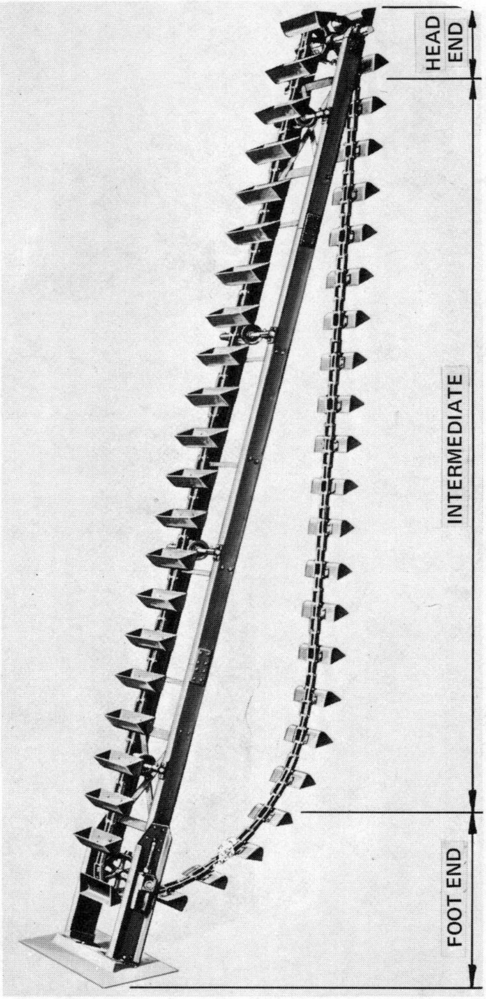

Figure 11-2 Open inclined elevator.

2. Positive or perfect discharge
3. Continuous bucket
4. Super capacity

These are shown in Figure 11-5 and are described in the following paragraphs.

Centrifugal Discharge Elevators

The centrifugal discharge elevator (Figs. 11-6 and 11-7) is designed so that the buckets scoop material from the boot and discharge it by centrifugal force as they pass over the head sprocket. It is, therefore, a comparatively high speed elevator. The average centrifugal discharge elevator headshaft speed varies with the diameter of the head sprocket and nature of the material, but is usually between 35 and 50 rpm.

Elevators handling grain, cement, and so on, with large-diameter head sprockets and high-speed buckets may run at much higher speeds.

Centrifugal discharge elevators work best with fine, loose material like ashes and crushed coal that can be thrown from the buckets and is not easily degraded.

Because the chain runs through the material in the boot when the buckets are loaded, steel bushed rollerless chain should be used. This is because chains with rollers often become loaded with dust between the rollers and bushings. If any moisture is present, this tends to cake the material and thus prevent the rollers from turning. The pitch of the chain will vary depending on the size of the buckets and the weight of the material, but since the speed must be relatively high, a chain pitch of 6 in. is used in most applications. A single strand of steel bushed rollerless chain will carry most loads on a centrifugal discharge elevator.

Attachments. When a single strand of chain is used, the buckets are bolted to

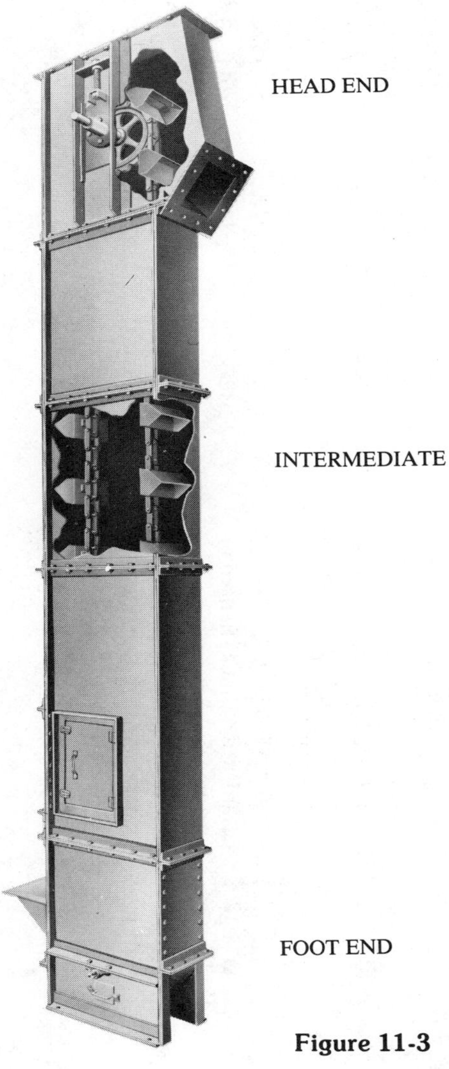

Figure 11-3 Enclosed vertical elevator.

Table 11-4 Typical Engineering Steel Chains for Elevators

Pitch (in.)	Description of chain	ANSI number	Typical working load (lb)	Average weights per foot (lb)		
				Plain chain	All K attachments	All G attachments[a]
2.609	Straight sidebar, steel bushed, roller-less; cottered or riveted	S188	2740	3.8	5.8	—
3.075	Same	S131	4450	7.4	10.2	—
4.00	Same	S102B	6300	6.9	9.0	—
4.760	Same	S111	8800	10.2	15.2	—
6.00	Same	S110	6300	6.3	8.6	—
6.050	Same	S150	15000	16.6	24.0	—
6	Same—various sizes	—	9200–21000	10.4–34.0	15.0–65.0	—
7	Same	—	21000	31.0	75.0	—
9	Hardened bushing; roller chain, rollers flanged or smooth; cottered or riveted	12-09-½	7200	12.0–12.5	16.0–16.5	13.0–27.0
9	Same—except flanged roller, only	14-09-2	9200	16.5	22.3	18.0–29.0
12	Hardened bushing; flanged or smooth rollers; cottered or riveted	12-12-½	7200	10.0–10.5	15.3	11.6–24.0
12	Same—flanged roller only	14-12-½	9200	14.0	20.3	15.4–25.0
12	Same	16-12-2	12300	21.5	30.4	23–32

[a]Weights of G attachment links depend on style of bracket.

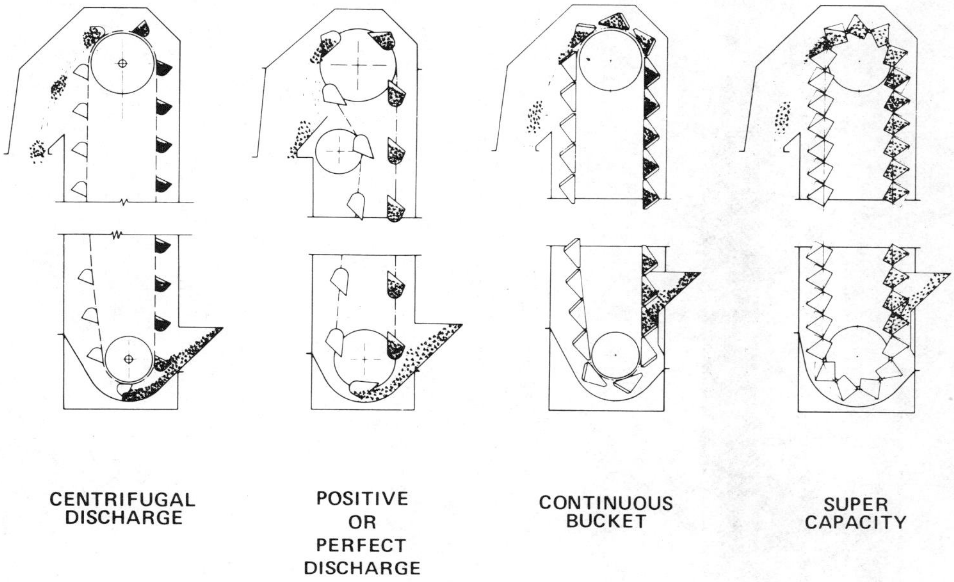

Figure 11-5 Types of elevators.

K-type attachments (see Fig. 11-8). The selection of the number of holes is dependent on the forces and impact developed. For the larger capacities, K attachments having more than two holes are used with buckets up to 24 in. wide. For elevators requiring larger chain and buckets, consult the chain manufacturer. Double-strand chain, centrifugal discharge elevators are not recommended.

Elevator Buckets. Buckets for centrifugal discharge elevators come in many styles, as shown in Figure 11-9. For heavy, moderately abrasive loads, style AA malleable iron or pearlitic malleable iron buckets are used. Specially shaped buckets are available for large capacities or for fine or wet material. Reinforced welded steel buckets are used for high-speed cement or grain elevators.

Style A cast elevator buckets are used when very small buckets are required for handling coal, cement, ashes, sand, stone, gravel, and similar materials. They are furnished in either malleable or pearlitic malleable iron.

Style AA cast elevator buckets, which have generally replaced style A in popularity, are designed similar to style A except that a heavy reinforcing band is cast along the front. Style AA buckets are furnished in either malleable or pearlitic malleable iron.

Style AA-RB cast elevator buckets are used for handling heavy abrasive products under severe service and impact conditions. Buckets are designed similar to style AA with reinforcing ribs on the front and with backs twice the normal thickness. They may be furnished in either malleable or pearlitic malleable iron.

Style AC cast buckets are designed specifically to increase elevator capacity by closer bucket spacing. Style AC buckets are used for handling material such as raw meal, crushed clinker, and finished cement. Air ports are provided in the bottom of

Figure 11-6 Centrifugal discharge elevator with foot-end takeup.

the bucket to act as a relief valve for faster loading and unloading of light, fluffy materials.

Style B cast elevator buckets are used on inclined elevators for handling coarsely broken materials such as coke, stone, ore, and so on. Clean discharge at relatively slow speeds is made possible by the low front design. Style B buckets are furnished in either malleable or pearlitic malleable iron.

Style C cast elevator buckets are designed to handle materials which tend to stick or pack in buckets such as clay, sugar, salt, wet grain, and finely pulverized wet ore. Style C buckets are furnished in either malleable or pearlitic malleable iron.

Style SC cast elevator buckets are designed primarily for handling foundry sand. SC buckets have an extra heavy body with a reinforcing band cast along the front edge and around the ends to resist abrasion, and are furnished in either malleable or pearlitic malleable iron.

Salem steel elevator buckets are used for powdered or granular free-flowing materials. They are formed from one piece of sheet metal and have a smooth round con-

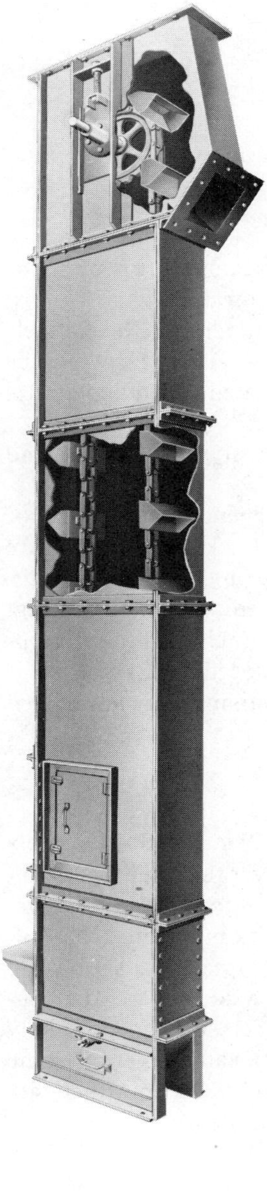

Figure 11-7 Centrifugal discharge elevator with head-end take-up.

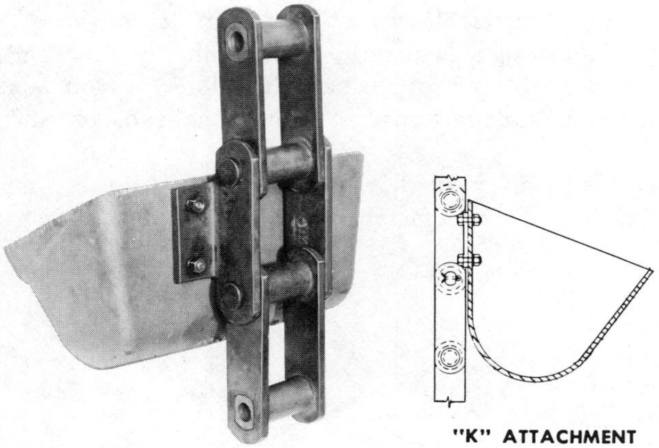

Figure 11-8 Bucket attached to single strand of chain with K attachment.

tour for clean handling and delivery of materials. There are no seams in front or on the ends of these buckets. Backs are reinforced for strength and durability.

Continuous elevator buckets are used for gritty, bulky material with relatively large lump size at slower speeds than centrifugal discharge elevators. They may be either regular or super capacity and may be either cast or fabricated depending on size or usage.

Buckets made of nonmetallic materials, such as high-density polyethylene and nylon, are also available for special handling requirements.

The manner in which buckets are spaced along the chain depends on the design of the particular installation. As a rule, for cast malleable iron buckets the spaces are 2-1/2 to 3 times the projection of the bucket, the principle being that the load discharged from one bucket should be thrown clear of the preceding bucket (see Fig. 11-10). Buckets are commonly attached to every third or fourth chain pitch. Exceptions to the above are for specially designed buckets or high-speed discharge.

The bucket length may be governed by the rule of thumb that it may not exceed five times the chain bushing length (see Fig. 11-11).

Positive Discharge Elevators

The positive discharge, or perfect discharge elevator (Fig. 11-12) moves slowly (rarely more than 120 ft/min). The buckets are completely inverted over the discharge spout by a pair of snub wheels. This deliberate method of discharge is intended for cohesive, powdery, and granular material and for sluggish loads that stick in the buckets. Pulverized coal, powdered feldspar, and wet sugar beet pulp are typical loads.

Positive discharge elevator buckets are end-hung between a double strand of Engineering Steel Chain. An A attachment with an A bucket wing is the most commonly used mount (see Fig. 11-13). Usually, the positive discharge elevator takes the same standard malleable iron buckets as the centrifugal discharge elevator, but there are also specifically contoured steel buckets for handling sticky materials.

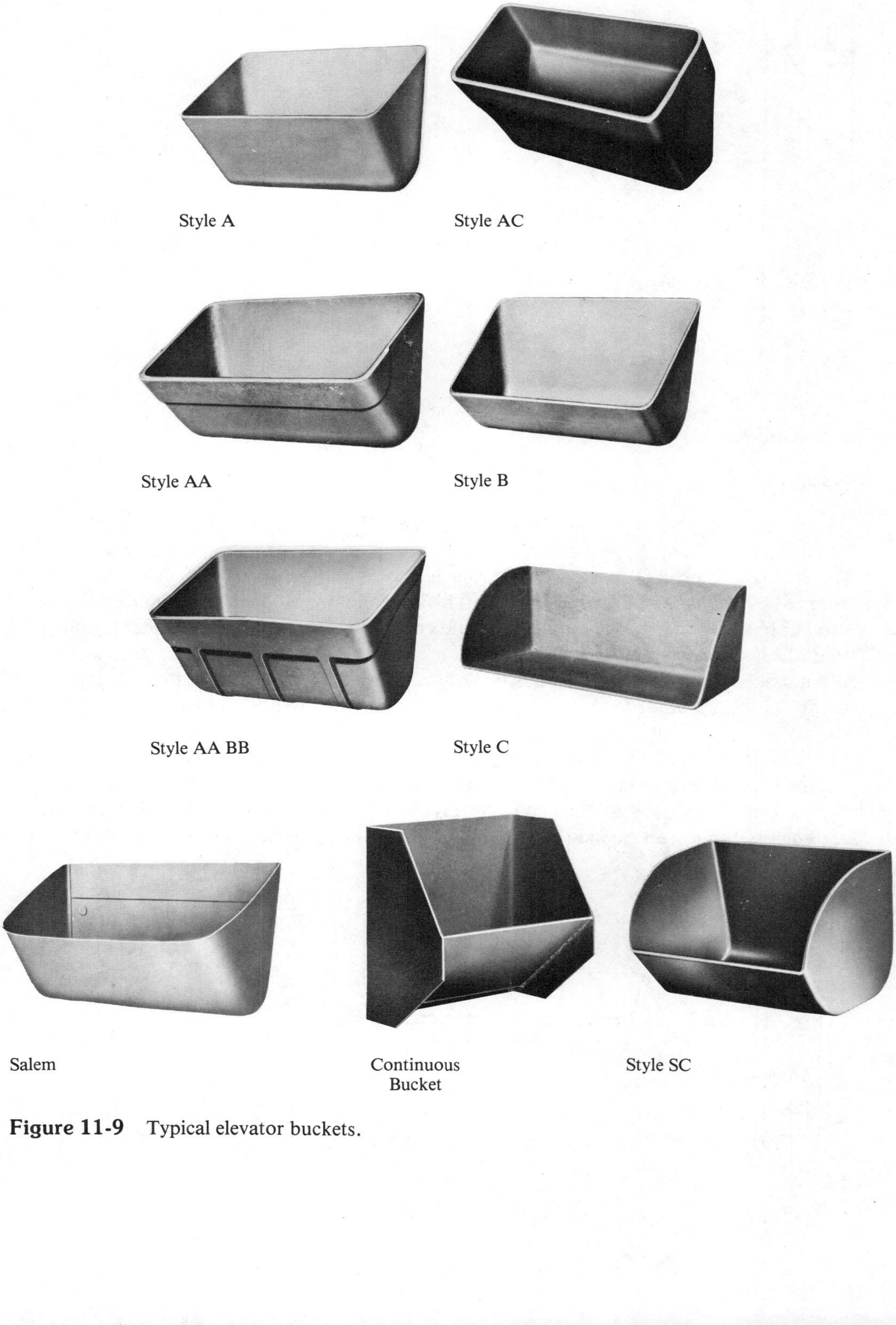

Figure 11-9 Typical elevator buckets.

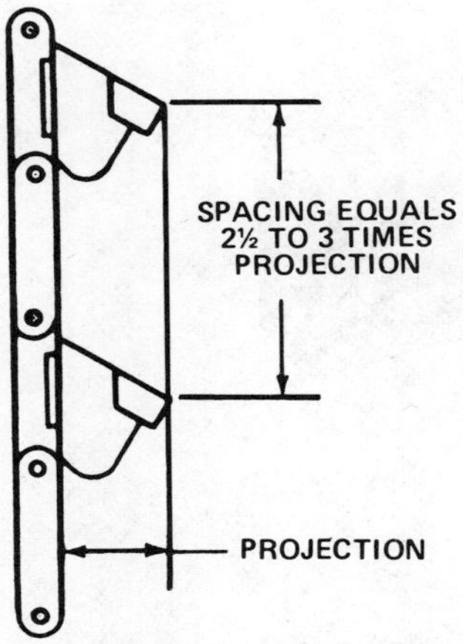

Figure 11-10 Bucket spacing.

The buckets must be spaced to keep the loads from being dumped onto the preceding bucket.

Continuous Bucket Elevators

The continuous bucket, or continuous discharge, elevator usually operates at speeds of 100 to 125 ft/min. It is unique in that the buckets are spaced continuously and have flat fronts and projecting sides so that each bucket acts as a chute for material discharged by the bucket directly behind it. This combination of continuous buckets and moderate speed makes the capacity of the continuous bucket elevator nearly twice that of the positive discharge elevator, but less than that of the high-speed centrifugal discharge elevator. The continuous buckets are usually fed through a loading leg or chute. This method of loading, together with the gentle method of discharge, makes the elevator especially suitable for material that is easily degraded.

Regular Capacity Continuous Bucket Elevators. Regular capacity vertical and inclined elevators operate at about the same speeds (100 to 125 ft/min, depending on the material), and use steel bushed rollerless chain for severe work. Vertical elevators are enclosed in a steel casing (Fig. 11-14), while inclined elevators usually work on a

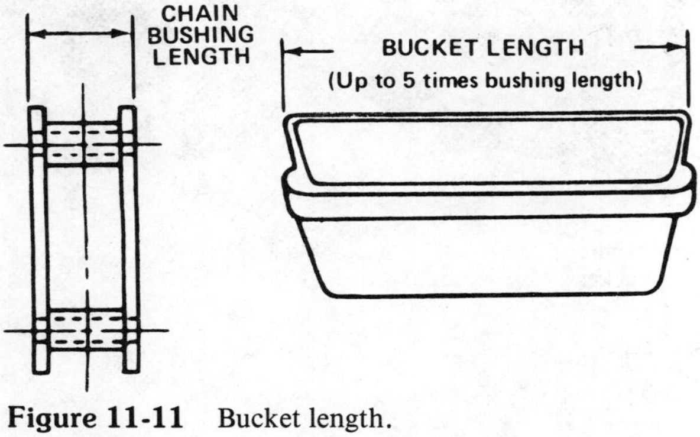

Figure 11-11 Bucket length.

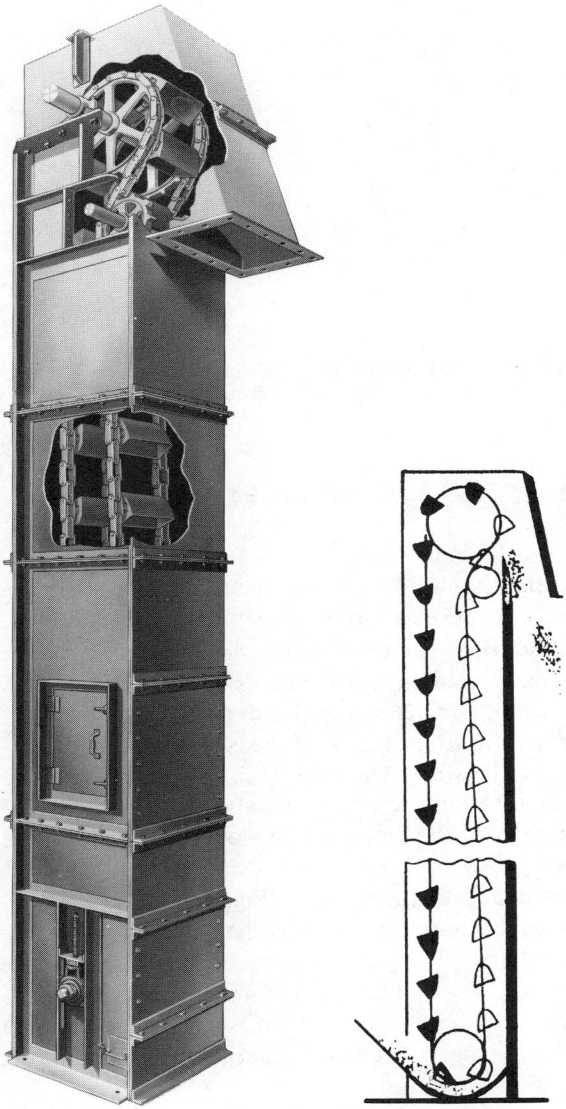

Figure 11-12 Positive or perfect discharge elevator.

simple frame and are used for outdoor work where it is not important to contain material within a casing or exclude outside contaminants.

Engineering Steel Chain may be installed in single or double strands, depending on the length of bucket and how much strength is needed. If a single strand is chosen, K attachments are used. For double strands, either A or G attachments are generally preferred. Continuous elevator buckets are made of welded steel and can be furnished in many standard and custom designs, all of them with projecting sides that may overlap in varying degrees depending on how smooth a chute is needed to discharge the

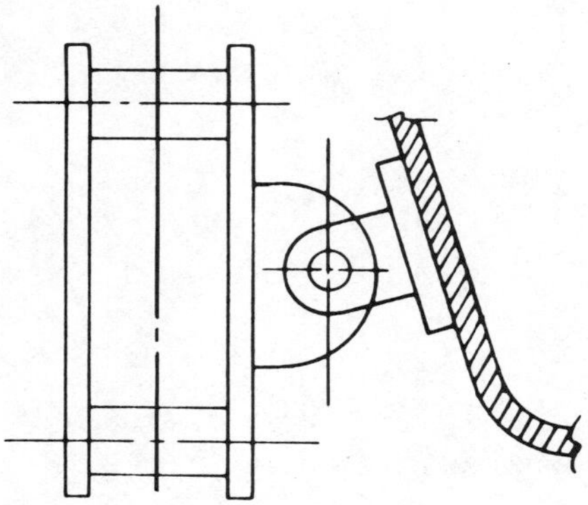

Figure 11-13 "A" chain attachment with bucket wing for end-mounting buckets.

material without spilling (see Fig. 11-15). As a rule, the finer the material, the deeper the overlap.

Super Capacity Elevators. A special kind of continuous bucket elevator is the super-capacity type (see Fig. 11-16). Its extra long buckets are end-hung between two strands of chain and carry heavy loads, from fine sand to large lumps of coal and ore.

Since the super capacity elevator carries very heavy loads and since a smooth run is necessary for effective discharge, long-pitch bushed straight sidebar roller chain is normally used. The rollers are between the sidebars, and this design allows the chain to roll smoothly between angle tracks fixed to the inside of the casing on vertical elevators, or part of the frame on inclined elevators. The material is fed directly into the buckets from a hopper so that it does not enter the chain joint and wear the inside diameter of the roller and outside diameter of the bushing as it would with scoop-type loading.

Super-capacity buckets are mounted to G attachments (Fig. 2-25), and their capacity is explained not only by their bucket length but also by the fact that the double strand of chain allows them to extend back of the chain centerline, as shown at the right in Figure 11-5. The sides of the bucket extend outward far more than those of the regular capacity continuous buckets and are shaped so that, when inverted, each bucket forms a deep trough to receive the material being discharged from the following bucket.

Takeups

Some type of takeup must always be used in elevator installations. Types of takeups are shown in Figure 11-17. They are used primarily to maintain sufficient tension on the chain to assure proper seating on the sprocket and to compensate for elongation of the chain due to wear. They are especially useful when installing chain in the elevator casing by allowing the reduction of shaft center distances.

In a standard basic elevator either foot-end or head-end takeups may be used. A foot-end installation, however, is preferable, but for elevating materials that require frequent clean-out or which tend to pack excessively, head-end takeups are recommended.

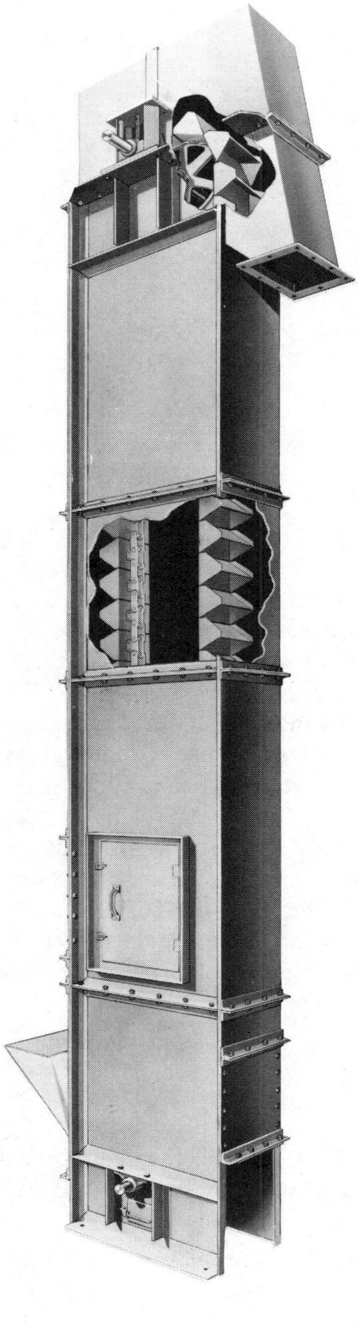

Figure 11-14 Regular capacity continuous bucket elevator.

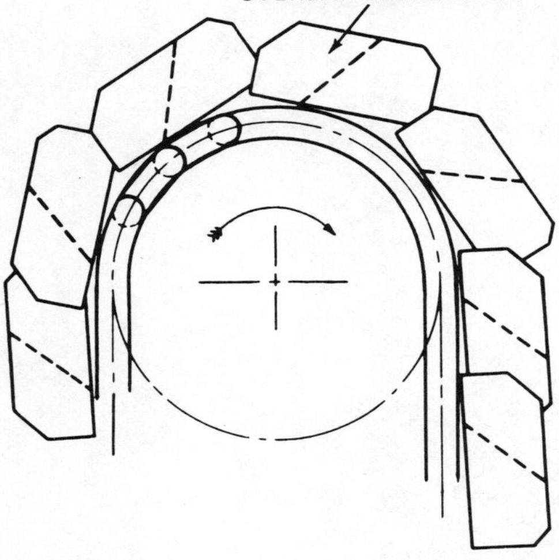

Figure 11-15 Typical overlapping bucket assembly at discharge.

Design and Selection of Chain and Bucket Elevators

Although the various manufacturers' standard elevators vary somewhat in type and ratings, Table 11-18 gives an approximate comparison of what the four basic types can do. Note that there is some overlapping. Generally, however, centrifugal discharge elevators are used for rapid conveying of free-flowing granular materials, and are popular for elevating coal, sand, cement, and so on. Positive discharge elevators are used for rather fine materials that tend to stick to the bucket or adhere to each other. Continuous bucket elevators may be used for some of the same materials as centrifugal elevators and are commonly used for gravel and similar mixed and lumpy materials. Continuous elevators are slower, have lower capacities, and require less maintenance when conveying heavy or variable loads. The super-capacity elevators are the workhorses of the group. They are used not only for conveying the materials shown in Table 11-18, but for conveying almost any heavy bulk material at high capacities over long centers.

Selecting Engineering Steel Elevator Chain

The following two rules of thumb apply in nearly all cases:

1. Use Engineering Steel bushed straight sidebar rollerless chain for elevators in which the chain hangs free and in which the chain is exposed to the conveyed material as it is scooped up during loading.
2. Use Engineering Steel bushed straight sidebar roller chain for elevators in which the chain rides on tracks, generally super capacity elevators.

Prior to selecting the chain and sprockets, the following should be known or determined:

Figure 11-16 Super-capacity vertical elevator.

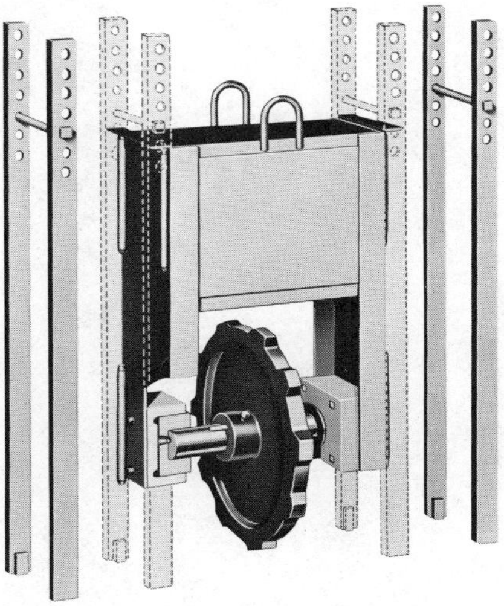

Internal gravity takeup

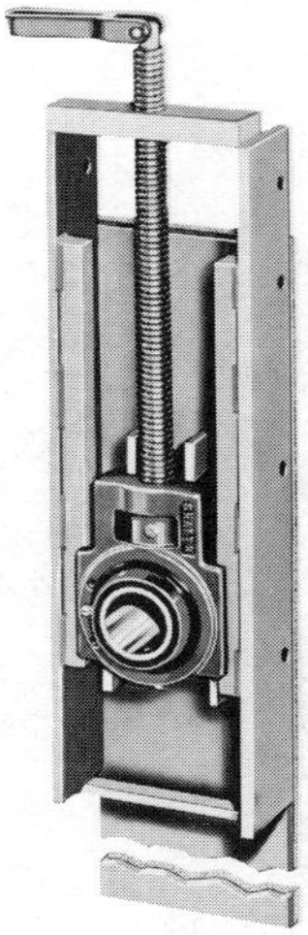

Screw take up

Internal takeup

External takeup

Figure 11-17 Types of takeups.

Table 11-18 Maximum Capabilities of Standard Chain and Bucket Elevators*

TYPE	CAPACITY IN TONS PER HOUR**	SPEED FPM†	MAXIMUM DISTANCE BETWEEN SHAFT CENTERS	MATERIAL TO BE HANDLED
CENTRIFUGAL DISCHARGE	350	300	80	††Cement (bulk/clinker), ashes, crushed bauxites, coal, soda ash (dense), many agricultural items — — wheat, coffee, rice, bran, etc.
POSITIVE DISCHARGE	70	120	80	Baking powder, carbon black, powdered coal, charcoal, wet sugar pulp, powdered feldspar.
CONTINUOUS BUCKET	120	140	80	Cement (bulk), fine coal, gravel, cinders, bauxite, phosphate rock.
SUPER CAPACITY	750	100	120	Crushed stone, cement clinkers, gravel, larger coals, crushed slate and gypsum slag.

*The figures included in this table are nominal and based on optimum standards relating to chain speeds, head sprocket diameter, and other criteria which reflect good general practices. Do not exceed these figures without contacting the chain manufacturer.

**At 100 pounds per cubic foot, except that super capacity is at 125 pounds per cubic foot.

†Higher speeds are permissable with light, fluffy materials.

††Special centrifugal discharge cement mill elevators have been designed to handle up to 900 TPH with centers up to 140 feet.

1. Required capacity in tons per hour
2. Elevator chain and bucket speed in feet per minute
3. Distance between sprocket centers
4. Size of elevator buckets
5. Weight of material per foot of elevator as determined by the equation

$$W = \frac{33.33T}{S} \tag{11-1}$$

where

T = capacity, tons/hr
S = chain and bucket speed, ft/min
W = weight of material per foot of elevator

6. Service conditions such as abrasiveness of material; clean, corrosive, hot or cold environment; degree and frequency of shock; operating hours per day

7. Aeration for pulverized materials and its effect on material density must be considered
8. Weight of moving parts (buckets, attachments, chain) per foot of elevator

Some of these items can only be approximations at this point. For example, the user does not know the weight of the chain until he or she selects it, so a back-check should be made when the selection is completed.

Selection Steps

Step 1. Determine chain pitch. In nearly all cases this is predetermined by bucket size and spacing. With continuous buckets and super capacity elevators, one bucket per pitch is normally used and should never exceed 2 pitches of chain when the chain pitch is 6 in. or less. In other types, one bucket every three or four pitches may be used if the spacing falls within the range of 2-1/2 to 3 times the projection for AA buckets, which are generally used. AC buckets are usually spaced every second pitch. In all cases choose a pitch length that will permit proper spacing to give the required capacity at the correct speed for the headsprocket diameter. Specific chains and pitch lengths are usually selected from styles and sizes that have become known through regular use as typical elevator chains (see Table 11-4).

Step 2. Tentatively determine the size of the sprocket in terms of number of teeth, chain pitch, and rpm by referring and checking to conform to limitations found in Tables 11-19 and 11-20. In general, the larger the number of teeth in the sprocket, the smoother the operation. When material is discharged by centrifugal force, its trajectory will be determined by the sprocket diameter and revolutions per minute. The basic equation for rpm is

$$\text{Minimum head shaft rpm} = \frac{44.3}{R} \tag{11-2}$$

where R is the distance in feet from center of rotation to the center of gravity of the discharged material.

Good elevator design requires proper balance between these factors. (For calculating sprocket pitch diameter from pitch and number of teeth, see Table 4-45.)

Table 11-19 Maximum RPM as a Function of Chain Pitch and Number of Sprocket Teeth

NUMBER TEETH IN SPROCKET	PITCH OF CHAIN															
	1.63	2.00	2.30	2.50	2.61	3.00	3.10	3.50	4.00	4.76	5.00	6.00	7.00	8.00	9.00	12.00
	RPM															
6	144	106	86	76	70	58	54	46	37	28	27	20	15	13	11	7
7	168	124	100	88	82	67	63	53	44	33	31	23	18	15	13	8
8	192	141	114	101	94	77	72	61	50	38	36	27	21	17	14	9
9	216	159	129	114	106	87	81	69	56	41	40	31	24	19	16	10
10	241	177	143	126	118	96	90	76	62	47	45	34	26	22	18	12
11	265	194	158	139	129	106	99	84	69	51	49	37	29	24	20	13
12	289	212	172	152	141	115	107	92	75	57	54	41	32	26	22	14
13	313	229	192	164	153	124	117	98	80	61	57	44	34	28	24	15
14	337	247	201	177	165	135	126	107	87	67	63	48	37	31	25	16
15	361	265	215	190	177	144	135	114	93	71	66	50	40	33	27	18
16	385	283	229	202	188	154	144	122	100	76	71	54	42	35	29	19
18	433	318	258	228	212	173	162	137	112	85	80	61	48	39	33	21
20	433	318	258	228	212	173	162	137	112	85	80	61	48	39	33	21
24	433	318	258	228	212	173	162	137	112	85	80	61	48	39	33	21

Table 11-20 Maximum Linear Speeds Ft/Min for Conveyor and Elevator Chains

NUMBER TEETH IN SPROCKET	PITCH OF CHAIN															
	1.63	2.00	2.30	2.50	2.61	3.00	3.10	3.50	4.00	4.76	5.00	6.00	7.00	8.00	9.00	12.00
	FPM															
6	117	106	99	95	92	87	83	80	74	67	67	61	55	52	50	43
7	160	144	134	128	125	117	113	108	102	93	90	83	76	72	68	58
8	209	187	174	168	164	154	148	142	133	121	120	108	99	94	88	76
9	264	238	222	213	207	195	188	181	168	148	147	139	127	119	112	97
10	327	294	274	262	256	240	232	221	206	188	187	170	155	147	138	120
11	395	355	333	318	310	291	281	269	253	223	220	203	187	178	167	146
12	471	423	395	379	369	345	331	321	300	274	270	246	228	212	198	174
13	552	495	478	445	434	405	393	374	350	315	313	286	263	248	234	204
14	640	576	539	516	502	472	455	436	405	372	367	336	304	289	272	236
15	735	662	617	593	577	540	523	499	466	427	416	381	351	332	311	271
16	836	754	702	673	657	616	595	569	533	482	473	432	400	337	355	308
18	1058	954	889	854	832	778	753	719	672	612	600	549	507	477	449	390
20	1176	1060	988	949	924	865	837	799	746	680	666	610	563	530	499	434
24	1411	1272	1186	1139	1109	1038	1004	958	895	816	799	732	676	636	599	520

Step 3. Calculate the maximum pull on the chain. In a vertical chain elevator the main elements of this pull are the weight of the chain and buckets, the weight of the material, the pull induced by the digging or loading action of the buckets, the load induced by the takeup, and the friction set up by the movement of the components (bearings, chain joints, etc.). If the elevator is inclined, an additional factor induced by the catenary effect of the return run (if hanging free) must be taken into consideration. Inclined elevators with a catenary return run will not be covered in this discussion. The maximum chain pull P can be calculated by

$$P = W_2 + W_3 + W_4 + Q + \frac{P_1}{2} \quad \text{(see Fig. 11-21)} \tag{11-3}$$

where

P = maximum chain pull
W_2 = total weight of either the carrying or the return strand of chain
W_3 = total weight of buckets on either the carrying or the return strand of chain
W_4 = total weight of material in buckets along the carrying strand of chain (anticipated surge conditions or bucket overfill must be considered)
Q = chain pull (tension) induced by digging or loading action in the boot and other miscellaneous elevator friction
P_1 = takeup tension

W_2, W_3, and W_4 can be readily found. For instance, the weight of the chain can be found by multiplying the weight per foot of the selected chain by the elevator centers. The weights of the buckets and the weight of the material can be found in a like manner.

The chain pull due to the digging action Q of the boot or the loading action of the carrying leg as well as the friction throughout the elevator can be calculated by the equation

$$Q = 12 \times \frac{W_5 \times J}{d} \times \text{boot sprocket diameter in inches} \tag{11-4}$$

where

W_5 = weight of material in one bucket, lb

d = bucket spacing, in.

J = an empirical corrective factor with the following values for stated conditions:

= 1.0 for centrifugal discharge elevators handling coarse lumpy material

= 0.67 for centrifugal discharge elevators handling fine free-flowing material

= 0.50 for continuous bucket elevators

Service factors are to be applied to find the design working load of the elevator chain (see step 5).

In setting the takeup in the elevator, care should be taken to prevent extra tension in the elevator chain. If such is necessary, an additional factor must be included in the basic chain pull equation. Should additional equipment be driven from the elevator foot shaft, such tensions must be calculated, converted to pounds, and included in the basic equation.

When vertical elevators decrease in sprocket center distances the Q value becomes relatively greater in importance. If adjustments for difficult conditions must be made, the J corrective factor can be adjusted to suit.

Chain Pull Equations

P_1 = takeup tension for screw-type takeups with proper adjustment tension will be as near zero as possible. P_1 for gravity takeups will be equal to the tail carriage machinery plus any resultant force from added weights (Fig. 11-21).

Tension P_2 can be determined by

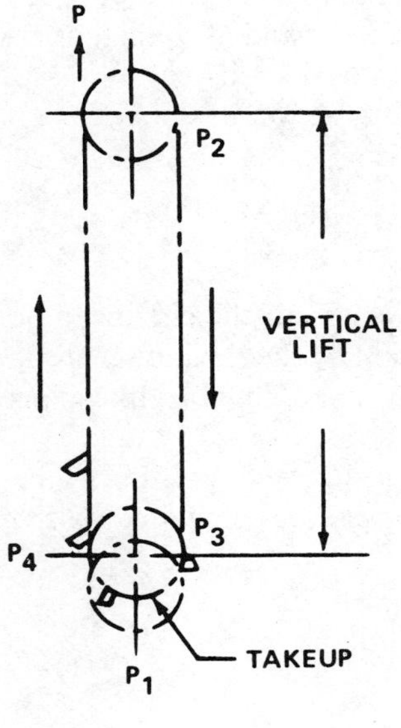

Figure 11-21 Diagram for maximum chain pull equation.

$$P_2 = W_2 + W_3 + \frac{P_1}{2} \tag{11-5}$$

P_3 is the point of least chain tension and theoretically should have a zero value.

$$P_3 = \frac{P_1}{2} \quad \text{(Fig. 11-21)} \tag{11-6}$$

Tension P_4 can be determined by the following equation, which has been explained previously:

$$P_4 = Q = \frac{12 \times W_5 \times J}{d} \times \text{boot sprocket diameter in inches} + \frac{P_1}{2} \tag{11-7}$$

Step 4. To the total pull P (Fig. 11-21), apply the factors for chain and speed and service conditions (shock, type, and frequency of operation, etc.) from Table 11-22. This gives the design working load on which chain selection is based. Service factors are multiplied.

Step 5. Select a chain whose maximum rated working load is equal to or greater than the design working load.

Table 11-4 lists such chains of the steel bushed rollerless and long-pitch roller conveyor types deemed suitable for bucket elevator service and listed in ANSI Standards B29-12 and B29.15. For these, Table 11-4 shows ANSI designations, working loads, and weights of plain chain and of a few styles of attachments.

Although Table 11-4 provides enough information to make a tentative chain selection, there may be many other factors involved, and it is really desirable to consult chain and bucket elevator catalogs before a final selection is made.

Step 6. Sprocket, traction wheel, hub, and shaft selection are given in Chapter 4.

Example: Elevator Chain Selection

A centrifugal discharge bucket elevator has been selected to handle 3/4-in. minus lump coal to a belt system feeding surge bins. Buckets and some components are already on hand and are desired to be used, if possible. Requirements are as follows:

1. Required capacity of bucket elevator: 60 tons/hr
2. Elevator chain and bucket speed: 240 ft/min
3. Distance between centers: 70 ft
4. a. Size of elevator buckets: 16×8 in.
   b. Capacity of buckets: 0.34 ft^3
   c. Weight of buckets: 21 lb each
5. Weight of material per foot of elevator, as per equation (11-1):

$$W = \frac{33.33T}{S} = \frac{33.33 \times 60}{240} = 8.33 \text{ lb/ft}$$

The coal has a bulk density of 50 lb/ft^3.

Table 11-22 Service Factors

CONDITIONS		SERVICE FACTORS
FREQUENCY OF SHOCK	Infrequent shock	1
	Frequent shock	1.2
CHARACTER OF CHAIN LOADING	Uniform or steady load	1
	Moderate shock load	1.2
	Heavy shock load	1.5
ATMOSPHERIC CONDITIONS	Moderately dirty and moderate temperature	1.2
	Exposed to weather, very dirty, abrasive, mildly corrosive and reasonably high temperatures	1.4
DAILY OPERATING RANGE	8–10 hours	1
	10–24 hours	1.2

6. a. Service conditions (see Table 11-22): (1) frequency of shock—occasional frozen lumps in winter; otherwise smooth; factor: 1.1; (2) shock is moderate; factor: 1.2
   b. Atmospheric conditions: exposed to weather; very dirty; material handled is mildly corrosive; quite corrosive when wet; factor: 1.3
   c. Daily operating range: averages 12 to 14 hr; factor: 1.2
7. Aeration, and so on, is not a factor.
8. Weight of moving parts: assume the standard 6-in.-pitch ANSI S110 chain (weighing 6.3 lb per foot for plain chain) will have adequate capacity. If it does not, then S111 chain, weighing 10.2 lb/ft, will have to be used (Table 11-4). The chain will have K-2 attachments for the buckets, spacing not yet established. For S110 chain, K-2 attachments weigh 8.6 lb/ft; for S111 chain they weigh 15.2 lb/ft.

Step 1. Determine chain pitch. The buckets to be used are 16×8 in., and have a rated capacity of 0.34 ft^3. The bulk density of the material is 50 lb/ft^3, so the buckets will carry $0.34 \times 50 = 17$ lb each.

To achieve a capacity of 60 tons/hr, the elevator has to carry an average of 8.33 lb/ft, so the maximum permissible bucket spacing is

$$\frac{17\times 12}{8.33}=24.49\text{ in.}$$

Therefore, buckets every fourth link of a 6-in. pitch chain will produce the rated capacity.

Step 2. Determine tentative sprocket size. On a centrifugal discharge bucket elevator, the shaft rpm, coupled with the radius of the bucket center, must be such that the bucket will discharge the material. Against this is that chains have limits of the speeds at which they may be operated.

From Tables 11-19 and 11-20, maximum allowable speeds of a 6-in. pitch chain in bucket elevator service is 37 rpm shaft speed and 203 ft/min linear chain speed over an 11-tooth sprocket. Over a 12-tooth wheel, these figures are 41 rpm and 246 ft/min and for a 13-tooth sprocket they are 44 rpm and 286 ft/min. On this basis, with a chain speed of 240 ft/min required to achieve capacity, a 12-tooth sprocket would just answer.

Next the minimum shaft rpm necessary that bucket discharge will be satisfactory must be determined. Calculating the pitch diameter of a 12-tooth sprocket for 6-in. pitch chain (securing the necessary angle cosecant from Table 4-45):

$$6\times 3.8637=23.18\text{ in. pitch diameter}$$

Then, assuming that the center of gravity of the material being discharged is approximately 5 in. outside the sprocket pitch radius, the R of centrifugal discharge is [equation (11-2)]

$$R=\left(\frac{23.18}{2}+5\text{ in.}\right)\div 12=1.38\text{ ft}$$

Minimum shaft rpm is thus

$$\frac{44.3}{1.38}=37.7\text{ rpm}$$

Thus, operating at 240 ft/min, the shaft rpm will be

$$\frac{240\times 2\text{ links per foot}}{12\text{ teeth per revolution}}=40\text{ rpm}$$

While a 12-tooth sprocket calculates to be adequate, many engineers would go to a larger sprocket both for better discharge, and to permit a higher allowable chain speed.

Step 3. Calculate the maximum chain pull. Using equation (11-3), chain pull P under full load may be calculated, following

$$P=W_2+W_3+W_4+Q+\frac{P}{2}$$

where W_2 is the total weight of the chain on either the carrying or the return strand.

With 70 ft between centers, 75% of the chain will be plain chain weighing 6.3 lb/ft and the rest will be attachments weighing 8.6 lb/ft:

$$\begin{aligned} 6.3 \times 70 \times 0.75 &= 330.75 \text{ lb} \\ 8.6 \times 70 \times 0.25 &= \underline{150.5 \ \ \text{lb}} \\ \text{Total} \quad & \ \ \ 481.25 \text{ lb} \ \text{ for the strand} \end{aligned}$$

W_3 is the weight of the buckets on either the carrying or the return strand. There is a bucket every other foot of chain, or 35 buckets. Total weight of buckets is, therefore:

$$35 \times 21 \text{ lb per bucket} = 735 \text{ lb per strand}$$

W_4 is the weight of the material on the carrying strand. Since there are 35 buckets with a capacity of 0.34 ft³ each, and the material weighs 50 lb/ft³.

$$W_4 = 35 \times 0.34 \times 50 = 595 \text{ lb}$$

This may be compared to the required weight of material per foot of elevator, or $8.33 \times 70 = 583$ lb.

Q is the digging load, and this, under normal conditions, will be obtained by equation (11-4), where W_5 is the weight of material per bucket:

$$Q = 12 \times \frac{W_5 \times J}{4 \times \text{the pitch}} \times \text{the sprocket diameter}$$

$$Q = 12 \times \frac{17 \times .85}{4 \times 6} \times 23.18 = 177 \text{ lb}$$

In the calculation above, J is estimated from the factors given on page 298, being for a material considered to have relatively small lumps with fair flow characteristics.

P_1 is assumed at 2000 lb, assuming a need for a fairly heavy gravity takeup that will "give" and yet resist the digging loads from the occasional wet caked or sometimes frozen lumps of crushed coal from a storage yard, which can momentarily be quite high. Then

$$P = 481 + 735 + 595 + 177 + \frac{2000}{2} = 2988 \text{ lb}$$

Referring to Figure 11-21, we have

$$P_2 = W_2 + W_3 + \frac{P_1}{2} = 481 + 735 + 1000 = 2257 \text{ lb}$$

$$P_3 = P_1 \div 2 = 1000 \text{ lb}$$

$$P_4 = Q + \frac{P_1}{2} = 177 + 1000 = 1177 \text{ lb}$$

Step 4. Find the design working load for the chain:

$W_1 = P \times F_1 \times F_2 \times F_3 \times F_4$ where F_1, F_2, . . . are the service factors determined earlier

$$W_1 = 2988 \times 1.1 \times 1.2 \times 1.3 \times 1.2 = 6153 \text{ lb}$$

Step 5. Select the chain. The chain tentatively selected, ANSI S110, with a rated working load of 6300 lb, is adequate. Again, some engineers would decide to use a heavier chain, depending on such factors as how important the installation might be to an operation, or from other factors not a part of these calculations.

Step 6. Select sprocket or traction wheel hubs and shafting. This step was covered in Chapter 4.

Roller Chain Equipped Bucket Elevators

Most bucket elevators using Roller Chain are used for unit handling or food or chemical bulk handling applications. Generally speaking, standard and extended pitch conveyor series (B29.4) Roller Chains with straight-edged link plates are normally used for bucket elevator applications.

Specification of Roller Chains for bucket elevators is done in exactly the same manner as for Engineering Steel chains, the necessary information concerning working loads, chain weights, and static or rolling friction (if needed) being taken from Tables 10-18, and 10-12 to 10-14.

Pivoted Bucket or Pan Conveyors/Bucket Elevators

Figure 11-23 shows a style of equipment widely used to handle materials vertically, this particular unit being often seen in food processing plants. The concept carries different names in various applications. In food handling applications, such as that of Figure 11-23, which commonly use Roller Chains, it is termed a bucket elevator. With Engineering Steel chains, and in a much larger version such as would be used to handle coal in a power plant (Fig. 11-24, a common and effective application), it is called a pivoted bucket conveyor. On long slopes, with no vertical run, the concept is sometimes called a pivoted pan conveyor, when the pivoting carrier idea has but two shafts.

Figure 11-25 is a diagram of a typical pivoted bucket conveyor path with six shafts, and Figure 11-26 shows the method of discharge in diagram form.

Operation Practices

The advice on the maintenance and lubrication of conveyor chains, as given for conveyor chains in Chapter 12, can be applied to chains used in bucket elevators as well, except where it is modified by the following remarks.

Elevator Stoppage. In normal operation, the elevator should always be emptied before it is stopped. If the elevator is stopped while it is loaded (as in a power failure, for instance), the chain will run backward. A backstop should be used to prevent this. One type of backstop is a wheel keyed either to the head or foot shaft that turns forward with the shaft. The slightest backward motion, however, brakes the wheel, and the elevator is immobilized until the forward motion is resumed. Figure 11-27 shows a typical backstop.

Inspection. The inspection door on the side of the elevator casing makes the

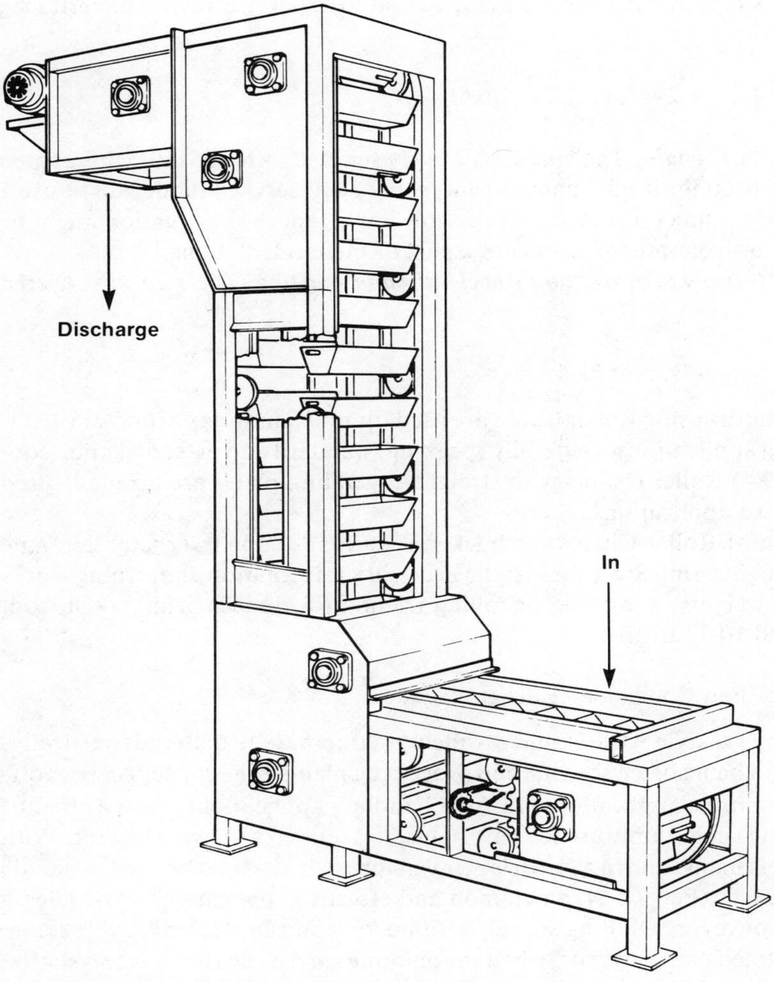

Figure 11-23 Pivoted bucket elevator using Roller Chain, largely of stainless steel, used for handling food, drugs, and other bulk materials requiring sanitized equipment.

chain accessible for inspection and repair. Check the chain, attachments, and buckets as you would those on a bulk handling conveyor. If the chain pitch has stretched appreciably, adjust the takeups.

Operation. If for any reason the elevator is out of service for a time, it should be run regularly to prevent freeze-up. As a rule, it should be run at least once a week, and be turned over every day if it is handling sticky material. Otherwise, the material may be set and the chain will be severely strained when the elevator is started up again.

Lubrication. Bushed roller and bushed rollerless chains, whether used on elevators or conveyors, are lubricated in the same manner. For details, see Chapter 12.

Figure 11-24 Discharge portion of pivoted bucket conveyor handling coal into hoppers over boilers of a power plant.

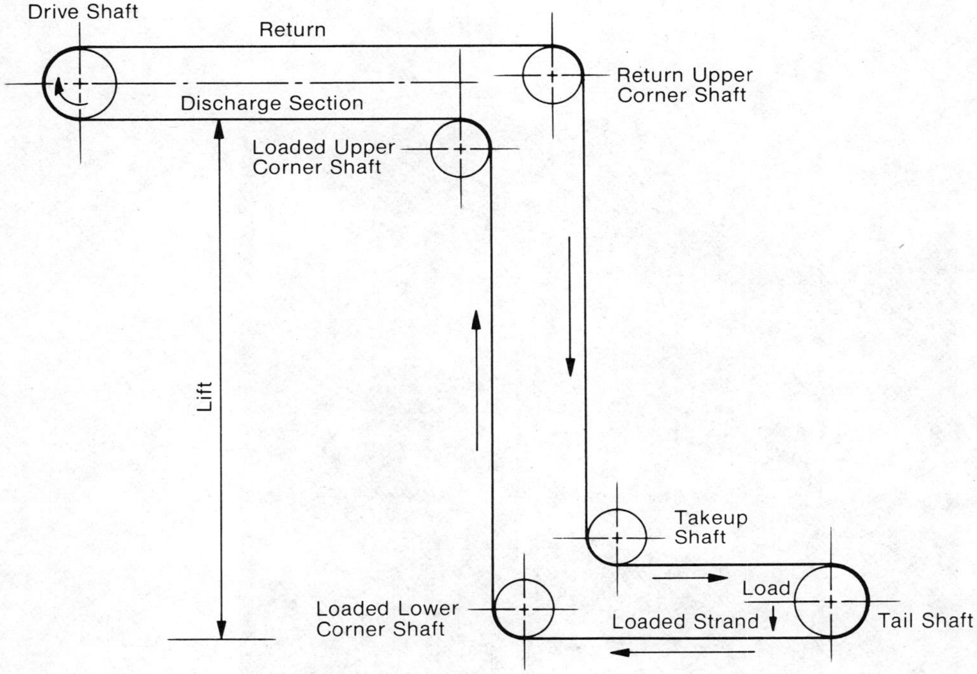

Figure 11-25 Diagram of the bucket path of the conveyor of Figure 11-24.

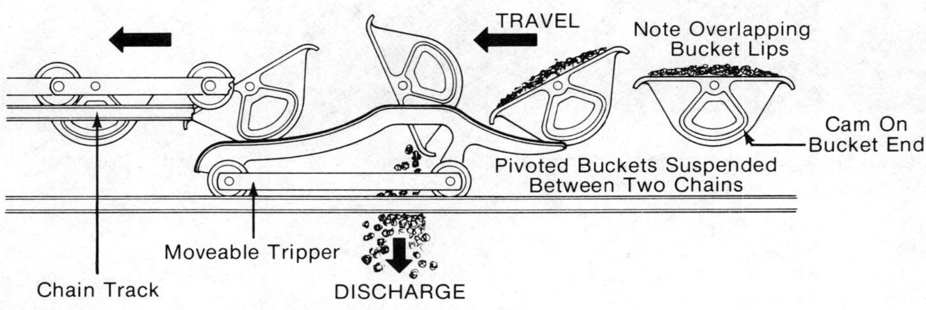

Figure 11-26 Diagram of discharge arrangement of pivoted buckets.

Figure 11-27 Typical backstop.

12

Installation, Lubrication, and Maintenance of Chains

As with any other machine element, certain minimum standards of operating procedures are assumed in designing, manufacturing, and specifying for application the chains and sprockets discussed in this book. These products will usually perform well beyond expectations of life and efficiency if reasonable attention is given to correct installation design and procedures and adequate lubrication and maintenance.

Unfortunately, these aspects of operation are often taken for granted when an installation is planned, sometimes with very unfortunate results. On the more positive side, even a little consideration of these needs will usually not only result in an installation that works adequately, but one that remains trouble-free for many years.

Much of the information required to design an installation properly using chains has been included in previous chapters, particularly those dealing with materials handling applications. Other information, procedures, and concepts are included in this chapter, mostly of a general nature and dealing with what happens at the installation.

The sections deal with installation, lubrication, and inspection and maintenance, and are set up in terms of generalities as applicable, followed by specifics dealing in turn with particular types of chains and applications. Instructions to connect and disconnect the various types of chains fall into the latter category.

Planning for Installation, Lubrication, and Maintenance

Certain procedures, usually considered routine, will pay real dividends if observed carefully in the process of laying out a drive or conveyor. Major areas to be considered are as follows:

1. Rigidity and support, including internal loads and deformation
2. Space and accessibility for installation and repair
3. Accessibility for periodic lubrication, inspection, and maintenance
4. Guards and protective equipment, relative not only to personnel safety but to

protect chains and other equipment from foreign material that would impair its operation

Although most of the above is generally obvious, it is often not fully considered or left as an on-the-site process.

Adequate supports for shafting and conveyor ways are important, not only in terms of strength but of potential deformation under load that can cause chain misalignments such as will lead to rapid wear. Internal loads—the tendency of a chain under load, whether in a drive or a conveyor, to pull the shafts toward one another, should be considered.

Space should be provided for chain sag in a drive and also in some conveyors. Means of takeup, such as sag as well as for mechanically adjusted takeups, should be made accessible. Clearance should be adequate around sprockets to clear chains and attachments when the chain rides up on the teeth as it wears.

Accessibility for routine lubrication, inspection, and maintenance should be provided. Service platforms are sometimes justified. In enclosed installations, access doors and lockout switches located in sight of those working on the installation should be provided.

Installation of Sprockets and Chain

Chain, regardless of what type, is usually received in coils of a convenient length, often approximately 10 ft. Extra pins, connecting links, and cotters are usually wired to the chain outside end, which is also wired to the coil to keep it from unrolling. Coils may be boxed as received, but will usually be stacked and strapped on skids for shipping. Measures to inhibit rust while in storage are usually provided, usually in the form of rust preventive dips or coatings unless such are not acceptable in the application. It is usually desirable to store new chain awaiting installation in an area protected from weather.

Chains received should be checked against specifications relative to quantities received, type and spacing of attachments, and to be sure that necessary connecting links, pins, cotters, and other such items are present.

Shaft and Sprocket Alignment

Shafting and sprockets should be carefully aligned before chains are installed. This is particularly important on drive installations, where spans are short and loads are high, but it is also important on conveyors and bucket elevators where more than one strand of chain is used, connected by through rods, flights, or buckets. Even for these, the shorter the span, the more important the shaft alignment.

Bearings for shafting should be such that they can be firmly locked into position and the shafting checked for alignment as shown in Figure 12-1. Bearings usually can be shimmed to move vertically; if not on takeups, hole clearances over bolts (the holes usually being slotted) will normally provide enough adjustment to compensate for misalignments of a welded structure.

Figure 12-1 shows shafting on short centers as used for a drive, and Figure 12-2 shows the sprockets in position for alignment.

Aligning Shafts (see Fig. 12-1). Carefully level each shaft, checking the adjust-

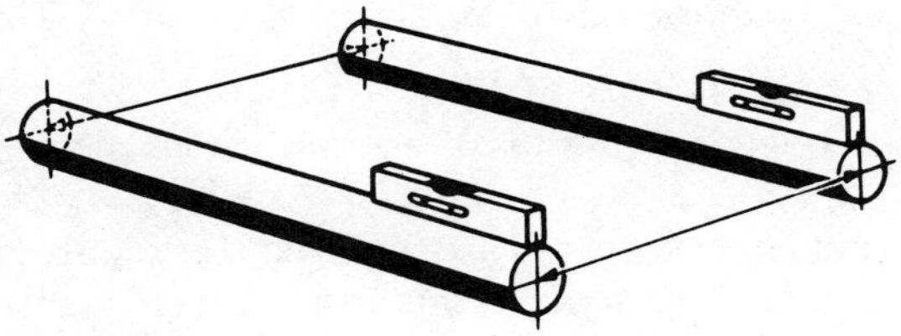

Figure 12-1 Aligning shafts.

ment with a machinist's spirit level applied directly to the shaft. (Where multiple width sprockets are used, the level may be applied across the teeth.)

Check the shafts for parallelism with a feeler bar. After adjustment, recheck the shaft levels; repeat the adjustments until both level and alignment are satisfactory.

Aligning Sprockets (see Fig. 12-2). Check the axial alignment of the sprockets with a straightedge applied to finished surfaces on the sides of the sprockets. A taut piano wire may be used if the center distance is too great for the available straightedge.

If a shaft has axial float, block the shaft in its running position before aligning the sprockets.

Be sure to secure the sprockets against axial displacement by tightening the set screws.

Alignment procedures as illustrated in Figures 12-1 and 12-2 will assure that: (1) shafts are parallel and in the same plane, and (2) sprockets are in line and not offset on the shafts.

For conveyors, normally having far greater shafting center distances than with drives, the important criteria are the distances or "spans" between pairs of sprockets on the shafting of each conveyor, as in Figure 12-3, and the alignment of the teeth of the pairs of opposite sprockets with one another. The latter situation is noted in Chapter 9 (see Fig. 9-50).

Although both of these qualities are normally established when sprockets and shafting are bought as units, accidents can happen in shipping, and inadvertent mixing of match-marked but unassembled sprockets is not unknown.

It is obvious that if pairs of sprockets on the various shafts of a conveyor or bucket elevator are not the same distance apart, and the chain strands are cross-connected

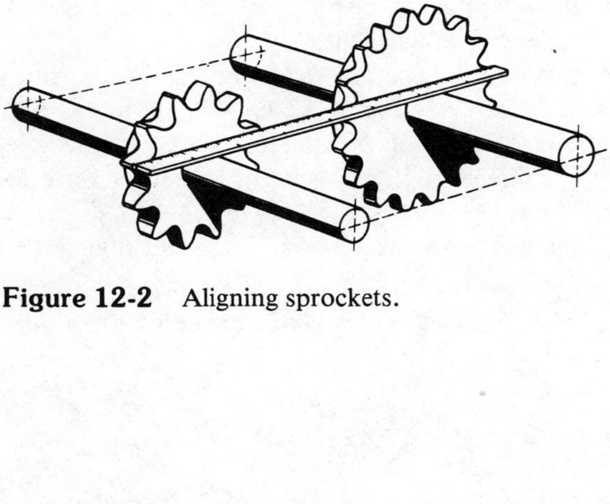

Figure 12-2 Aligning sprockets.

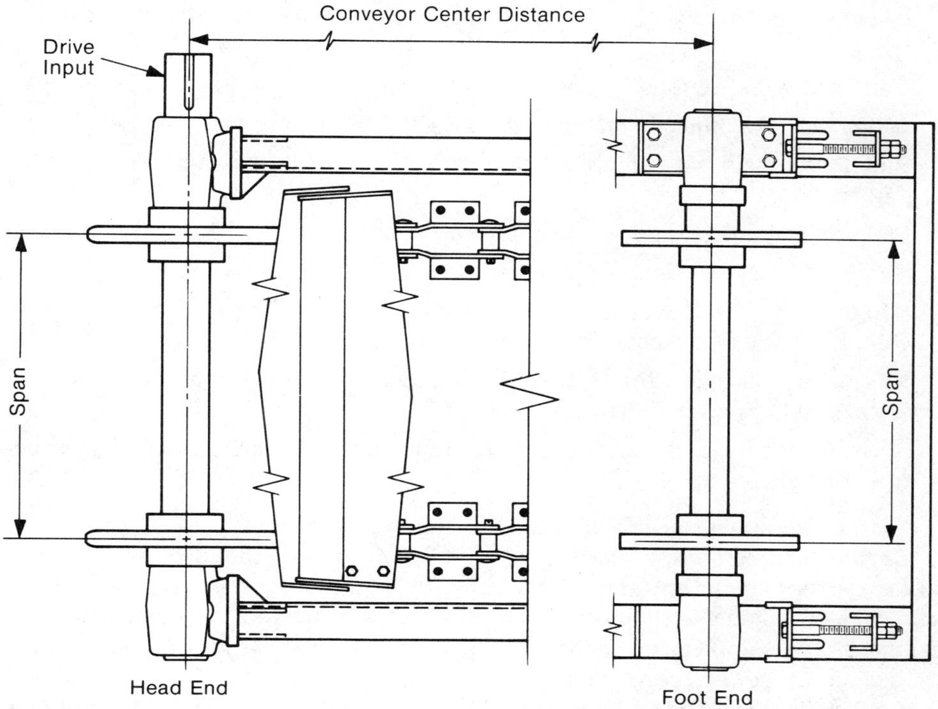

Figure 12-3 Alignment and spacing of sprockets on a double-strand conveyor.

in any way, the conveyor may not run at all on startup or the chains may be gradually destroyed.

Although shafting for such conveyors is normally keyseated in line, and sprockets keyseated in matched pairs with matchmarks provided, these factors should never be taken for granted. Sprocket spacing and alignment of shafting should always be checked after installation but before installing any chain.

For conveyors, tail shaft assemblies are frequently furnished with only one sprocket keyed to the shaft, the other being free to turn on the shaft and held in place with set collars. It is important that these set collars be firmly fixed in place, often done by "spotting" the shaft to assure correct location of set screws on subsequent adjustments. The purpose of this is to provide latitude for the sprockets to adjust if wear is uneven between the two strands. However, the distance between the pair of sprockets still remains a critical dimension.

After rechecking all dimensions and assuring that clearances around the sprockets are adequate, chain can be installed.

Chain Installation

Generally speaking, installation of chain is undertaken in several basic steps, regardless of the type of chain or application, as follows:

1. Uncoiling and laying out the sections of chain on a flat surface to arrange and check.
2. Connecting the sections.
3. Installing interconnecting members, as flights, through rods, as applicable.
4. Pulling the chain onto the sprockets and "threading" the chain into its intended path.
5. Make final connection or connections of the chains.
6. Thoroughly inspect the installation.
7. "Jog" start, manually, if possible, or with power, if necessary, for at least one complete cycle of the chain or chains through the intended path.
8. Run, with light loads, for several minutes, and with loads for an hour or so, adjusting takeups or chain tensions as needed.
9. Recheck all nuts, bolts, and cotters, including those of bearing blocks and structure for tightness.

Most of these steps need elaboration as follows.

Laying out the chains. The purpose of this is to establish the correct direction of travel for offset chains and those used in multiple strands with nonsymmetrical attachments. Attachments should be checked for correct position, arrangement, and alignment. Care should be taken that matched strands, for conveyors or bucket elevators, remain in the correct relationship.

Connecting the sections. Different kinds of chains vary somewhat in methods of connecting. These are covered in detail in the ANSI and American Chain Association publications, with a good many specific instructions that can only be summarized here. The variations stem from two basic factors:

1. Whether the chain is straight or offset link, which can establish whether connecting link subassemblies or simply individual pins are used to make the connection
2. Differences in pin configuration and the shapes of the holes in mating parts in which the pins articulate or must be locked to prevent rotation

Generally speaking, it may be seen from Chapter 3 that straight link Roller Chain, while having special connecting links, often uses standard pin links with cotters to connect chain being installed. The difference between the two is that the special connecting link, intended to facilitate replacement and repair, has either a very light or no interference fit at all between pins and link plates, whereas the standard link has the usual degree of interference and may require considerable pressure to assemble. Figure 12-4 shows the correct method of assembling sections of chain into a strand before placing it on the sprockets. A connecting link, consisting of a pin link plate with two pins in place, is used to connect the sections, with the link plate down and the pins backed up or "bucked" so they cannot move in the lower link plate. The loose link plate is then placed on the pins and driven down over them, with alternate blows on the link plate at each end between the pins (or with a hollow punch over the pin). The cotter is then inserted and spread. The "incorrect assembly" shown in Figure 12-4 causes problems in that the pins are not straight in the holes, necessarily, and galled interference fits and bent link plates may result.

Connecting of Silent Chain is basically as follows:

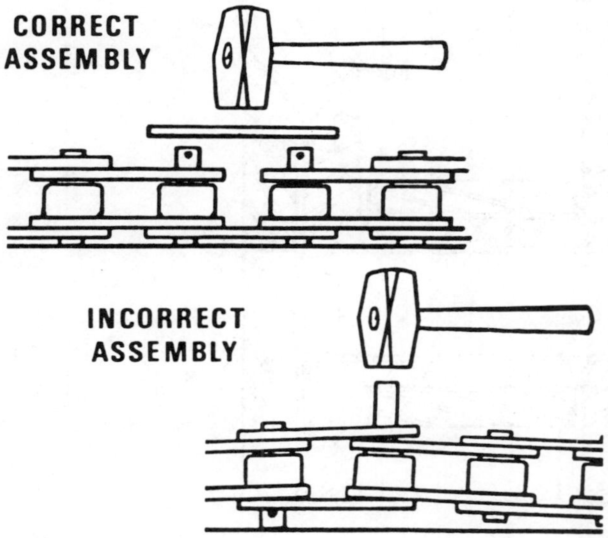

Figure 12-4 Installation of connecting links of Roller Chain, prior to placing on sprockets.

1. Lace the ends together, aligning the holes.
2. Insert any bushing segments, drifts, or rockers, as applicable.
3. Insert pins, and provide washers or cotters and/or rivet, as applicable.

As noted in previous chapters, Silent Chains differ in details of joint construction for all manufacturers, and parts are neither interchangeable nor can chains of different make be coupled.

A brochure entitled "Connect and Disconnect Instructions for Silent Chain," published by and available from the American Chain Association, gives details of connecting or disconnecting the standard chains of the four manufacturers of Silent Chain.

Previous chapters have indicated the wide variations in design among pins for Engineering Steel Chains, but there is one common similarity that greatly enhances ease of assembly. This is that most such pins have an under-the-head shoulder larger than the body diameter of the pin. This allows the pin body to clear the head end sidebar hole as the pin is pushed into the hole (see Fig. 12-5). Only when the cotter end of the pin reaches the opposite side of the chain and the shoulder of the pin starts to seat in its sidebar hole is any interference fit encountered requiring heavy pressure.

Figure 12-6, taken partially from a proposed standard for Engineering Steel chain, "Connect-Disconnect Instructions," illustrates the wide variety of pin designs used in Engineering Steel Chains, the variations being mostly in shoulder and shank end proportions and shapes.

Many Engineering Steel Chains are quite large, and interference fits of pins in sidebars very heavy. Assembly of such chains into strands for installation should be carefully planned, not only for safety reasons but to avoid damaging the chains themselves. The forces necessary to install the pins are such that pressing equipment is pre-

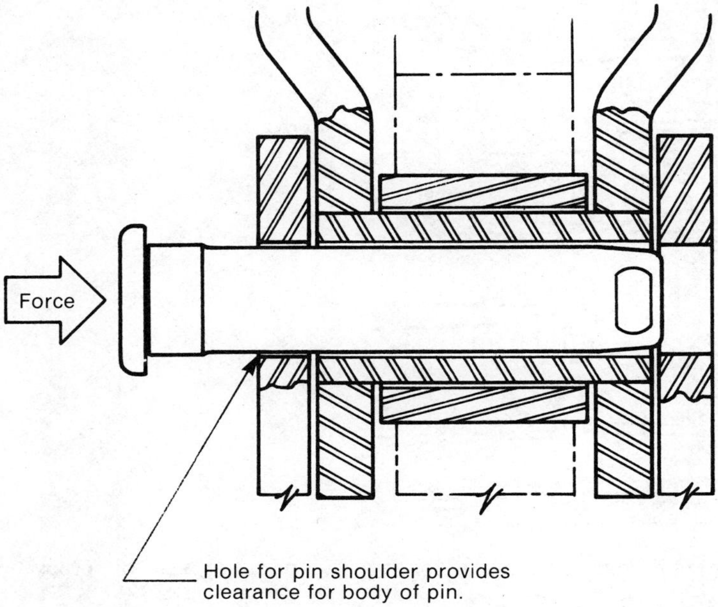

Figure 12-5 Assembly of chain having shouldered pins.

ferred for assembly; however, if not available, a large hammer and appropriate drifts must be used. A large vise or a similar type of holding device or clamps is recommended to secure chains being assembled in place.

The following are commonly used tools in connecting chains prior to installing (after giving first priority to safety equipment for eyes, etc.):

1. Pressing equipment
   a. Hydraulic press specifically designed for use with chain
   b. Jack (mechanical or hydraulic)

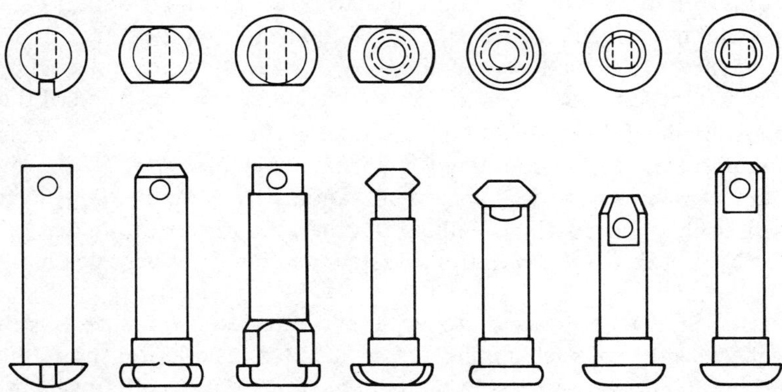

Figure 12-6 Some of the many styles of pins used in Engineering Steel Chains.

  c. Arbor press (hydraulic or mechanical)
2. Heavy hammer, and appropriate drifts or punches
3. Support fixtures
   a. Chain vise
   b. C-clamps
   c. Parallel bar spacers, "doughnuts," or short sections of heavy tubular shapes or shaped washers
4. Block and tackle or other suitable lifting and pulling devices
5. Cutting tools (mostly for disconnect usage)
   a. Pliers, or adjustable wrenches (for cotter removal)
   b. Grinder (for disassembly only)
   c. Hacksaw
   d. Chisel

Figures 12-7 and 12-8 show the methods of field connecting used for offset and straight link chains, which vary only in that offset link chain requires but one pin to be installed. Two pins are installed to connect straight link chain. This applies only to chains with pins having shoulders for straight link chains; otherwise, the method used for Roller Chain applies.

The steps in making the connection are as follows:

1. Draw the ends of the chain together so that the pin holes and barrel or bushing I.D. in the mating link aligns.
2. Insert the chain pin or rivet through the mating links, aligning the pin rotationally so that any mechanical locks engage.
3. Support the opposite side of the chain so that there is enough space below for complete insertion of the chain pin or rivet.
4. Press or drive the chain pin or rivet into place.
5. Check for free articulation of the chain joint prior to riveting; it is sometimes necessary to strike the tip of the rivet pin a sharp blow to free the joint.
6. Assemble the cotter in the chain pin or rivet over the head of the chain rivet pin to form a head.
7. Check for free articulation of the chain joint.

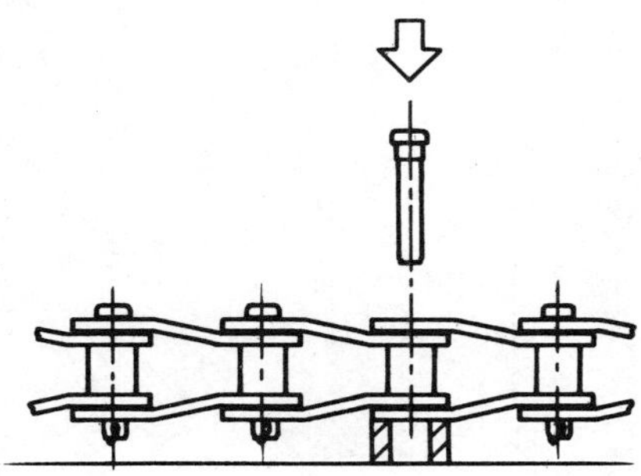

Figure 12-7 Assembly of Engineering Steel offset chain.

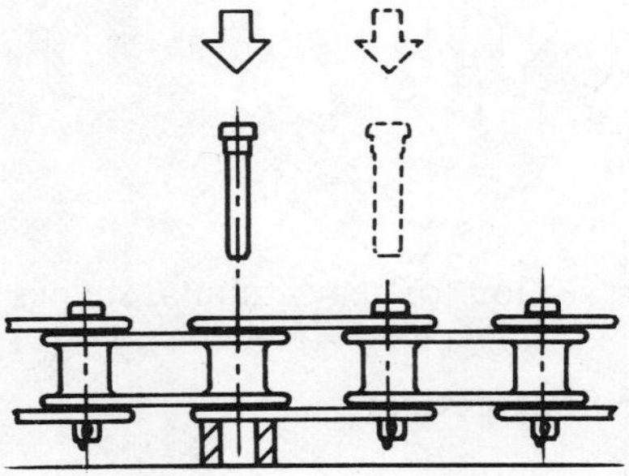

Figure 12-8 Assembly of Engineering Steel straight link chain.

Figures 12-9 to 12-11 show typical pins and also the assembly of bar-link chains and other chains of the steel bushed rollerless and heavy duty roller conveyor types having pins with a reduced-size shoulder at either end.

Here the process is essentially the same as for Roller Chain, a subassembly being formed by assembling two pins into an outside link and putting this subassembly in place connecting the sections. The remaining sidebar is then driven on from the top and cotters installed or the pin end riveted.

It should be noted that attempting to install pins without shoulders, such as the one shown at the extreme left in Figure 12-9, by driving clear through the upper sidebar is not recommended. Such a procedure may destroy the interference fit between the pin and the sidebar hole. Therefore, such pins should be made into a subassembly and assembled in the manner shown.

Installing Interconnecting Members Between Chains Used in Two or More Strands. Conveyors such as apron flight conveyors are usually furnished in short sections with flights and other parts fully assembled. In such cases, due to the weight, only a few sections at a time will be floor connected, before pulling onto the sprockets.

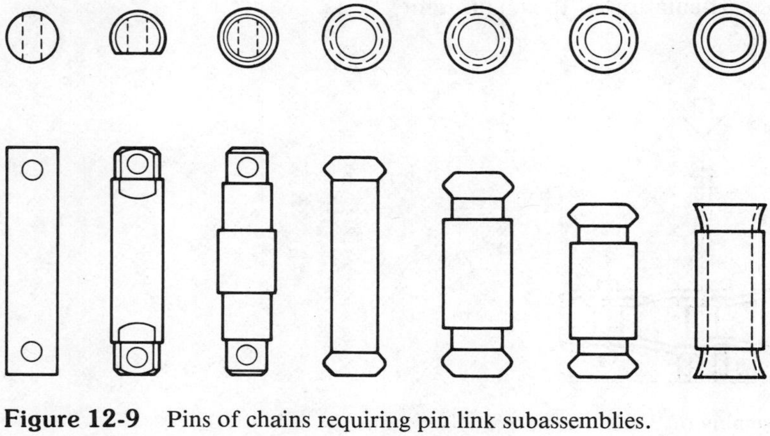

Figure 12-9 Pins of chains requiring pin link subassemblies.

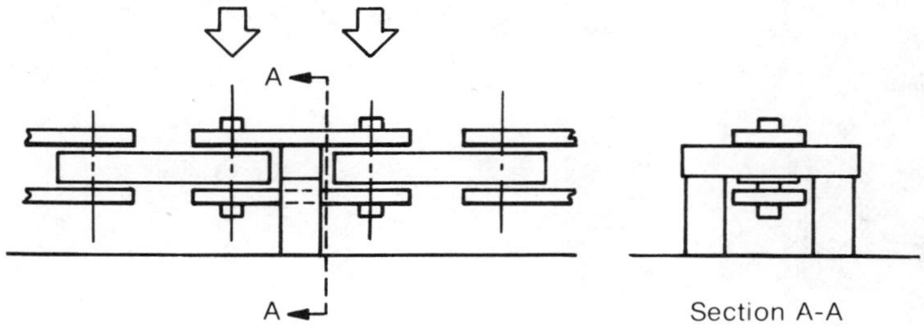

Figure 12-10 Assembly of bar link chains.

Flighting assembled with two or more chains should be floor assembled as far as possible, so that alignment and other details can be checked.

Placing Chain on Sprockets and into Chain Path. Several factors must be observed in this process. Safety, always paramount in handling chain (see the last section of this chapter), is of particular concern here. From the standpoint of securing a satisfactory installation, a great deal of care is necessary in handling chains used in multiple strands in bucket elevators or conveyors, and with buckets or flighting already assembled. The conveyor should be pulled straight into its intended path or onto its sprockets. Usually, the strand leading to the head sprockets is installed first using a "come-along" or block and tackle to pull the chain to the head shaft. Some installers, at that point, will use conveyor power to jog the chain with care along its intended circuit. It is usually desirable to thread the conveyor so that the final connections can be made at a sprocket and where minimum tension is on the chain.

Making Final Chain Connections. Final connection is made in exactly the same way that a floor connection of sections is made, except that the chain is usually not in a position that the connection can be so easily made. The sprocket can often be used as an "anvil" to "buck" when pins are driven in or sidebars driven on, but care should be taken that the right parts are bucked. A general rule is that when pins are installed, sidebars or link plates must be bucked; when sidebars are driven on, both pins must be bucked.

Figure 12-12 shows the final connection on a Roller Chain drive. Careful atten-

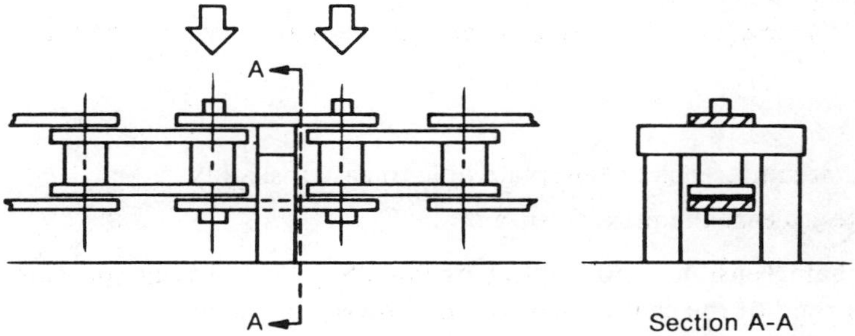

Figure 12-11 Assembly of steel bushed rollerless chains or other Engineering Steel Chains requiring connection with pin link subassemblies.

Figure 12-12 Connecting drive chains.

tion to details must be given in such installations so that the chain will give maximum service.

The pins of a connecting or pin link are inserted in the free ends of the strand, held in place by sprocket teeth and, if necessary, with C-clamps. The free link plate is driven on, and retainers—cotters or spring clips or other fasteners—are installed. After the fasteners have been inserted, it is important that the ends of the chain pins be pressed back so that the fasteners come up snugly against the outside of the connecting link plate. By doing this, three important things are accomplished:

1. The designed clearances between link plates across the chain width are maintained. A connecting link plate driven down too far on the pins "squeezes" the chain joint so that no oil can get down between the link plates.
2. This squeezing prevents joints from flexing freely as they go around the sprocket. Proper assembly of the connector will assure smoother chain action with a minimum of whipping of the chain.
3. It should be noted that maximum fastener life is achieved when it fits snugly against the link plate.

The use of offset connecting links should be avoided wherever possible for best performance of straight link chains.

If an offset link is necessary, it should be assembled permanently into the chain with press-fit pins.

Precautions.

Do not grind the pin circumference or link plate hole to ease assembly.

Do not heat link plates to ease the press fit of a pin.

Adjustment of Chain Tension. For a check of chain tension, turn one sprocket to tighten the lower strand of the chain; then measure the sag of the upper strand as shown in Figure 12-13. This sag, measured at the midpoint, should be approximately 2 to 3% of the length of the tangent to the sprockets.

An inclined drive should have less slack than a horizontal drive.

For a vertical drive, provision for adjustment of chain is desirable.

Inspection. Before any power is applied to the application, a thorough inspection of a newly installed drive, conveyor, or bucket elevator should be made. This should include but not necessarily be limited to the following:

All cotters or fasteners should be in place and bent or split, or head riveted as required.

All set screws should be in place and be tight.

All bolts and nuts should be in place and nuts tight with lockwashers or other fixation as needed.

Chains should be in the proper position on the sprockets with all C-clamps removed.

Direction of travel of the chain, and direction of rotation of the drive, must be checked out as being in agreement.

All obstructions such as pry bars, blocks, tools, and debris from construction should be removed, with particular attention given to chain guides, tracks, and ways. Gravity takeups should be free.

The chains, including all outboard rollers, should be lubricated as specified. Bearings and drives associated with the installation should be checked for proper lubrication.

Jog-Start. On a drive, jog starting with power is usually appropriate. On a conveyor or bucket elevator, this is particularly important to make sure that all obstructions are cleared and that everything operates properly. The chain or conveyor should be jogged for at least one full cycle through its path.

Operate at No Load. The installation should be operated (after verifying that it will operate smoothly, and adjusting takeups, sag, and tension) for several hours at no load or a very light load. Lubrication should be fully established during this period.

After several hours of operation at no load, loads may be applied.

Inspect and Retighten as required. After the installation has been operating a few days or weeks, all bolts, nuts, set screws, and so on, should be checked for tightness; cotters should be examined to assure that none have dropped out, and chain sag and tension or, in a bucket elevator or conveyor, takeups should be adjusted.

If all these steps are followed carefully, and no problems have been encountered, the new installation should operate satisfactorily for its design life or well beyond, assuming adequate lubrication and maintenance.

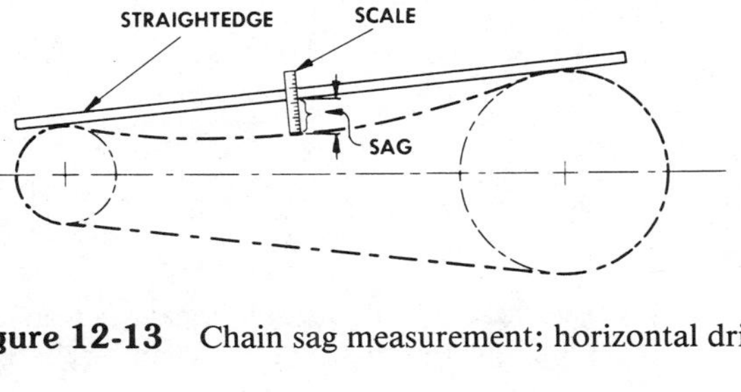

Figure 12-13 Chain sag measurement; horizontal drive.

Lubrication of Chains

Lubrication is an important factor in obtaining the desired life from a chain. In drives where chain is operated at established ratings, lubrication of the proper type is as much a part of the rating as the load and speed promised by the rating. This applies to and is clearly indicated in the horsepower tables for Roller Chain and Silent Chain, given in Chapters 5 and 7, and also applies to heavy duty offset Engineering Steel drive chains where horsepower tables are used.

Obtaining effective lubrication for a chain is a matter of correct application in terms of lubricant used and how and where in the strand it is provided. The basic problem to be solved in satisfactorily lubricating any chain is in getting clean lubricant to the bearing surfaces of pins, bushings, and rollers, as applicable. The problem is compounded by the fact that chains are moving in operation and direct connections for lubrication are nearly always impracticable if not totally impossible.

Figure 12-14 shows a cross section through an Engineering Steel drive chain with clearances exaggerated to show the paths that lubricant must follow to reach bearing areas. Figure 12-15 indicates the desired point of application—on the slack side so that clearances will be more open for oil to flow through them, and inside so that gravity (and centrifugal force, if of moment) will cause the lubricant to flow into the chain and not simply drip off it. These factors apply to nearly all chains, but other details can differ widely relative to the types of application and kind of chain involved and are better reviewed from that standpoint.

Lubrication of Drive Chains

Roller Chain and other chains used in drives have essentially the same lubrication needs, as indicated in the horsepower tables of Chapters 5 and 7. Three classifications of lubrication—manual, oil bath, and oil stream lubrication—are involved. These classifications, as an established variable of the rating function, are also included in the horsepower charts and tables for heavy duty offset Engineering Steel drive chains

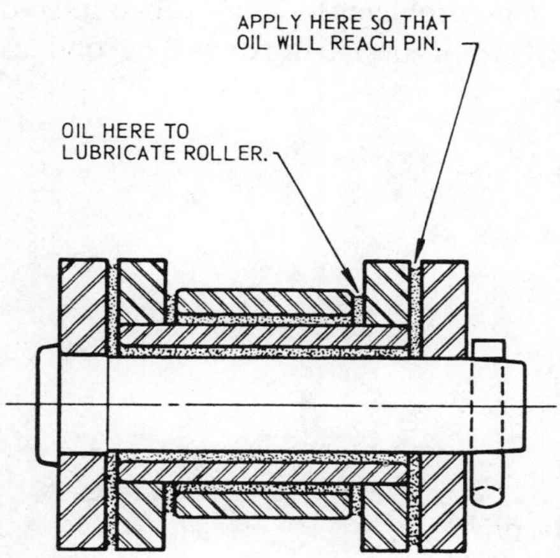

Figure 12-14 Paths of lubricant in reaching bearing surfaces of steel roller chain.

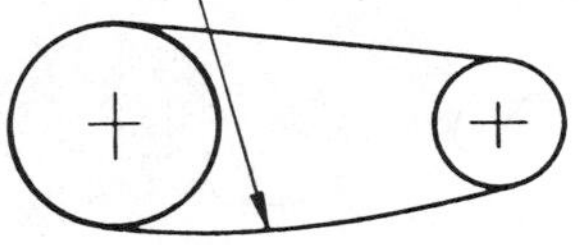

Figure 12-15 Application of lubricant to chains.

given in Chapter 6. Slow-speed drives using these chains, not always possible to lubricate as recommended, will be reviewed separately relative to lubrication.

Lubrication Classes as a Drive Chain Horsepower Rating Factor. A drive chain is a power transmission device consisting of a series of connected parts that move relative to one another under load. As is the case with any such device, the chain should be lubricated if it is to perform satisfactorily. It has been demonstrated that in the presence of lubricant, articulation between the pin and bushing of a roller pin joint establishes a hydrodynamic oil film in a manner analogous to that formed in journal bearings. Fluid lubricant must be applied to establish this oil film and minimize metal-to-metal contact. If supplied in sufficient volume, a lubricant also provides effective cooling and impact damping at high operating speeds. The amount of lubricant and the method of application required for adequate lubrication varies with the operating conditions of the drive. The horsepower rating tables in Chapters 5 and 7 are marked to indicate the type of lubrication required under various operating conditions. These markings are merely guidelines and under some conditions may be altered substantially to meet the requirements of a particular application. When deviating from the normal lubrication requirements, it is advisable to seek the recommendation of the chain manufacturer.

Lubricant Flow. Figure 12-16 shows a cross section through a Roller Chain, with clearances greatly exaggerated, to indicate the flow of the lubricant from spray nozzle to bearing surfaces. This is the style of lubrication required for maximum performance of such chains in drive applications. As may be seen, two different bearing pairs are involved, the roller and outside surface of the bushing forming one pair.

Most important is the lubrication of the pin and bushing surfaces, which articulate with each other while the chain is under full load. To reach all these surfaces, the lubricant is applied to the upper edges of the link plates or sidebars on the slack strand of the chain shortly before the chain engages a sprocket. Then, as the chain travels around the sprocket, the lubricant is carried by gravity or centrifugal force into the clearances between the pins and the bushings. Spillage over the link plates supplies lubricant to the interior and the end surfaces of the rollers.

Lubricants for Roller and Silent Chain Drives. Lubricants for continuous or intermittent lubrication of chains used on drives should have the following characteristics:

1. Sufficiently low viscosity to reach the internal surfaces
2. Sufficient body to maintain the lubricating film under the bearing pressures
3. Freedom from corrosive ingredients

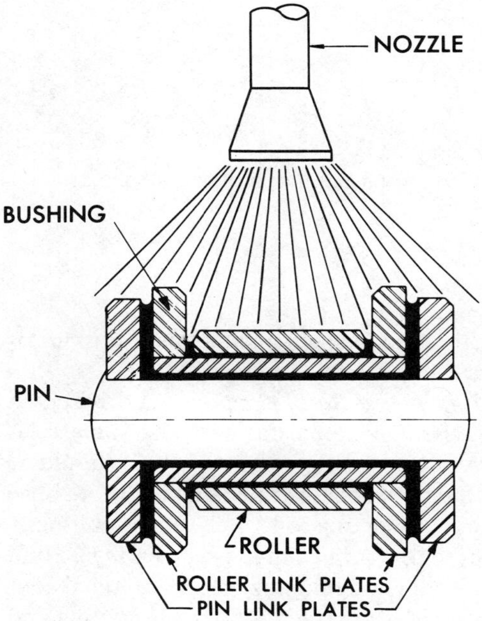

Figure 12-16 Chain lubricant distribution with oil stream lubrication.

4. Ability to maintain lubricating qualities under operating conditions of temperature, moisture, and so on.

Normally, these specifications are met by a good grade of nondetergent petroleum-base oil.

In general, detergent oils are not required but oils with antifoam, antirust, or film-strength-improving additives may be used.

Low-grade or impure oils should be avoided. Such oils may leave deposits in the chain joints which will prevent subsequent effective lubrication, and if not acid-free, they may damage the chain irreparably. Heavy oils and greases are also not recommended for the lubrication of Roller and Silent Chain, or heavy duty offset Engineering Steel drive chains selected for use from horsepower tables for continuous or frequent intermittent operation.

Viscosity Requirements. The proper lubricant viscosity for various surrounding temperatures to relubricate drive chains in continuous or frequent service is shown in Table 12-17.

Types of Lubrication for Drive Chains Selected from Horsepower Tables. There are three basic types of lubrication for drive chain service as indicated in horsepower charts and tables. These are:

Type I: Manual or drip lubrication

Type II: Oil bath or disk lubrication

Type III: Oil stream or pressure spray lubrication

The recommended type shown in the horsepower rating tables is influenced by chain speed and the amount of power transmitted. The types indicated are minimum lubrication requirements, and the use of a better type (for example, type III instead of type II) is acceptable and may be required by operating conditions other than speed and load. Chain life can vary appreciably depending on the way the drive is lubricated. The better the lubrication, the longer the chain wear life. For this reason it is important that the lubrication recommendations be followed when using the ratings given in the horsepower tables.

Consult chain manufacturers when it appears desirable to use a type of lubrication inferior to that recommended.

Manual Lubrication. For manual lubrication, oil is applied periodically with a brush (Fig. 12-18) or a spout can, preferably at least once every 8 hr of operation.

Drip Feed Lubrication. For drip lubrication, oil drops are directed between the edges of link plates or sidebars from a drip lubricator.

For multiple strand chains, a distribution pipe must be extended to feed oil to all link plates (Fig. 12-19). A wick packing is normally required to feed the oil uniformly to the holes in the distributing pipe.

Precautions must be taken against misdirection of the drops of oil due to windage. The volume of oil applied and frequency of oil application should be sufficient to prevent discoloration of the lubricant in the chain joints. This discoloration usually indicates the presence of products of fretting corrosion (red rust), which is an indication of inadquate lubrication.

Bath or Disk Lubrication. With bath lubrication the lower strand of chain runs through a sump of oil in the drive housing (Fig. 12-20). The oil level should reach the pitch line of the chain at its lowest point while operating.

With disk lubrication, the chain operates above the oil level. The disk picks up oil from the sump and deposits it onto the chain (Fig. 12-21), usually by means of a trough or collector plate. The diameter of the disk should be such as to produce rim speeds between 600 ft/min minimum and 8000 ft/min maximum.

In the case of either bath or disk lubrication, the temperature of the bath and the chain should not exceed 180°F. The volume of oil delivered from the slinger disk should be sufficient to prevent discoloration of lubricant in the chain joints.

Oil Stream Lubrication. The type III oil stream or pressure spray lubrication system is the most satisfactory of the three types of lubrication for chain drives run-

Table 12-17 Recommended Oil Viscosities for Various Temperatures

Temperature Degrees F	Recommended Lubricant
−20-+20	SAE 10
20- 40	SAE 20
40-100	SAE 30
100-120	SAE 40
120-140	SAE 50

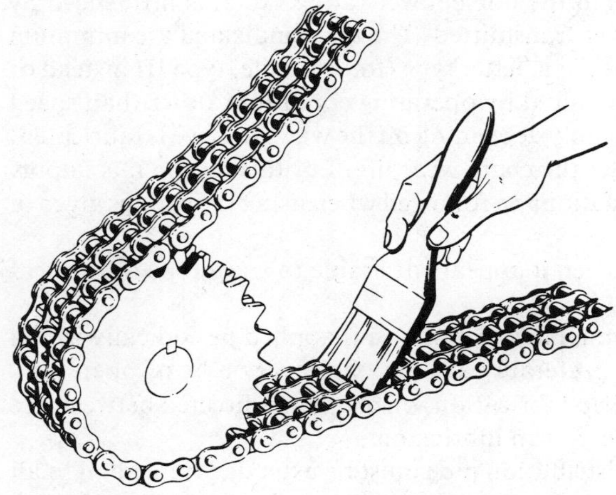

Figure 12-18 Manual lubrication.

ning at relatively high speeds and loads. It is an absolute requirement for Roller Chain drives operating in the indicated galling region for any extended period of time.

Figure 12-22 shows a typical oil stream system. A pump, driven from one of the main drive shafts, or by a separate motor, delivers oil under pressure to nozzles that direct an oil spray onto the chain. The oil should be applied inside the chain loop evenly across the chain width, and be directed at the slack strand. Excess oil collects in the bottom of the casing and is returned to the pump suction reservoir. A pressure-regulating valve bypasses to the reservoir any excess pump discharge. Cooling is accomplished by radiation from the external surfaces of the reservoir.

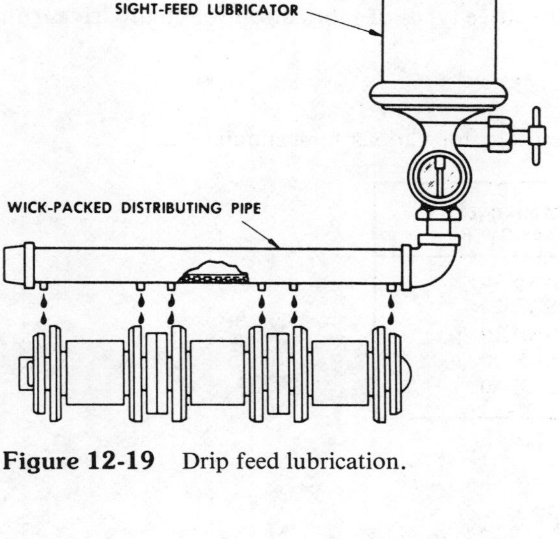

Figure 12-19 Drip feed lubrication.

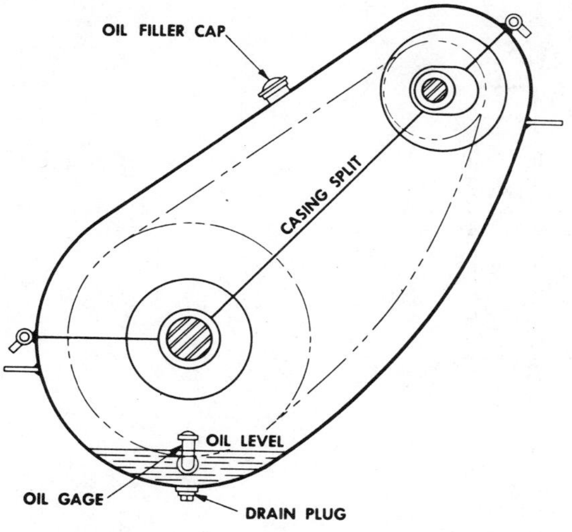

Figure 12-20 Shallow oil bath lubrication.

Lubricating oil in an oil stream lubrication system serves not only to lubricate but to carry away the wear products and heat of friction from a drive chain being operated at or near rated capacity.

The case should be designed large enough to accommodate ample oil. Table 12-23 gives minimum amounts of oil required to be circulated for various horsepowers, and these data, correlated with pump capacity, should be considered in establishing the case design.

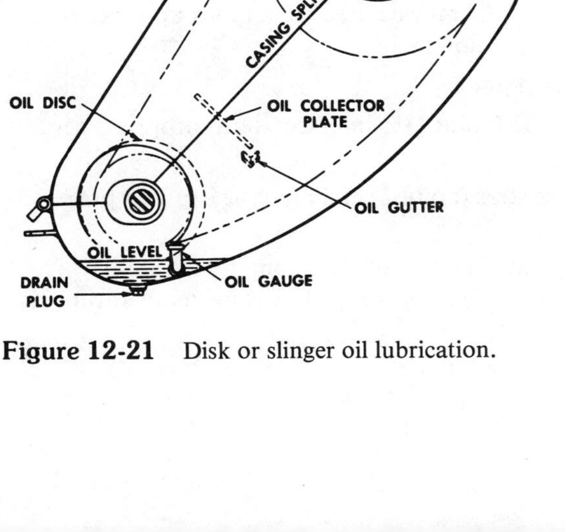

Figure 12-21 Disk or slinger oil lubrication.

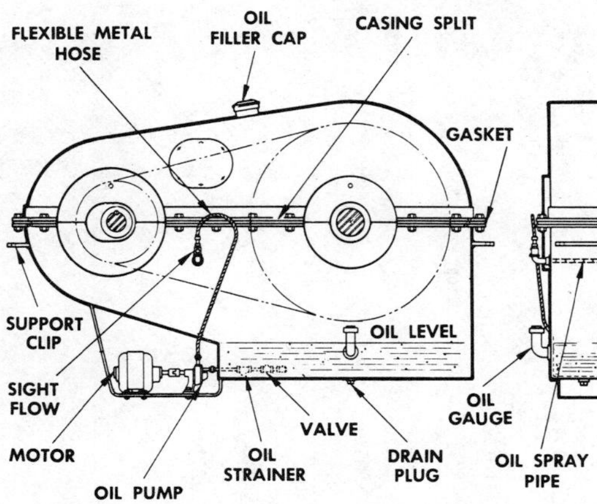

Figure 12-22 Oil stream lubrication.

To estimate the probable temperature rise of a chain case, the following formula may be used:

$$T = \frac{50.9HP}{AK} = {}^{\circ}\text{F above ambient}$$

where

T = temperature rise, °F
HP = transmitted horsepower
A = casing area exposed to air circulation, ft^2
K = radiation constant, Btu per square foot per hour per degree Fahrenheit temperature difference
K = 2.0 for still air
= 2.7 for normal free air circulation
= 4.5 for rapid air circulation

Good practice limits the allowable operating temperature to approximately 180 °F (temperature rise plus ambient). If the calculated temperature is greater than this value, a larger casing could be used or an oil cooler added to reduce the operating temperature to allowable limits. Figure 12-24 may be used for a quick approximation of possible temperature rise. Proceed as follows:

1. Compute the value of X and plot point 1.
2. Draw a vertical line from the X value (point 1) to intersect the appropriate centers (point 2).
3. Draw a horizontal line from the centers (point 2) and read the exposed projected casing area (point 3).
4. At the intersection of the appropriate HP and the horizontal line (point 4) from step 3, draw a vertical line and read the approximate casing temperature rise (point 5).

The value of X are as follows:

Standard casing

$$X = \frac{P}{6}(t + T) + W_c + 9$$

Oversize casing

$$X = R_1 + R_2 + W$$

where

P = chain pitch, in.
t = number of teeth small sprocket
W_c = chain width, in.
R_1 = casing radius, small end, in.
R_2 = casing radius, large end, in.
W = casing width in.
HP = horsepower
T = number of teeth large sprocket
A = area, ft^2

Case dimensions in the legend are as shown in Figure 12-25.

Example: Sizing Casing to Control Drive Chain Temperatures. Figure 12-26 gives the dimensions of a typical drive. Other data:

Horsepower: 97 hp needed

Chain used: No. 100-2 (double strand) Roller Chain, of 1.25-in. pitch

Chain rating: No. 100 Roller Chain at 700 rpm of a 17-tooth sprocket = 59.4 × 1.7 = 100.98 hp

Chain width W_c: approximately 3 in.

Table 12-23 Flow of Oil Required for Chain Drives

Horsepower Transmitted	Minimum G.P.M. of Oil Required
50	¼
100	½
150	¾
200	1
250	1¼
300	1½
400	2
500	2½
600	3
700	3¼
800	3¾
900	4¼
1,000	4¾
1,500	7
2,000	10

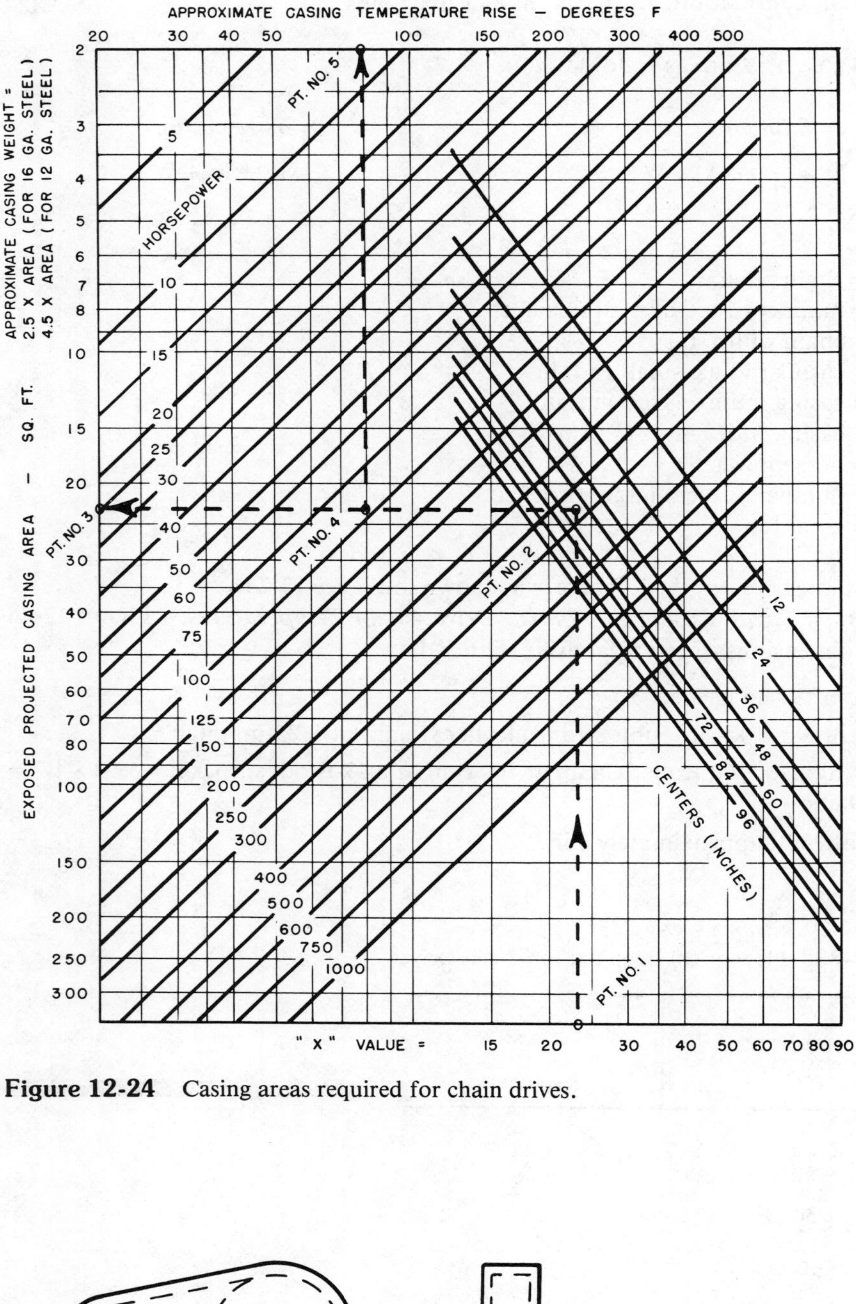

Figure 12-24 Casing areas required for chain drives.

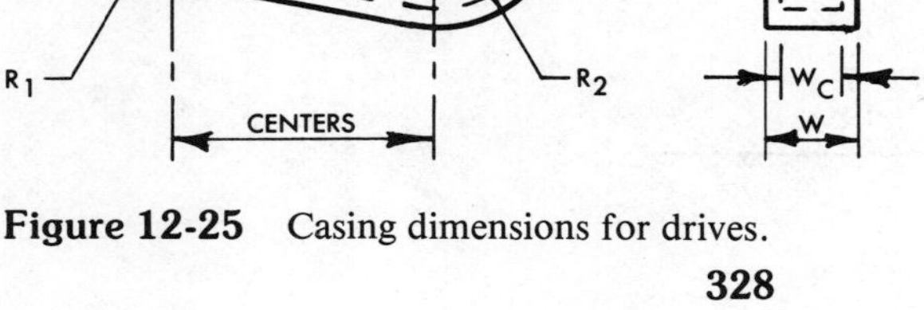

Figure 12-25 Casing dimensions for drives.

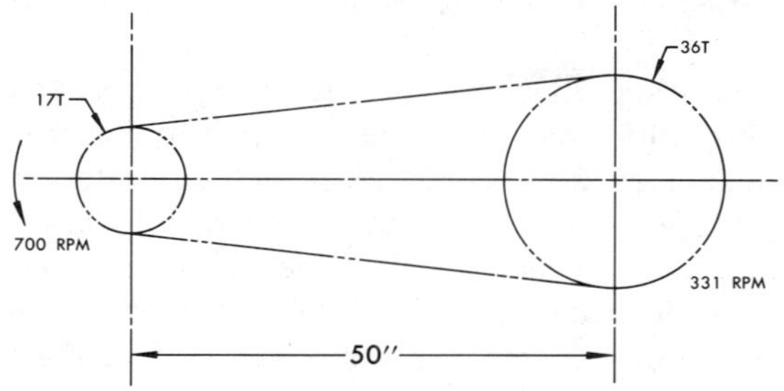

Figure 12-26 Dimensions of drive for casing design example.

$$X = \frac{P}{6}(t + T) + W_c + 9 = \frac{1.25}{6}(17 + 36) + 3 + 9 = 23$$

In Figure 12-24, at X = 23 (point 1) at the intersection with 50-in. center distance (point 2), draw a horizontal line to find the *exposed projected casing area* equal to 23 ft^2 (point 3).

Where this horizontal line intersects 97 hp (point 4), draw a vertical line to find the *approximate casing temperature rise* equal to 80 °F (point 5).

This means that the ambient temperature of the casing should not exceed 180 °F when the surrounding temperature is 100 °F. If the surrounding ambient temperature will usually exceed 100 °F, the chain center distance should be extended to increase the exposed projected casing area (square feet), thus reducing the temperature rise. If this cannot be done, an oil cooler must be provided or the casing made oversize.

Maintenance of the Lubrication System. Chain drives should be inspected at intervals to assure that the lubricating system is functioning properly. For all types of drive lubrication systems, check the quality and grade of the lubricant. If the chains have not been lubricated properly, and temperatures are noticeably above ambient, the joints will have a brownish, rusty color, and the pins of the connecting-link of the chain will be discolored a light or dark brown. Also, the pins will be roughened, grooved, or galled.

Properly lubricated chains will not show color in the joints, and the pins of connecting links will be brightly polished with a very high luster.

For manual lubrication systems, be sure that the lubrication schedule is being followed and oil properly applied. For bath or disk systems, and for oil stream systems, inspect oil levels and check that all sludge is removed. Such systems should be drained, flushed, and new oil provided at least once a year.

For oil stream or pressure spray lubrication systems, pump drive and pump delivery volume should be checked. Piping, nozzles, screens, and any filters should be checked for clogging.

Lubrication of Exposed Drive Chain on Movable Equipment. Many chain drives are used on equipment where only manual lubrication can be provided and where the chains are exposed to weather and other elements. Although Roller Chain is often found in such situations, most drives using Engineering Steel Chains are of this type.

Welded steel mill chains used to drive sanitary sludge collectors and heavy duty offset Engineering Steel drive chains used for propel drives on power shovels and cranes are two typical examples.

The chains intended for such installations, particularly the latter, are often packed with extreme pressure grease at the factory, intended to provide satisfactory lubrication during break-in periods. Subsequent lubrication is difficult from the standpoint of being effective, because the chains are operated in environments where lubrication is difficult or impractical. For example, the chains are often immersed in mud and water for extended periods without being cleaned or lubricated. Obviously, rust will occur rapidly under such conditions. Not so obvious is the fact that internal rust in the chain, plus accumulated dirt, may reduce chain clearances to a point where no lubricant can enter if provided. In addition, flexure of joints may be impaired so that poor sprocket action and reduced chain life will result.

For this reason, provisions should be made for such chains to be cleaned and relubricated regularly. If this is done, wear life will greatly increase and chain failures from poor sprocket-chain interaction will be avoided.

Most good machine oils will work if light enough to penetrate into sidebar clearances, with consideration given to temperature and environment. Application points on the slack side are recommended, as joints are more easily penetrated and gravity will carry oil into the chain. Oil should always be applied to sidebar edges, as discussed previously.

Drive chains of this type are sometimes operated in environments involving abrasive materials. Lubrication is always desirable even though the abrasive dust plus the lubricating oil may form what is essentially a lapping compound. The reason lubrication is recommended is that the effect of no lubrication—galling, seizing, and scoring—will destroy a chain in a short time. Occasional cleaning of drive chains is good practice and this is recommended strongly where mud and dirt may accumulate; otherwise, caking may seal joints, preventing effective lubrication and leading to rapid corrosive wear.

Lubrication of Conveyor, Bucket Elevator, and Tension Linkage Chains

Lubrication of chains on materials handling equipment or tension linkages, such as draw benches, depends to a great degree on location, environment, and accessibility. Conveyors or bucket elevators not exposed to weather or gritty materials or in locations where cleanliness is maintained, many of which use roller chains, may be lubricated with a good grade of vegetable or petroleum oil. With stainless steel chain, water-soluble lubricants and soaps may be used. Lubrication must be applied to sidebar edges to be effective (see Fig. 12-14).

Special lubricants are required for chains operating at extremely high temperatures or low temperatures where petroleum or vegetable oils will not flow. Protection against abrasive material can sometimes be obtained by using lubricants that become fluid at slightly elevated temperatures, but gel at operating temperatures to retard entry of abrasive particles between link plate or sidebar edges.

Heavy conveyors and bucket elevators using heavy duty Engineering Steel Chains are often lubricated with drip lubrication. Care should be taken that drip lubrication on roller-type conveyor chains, with gritty or abrasive materials present, will not cause rollers to stick and slide rather than roll. For such situations, chains with means of lubricating the chains internally on a scheduled manual program are available. Several methods of providing lubrication fittings exist.

Figure 12-27 shows a flanged cast roller, equipped with an oil can fitting. Such rollers can be provided with internal felt components, so as to be self-oiling, with lubricant renewed on a schedule, or can be fitted with flush-type pressure grease fittings.

Figure 12-28 shows cross sections through chain joints where lubrication is accomplished with grease pressure fittings in the ends of pins with grease passages drilled in them. The roller at the left (the method applies equally well to chains with solid or steel rollers, as the roller does not have to be hollow to serve as a grease reservoir) and bushing are lubricated through the pressure fitting in the head of the pin. The joint at the right has a through rod with the pressure fitting in its end.

It is recommended that pins and through rods less than 5/8 in. in diameter not be drilled in this manner.

The major advantage of lubricating chains in this way is that the lubricating process does not carry dirt, grit, and debris into the internal clearances of the chain. On the contrary, such foreign materials are flushed out by the grease from the inside, and the grease and dirt usually cake to form a solid barrier against the foreign materials entering. The disadvantage is cost and the need to lubricate each joint in each chain individually, as shown in Figure 12-29.

Heavy duty chains are sometimes equipped with rollers having antifriction bearings, as in Figure 12-30. Such rollers may have removable caps for grease packing or may be equipped with grease pressure fittings.

Rollers with antifriction bearings are normally used as outboard rollers on heavy-duty chains so as not to be subject to sprocket impact, or for other reasons.

Lubrication of chain is often a difficult process, because of lack of accessibility, the need to grease or oil every joint in a chain individually, and because attempts at "mass" lubrication, by dripping oil on a conveyor or bucket elevator chain, is often messy and unacceptable. Just the same, adequate lubrication, on a scheduled basis, can extend the life of such a chain many times over. A decision that a chain should not require lubrication cannot be based on economic considerations.

Inspection and Maintenance

Maintenance and inspection go hand in hand. Maintenance should be a product of inspection rather than breakdown. It should be a matter of routine adjustment and scheduled lubrication and inspection rather than on an emergency basis. Most chain

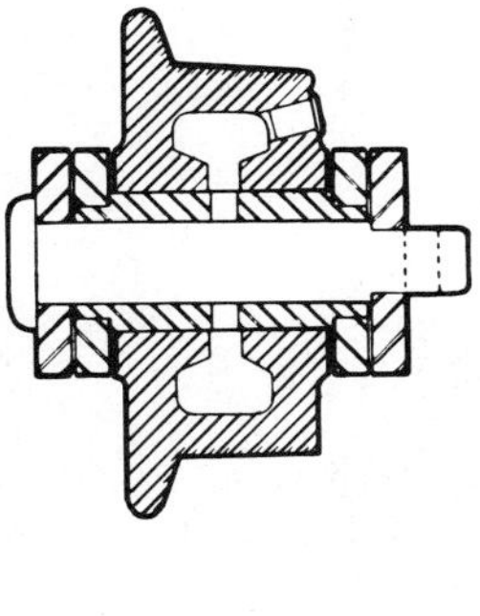

Figure 12-27 Conveyor chain joint showing roller with oil fitting. The hollow, cored center of the cast roller, filled with oil-absorbent material, serves as an oil reservoir.

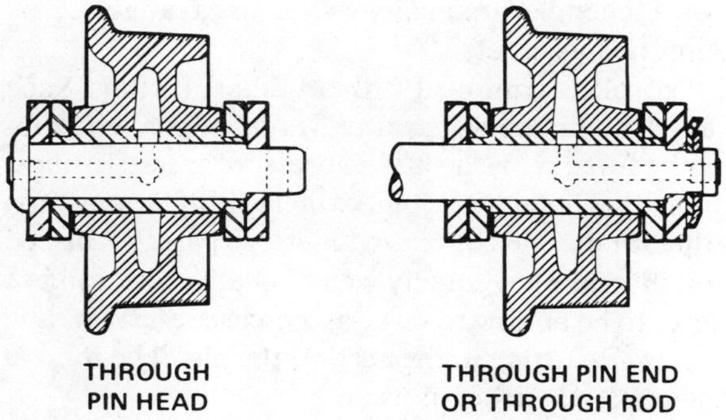

Figure 12-28 Chain joints with pins and through rods equipped with fittings for pressure lubrication.

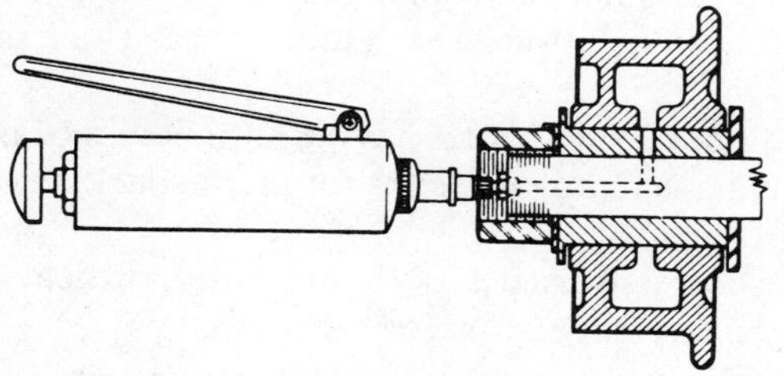

Figure 12-29 Chain joints must be regreased on an individual basis periodically. Annual regreasing is usually adequate.

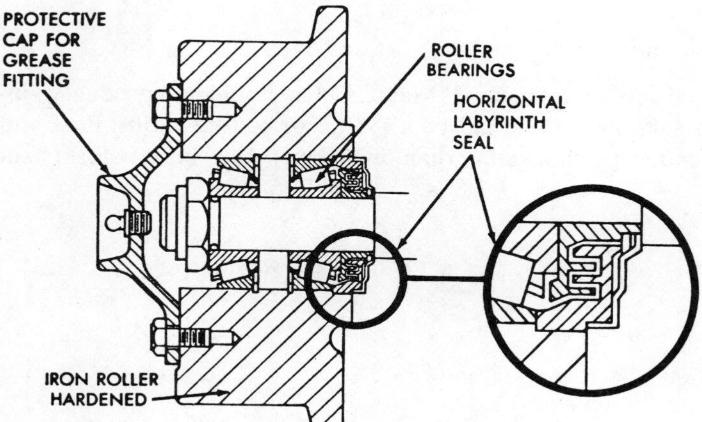

Figure 12-30 Chain roller equipped with antifriction bearings. Such rollers must be provided with grease fittings to carry clean grease into the bearings, which should have reliable closures. Regreasing on an annual basis or less frequently should be adequate.

installations will show signs of being in trouble long before emergencies take place.

A practical, recorded (checklists work well) inspection program should be established and then followed. Regardless of the type of installation involved, or the style of chain it uses, there are several criteria that should always be checked:

1. Check for abnormal wear on the sides of the sprocket teeth or the insides of the link plates or sidebars. Such wear can indicate a misalignment of the sprockets, or, in a two-stranded conveyor or bucket elevator, lateral slippage or displacement of sprockets or abnormal elongation of one of the chain strands.
2. Chain elongation will occur to a considerable degree when a new chain is first placed under load. This is simply a matter of pins and bushings settling in a place plus break-in wear. Elongation after break-in should be very gradual.

   Elongation is most easily evaluated in terms of slack or sag on the slack side of drives or at the foot sprockets of conveyors and bucket elevators. Sag on slack-side or takeup positions should be recorded periodically to detect changes more easily.
3. Lubrication should be a routine matter and is often handled as a part of the inspection procedure.

Suggested procedures for applications using Roller, Silent, or Engineering Steel drive chains may be somewhat different from those for conveyors or bucket elevators. These are reviewed in detail separately in the following sections.

Maintenance and Inspection of Drive Chains

Wear on link plates or sidebars and sides of sprocket teeth usually indicates misalignment in the drive.

Shaft and sprocket misalignment can sometimes occur by slippage of bearing blocks or sprockets on shafts and can be detected by inspection before becoming apparent in terms of damage to the chain.

Wear on working faces of sprocket teeth, such as a bright, polished surface, should appear as the drive runs-in. Scratches, grooves, galling, or visible changes in tooth form are a sign of trouble, the most common of which is sluggish roller action due to inadequate lubrication.

Chain elongation of as much as 3% of the assembled pitch indicates that the chain is riding high near its allowable limit on the sprocket teeth of Roller or Silent Chain sprockets.

A gradual increase in chain length is the result of normal wear. A sudden increase in slack indicates one or more of the following:

1. Lubrication failure
2. Excessive overloading or shocks
3. Displacement of bearings
4. Displacement or failure of takeups

An indication of the magnitude of chain elongation is readily available by maintaining records and comparing the chain sag.

Cleanliness in terms of accumulation of dirt or foreign materials is important. Dirt may often be evidence of corrosion. Foreign material packed into the chain or the sprocket teeth may cause chain or sprocket breakage.

Lubrication of all types should include checking the quality and grade of the lubricant.

For *manual* lubrication, make sure that the lubrication schedule is being followed and that the oil is being properly applied.

For *drip* lubrication, inspect the filling of the oiler cups and the rate of feed; check that the feed pipes are not clogged.

For *bath* or *disk* systems, inspect the oil level and check that there is no sludge. Drain, flush, and refill the system at least once a year.

For *oil stream* systems, inspect the oil level in the reservoir and check the pump drive and the delivery pressure; check that there is no clogging of the piping or nozzles. Drain, flush, and refill the reservoir at least once a year.

Maintenance and Inspection of Conveyor, Bucket Elevator, and Tension Linkage Chains

Generally speaking, inspection procedures for chains used for conveyors, bucket elevators, and tension linkages differ only in details from those for drives. The questions of alignment, chain elongation, and lubrication are at least as important, but other important details also need to be checked. Most of these are concerned with attachments, flights, or buckets, the presence of the materials being handled and other foreign materials, and the length of and use of two or more strands of chain in the installation. Because of the chain strand length, and also because conveyors and bucket elevators frequently run at much lower speeds than drives, passage over sprockets of individual joints is much less frequent. Loads may vary far more and shock loads be more frequent.

These added factors should be a matter of inspection and preventive maintenance, in addition to the elements previously discussed:

1. Guides, tracks, and areas below the conveyor should be kept clean and checked frequently.
2. Exit and entry points where chains enter or leave tracks or guides should be checked for deformation or buildup of material that would cause excessive pull on the chain or interference with attachments.
3. Overlubrication of tracks for roller conveyor chains with small-diameter rollers can cause such rollers to slide rather than turn and thus develop flats. All rollers of a conveyor or bucket elevator chain should be checked for rotation.
4. Damage of foreign materials between attachments such as apron flights or slats can cause bending or breakage. This can lead not only to material leakage, but successively larger materials entering the "gaps" thus formed can cause conveyor wrecks.
5. Conveyor or bucket elevator surging is sometimes caused by a stick-slip condition on long, very slow conveyors. Better lubrication or a slight increase in speed are possible remedies.
6. Chains handling gummy materials may need frequent cleaning or gums may set and harden and cause improper interaction with sprockets that can lead to chain breakage. Such conditions should be a matter of periodic recorded inspection and evaluation.

7. Buckets of bucket elevators digging in certain materials can be worn and also distorted so as to cause hang-ups. In this connection, inspection of both conveyors and bucket elevators, including a full cycle of inspection of each chain, should be standard practice.
8. Sag of tension linkages such as draw bench chains is often a key factor in maintaining adequate slack-side tension. Excessive wear, causing chain elongation such as will place the chain in contact with adjacent conveyor supports, can relieve slack-side tension and cause the chain to jump teeth. Sag of slack sections over nearby objects should be a matter of inspection and routine record.

Repair and Replacement of Chains

Repair and replacement of chains should be done with consideration of two important rules, abuse of which causes a substantial percentage of chain problems:

1. New links should never be used in badly worn chain. This is particularly important for drive chains or other short-pitch chains, but can also cause problems on other chains. New links are shorter than worn links and a shock is created every time the new link passes over a sprocket. In chains used with more than one strand, new links can cause strand mismatch. In both cases, the new link will not seat the same as the worn links on sprocket teeth, and chain breakage or sprocket damage can result.
2. New chain should never be used on badly worn sprockets. The resulting improper sprocket action can severely damage the chain in short order.

Replacement starts with disconnecting and removing a link so that the chain may be removed from the sprockets. This process is best done at a sprocket since the sprocket can serve to "buck" pressure or blows required to remove pins.

Regardless of the type of chain, there are several recommended procedures and precautions to be taken that apply to all types of chain.

1. Chain should always be disconnected when slack.
2. Field repair should be limited to replacement of complete links or sections only. Replacement of individual parts in the field is not recommended.
3. On chains of riveted construction, grinding away of pin heads is necessary to disconnect a strand of chain. It is usually necessary to grind into the chain link plate or sidebar to remove the swelled or upset portion of the riveted head. Since heat is usually generated from the grinding, the entire link should be discarded.
4. Chains that have been subjected to corrosion and having heavy interference fits are often parted by burning through with a torch. Adjacent links that have been exposed to heat should always be discarded, as the heat is usually sufficient to damage the parts.

Disassembly of chain is basically the reversal of the assembly process. Chains should be disconnected and removed from sprockets, if at all feasible, rather than being worked on in place. For disconnecting, straight link chains are best taken apart by driving both pins of an outside or pin link back through the near link plate or sidebar, and then removing the subassembly of two pins and the "far" link plate to disconnect the chain. Offset chains are readily disassembled by removing one pin, the opposite or far sidebar needing to be "bucked" by a bucking bar with a hole to clear the pin head.

Chain presses and screw-type disassembly tools are available for Roller Chains, but field disconnecting of heavy Engineering Steel Chains is usually done with a heavy hammer and drifts. When chains need to be disconnected and such is known a few days in advance, it is often desirable to mark a pin or pins for disconnection, and apply a penetrating oil to "unfreeze" the pin. When removal is done with a heavy hammer, care should be taken that a drift be used, as heavy blows directly on the pin head can "mushroom" and rivet the end (Fig. 12-31) if struck only slightly other than square.

Pins being removed should be kept as nearly square with sidebars or link plates as possible; otherwise, they may seize and be very difficult to remove, with the press fit destroyed.

Removal of split cotters is usually done with pliers. Solid, bent cotters are often straightened for removal with an adjustable wrench properly gapped.

Reconnecting New Sections or Links into Existing Chains. Reconnecting repaired or replacement chain, or new links installed to replace damaged parts, is done in the same manner as for new installations.

When replacement links or sections are provided in a chain in otherwise good conditions but with a few damaged links, links from the manufacturer of the original chain should be used. It is especially important that pins made by the manufacturer that supplied the sidebars be used in all cases involving either Silent Chain or Engineering Steel Chains and is highly desirable in all other cases. Shoulder and shank diameters of Engineering Steel Chains complying with ANSI standards are not necessarily the same from all manufacturers, and no Silent Chain parts made by any one manufacturer are interchangeable with chain parts made by another manufacturer.

Pins should never be ground smaller to eliminate interference fit to facilitate assembly. So doing will cause pins to work loose and rotate in sidebars or link plates, with resulting damage to the chain.

Sprocket Repair and Replacement. Sprockets with worn teeth should never be built up and ground down in an attempt to restore tooth profiles. It is nearly impossible to achieve correct tooth profiles with necessary tooth spacing by this process and such sprockets will often cause severe damage to the chain.

The life of a worn sprocket can sometimes be extended by reversing the sprocket on the shaft, provided that it is symmetrical relative to the hub, or there is space to do so. With double strand conveyors, reversing sprockets will usually require that the wheels also be switched relative to the strands to maintain alignment and match of teeth.

Cleaning of Chains. Even under the best operating conditions, periodic cleaning and relubrication of chains is good economics. Gummed lubricant and the products of normal wear can cause abnormally rapid pin and bushing wear. A drive chain exposed to dust or surroundings subject to other foreign material requires more frequent cleaning.

Drive chains or short strands in other service may be cleaned as follows:

1. Remove the chain from the sprockets.
2. Wash the chain in kerosene. If the chain is badly gummed, soak it for several hours in the cleaning fluid, and then rewash it in fresh fluid.

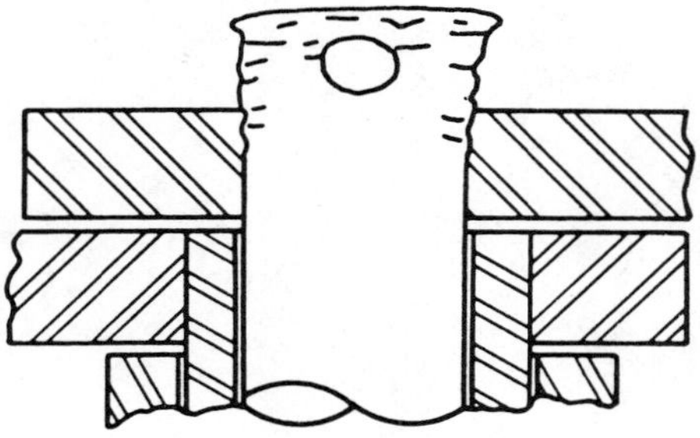

Figure 12-31 Angled or off-center blows of a hammer on a pin end, attempting to remove the pin, can "mushroom" the end so that it cannot be removed without damage to the holes in the sidebar or side plate.

3. After draining off the cleaning fluid, soak the chain in oil to restore the internal lubrication.
4. Hang the chain over a rod to drain off the excess lubricant.
5. Inspect the chain for wear or corrosion.

While the chain is off the sprockets, clean the sprockets with kerosene, and inspect them for wear or corrosion.

Storage of Idle Drive Chains and Sprockets. Unless properly protected, the components of a chain drive will deteriorate during long periods of idleness. If a chain is to be stored, remove it from the sprockets, clean and inspect it, and cover it with heavy grease. Then wrap it in heavy, grease-resistant paper. Store the chain where it will be protected from moisture, corrosive atmosphere, and mechanical injury.

The sprockets may be left in place on the shafts. Cover each with heavy grease and protect them from mechanical injury.

Before placing the drive in service again, thoroughly clean the chain and sprockets to remove the protective grease; then relubricate the chain.

Protection of Idle Conveyor Chains and Sprockets. Unless properly protected, the components of a conveyor system deteriorate during long periods of idleness, such as seasonal shutdowns. If a chain is to be stored, it can be removed from sprockets, broken down into sections that can be handled, cleaned, reoiled, coiled, and a rust preventive coating applied if storage is indefinite. Storage should be protected from moisture and mechanical injury.

Sprockets may be left on the shaft, protected with heavy grease.

Safety in Handling, Installing, and Maintaining Chains

The flexibility, use of interference fits, and loads involved and movement of chain and chain-operated equipment make safe practices an absolute "must" in working with such equipment. The procedures of connecting and disconnecting chains, when pins and link plates or sidebars are driven into place, require particular care. The following set of safety rules should always be followed, in addition to other local safety rules:

CAUTION—USE CARE TO PREVENT INJURY

Safety Guidelines For Connecting or Disconnecting Chain

1. Eye protection is required. Wear safety glasses, protective clothing, gloves, and safety shoes.
2. Lock out equipment power switch before installing or removing chain.
3. Determine correct direction for Pin/Rivet removal or insertion.
4. Support the chain or prevent uncontrolled movements of chain and parts.
5. Pressing equipment is recommended. Tools must be in good condition and properly used.

Glossary of Selected Terms

This section includes the "Glossary" of the Roller Chain *Design Manual* together with a somewhat similar section from the Engineering Steel Chain *Application Handbook,* called "Definitions of Selected Terms." Both those sections could be considered authoritative relative to the intent of the industry to standardize terminology.

The Glossary herein also includes certain other terms, used in chain catalogs in the past, and in some cases still used, which are not standard or even not approved on an industry-wide basis, for chains of the types discussed in this manual. Some of the terms are controversial; others are simply old or superseded. In this Glossary, these terms are identified as "old" or "former" or "manufacturers' " terms, meaning simply that these terms are still used where they will frequently be encountered. The purpose of including any of these terms is simply to explain its meaning, not to endorse the usage.

The purpose of a handbook such as this is to provide a potential user of chain, whether a student or an accomplished equipment designer, with a basis for a good understanding of chain use. The Glossary therefore includes not only the standard definitions but also those terms which the user is likely to encounter in catalogs, drawings, and specifications, whether old or new.

This Glossary is more representative than all-inclusive, but many terms not defined or given here are defined in the text and listed in the index.

Aeration Aeration is a property of many bulk materials of small particle size whereby, through agitation, the particles are separated by an adsorbed cushion of air. The bulk density is lowered and the mixture of material particles and air temporarily takes on many of the properties of a fluid.

Angle of flex The total angle of chain joint articulation as a chain enters or leaves a sprocket or wheel. The angle is equal to 360° divided by the number of teeth in the sprocket.

Articulation (v, "articulate") The action of a chain joint in flexing from the straight to an angle and back to the straight, as the joint passes around a sprocket or other path, causing it to flex.

Backing An old and now seldom-used term, usually applied to cast chains and always to chains not having rollers, for the dimension from the chain joint centerline to the contact point of the joint with the sprocket bottom diameter.

Backlash Movement (if any) of the chain along the pitch line of the sprocket when the direction of chain travel is reversed.

Block chain An alternative name used by some manufacturers for bar-link chain or for certain styles of leaf chains.

Block link An inside link, usually used with cast chain links but sometimes applied to Engineering Steel Chain inside links.

Bottom diameter The diameter of a circle tangent to the "pocket" configuration at the bottom of a sprocket tooth gap. *See also* Root diameter.

Caliper diameter The distance measured across the bottoms of two opposite tooth gaps in a sprocket having an even number of teeth, and measured between one tooth gap and the nearest opposite gap for a sprocket with an odd number of teeth.

Center distance The distance between the centers of the shafts of a chain drive.

Chain casing An oil-retaining sheet metal enclosure around a chain drive.

Chain guard An open guard of sheet metal, expanded metal, or similar construction around a chain drive.

Chain length The distance between the joint centers at each end of a taut chain strand. This distance is usually expressed in feet and/or inches or in pitches.

Chain pitch The average distance between joints (except for staggered pitch chains) of an assembled chain. In some cases, "joint," as defined here, will be a center of flexure not specifically identified with individual parts of the chain.

Chain pitch elongation or chain elongation Increase in measured length due to wear or excessive load. Normally expressed in percent increase in length.

Chain takeup An idler sprocket or similar device, mounted on an adjustable bracket to adjust the slack in a chain installation.

Chain width Defined somewhat differently for various chains, but usually the inside width of the chain, between roller link plates for Roller Chain or sidebars of inside links for Engineering Steel Chains. (Note that this is not the overall or clearance width of the chain.) The width of Silent Chain is the overall width, although the width between guides will also be given when appropriate.

Chills Cast or fabricated metal segments or rings which form the tooth contour in the mold of a cast sprocket. The chill cools the poured iron rapidly so that a hard, wear-resistant surface is produced in iron of the correct chemistry.

Chordal effect (chordal action) The effect produced by the chain joint centers being forced to follow arcs instead of chords of the sprocket pitch circle (see Chapter 4).

Chordal pitch The length of one side of the polygon formed by the lines between the joint centers as the chain is wrapped on the sprocket. It is a chord of the sprocket pitch circle, and is equal to the chain pitch.

Circumferential stress Stresses that act around the circumference of a bushing or cylinder.

Compressive stresses Stresses that act to compress a material and place the fibers in compression.

Connecting link For a Roller Chain, a pin link made with one link plate easily detachable to facilitate connecting or disconnecting the chain.

Crotch height For a Silent Chain, the height of the link crotch above the pitch line of the link.

Cutter A special cutting tool which machines the necessary tooth and tooth gap contours in a blank for a cut-tooth sprocket.

Cycle Change in load level as a chain passes around a system. Usually the change is from negligible load to a load peak on a regular basis as the chain undergoes a complete cycle of operation.

Design horsepower The specified horsepower for a chain drive multiplied by a service factor. It is the value used to select the chain size for the drive.

Digging or loading action In calculating the chain pull for a bucket elevator or a conveyor, a factor is added to the calculated load to account for forces generated by digging or loading.

Double-cut sprocket For double-pitch Roller Chains, a sprocket having two sets of effective teeth (*see* Effective teeth). Tooth spaces for the second set are located midway between those of the first set (see Chapter 4). Such teeth make contact with the chain only on alternate revolutions of the sprocket.

Double-pitch roller chain A Roller Chain having double the pitch of a standard Roller Chain, but otherwise having standard pins and bushings and standard or oversize rollers.

Effective teeth The number of sprocket teeth that engage the chain rollers during one revolution of the sprocket. The term normally applies to sprockets for double-pitch Roller Chain. *See* Hunting tooth sprocket.

Elastic limit The highest load a part (i.e., chain strand) can sustain without incurring a permanent change in length.

Embrittlement Reduction in ductility of materials due to exposure to certain environments or temperatures.

Engineering (Roller) Chain Manufacturers' term for nonstandard Roller Chain, usually double-pitch style but an odd pitch, intended for a specific special application.

Galling Developing a condition on the live bearing surface of a pin or bushing of a chain where excessive friction between high spots result in localized welding with subsequent tearing and a further roughening of the contact surfaces.

Hub and groove diameter The outside diameter of the hub, or the diameter at the base of a groove cut in the hub to provide clearance for link plates or sidebars.

Hunting tooth sprocket A sprocket with two sets of effective teeth (similar to double-cut sprockets; *see* Double-cut sprocket) arranged so that each set makes contact with the chain on alternate revolutions of the wheel. Applies to sprockets with cast teeth, intended for Engineering Steel Chains.

Included angle For Silent Chains, the angle included between the outer surfaces of the link plate contours. This angle affects the layout of the sprocket tooth form.

Knuckle Manufacturers' term for bushing of a steel bushed rollerless chain. Such bushings are usually of a shouldered design.

Knuckle chain Manufacturers' catalog term for steel bushed rollerless chain.

Link The portion of a chain connecting adjacent pitch centers.

Link plate One of the side plates of either a pin link or a roller link in a Roller Chain. For a Silent Chain, any one of the plates of which an assembled chain is composed.

Load classification A classification of drive loads based on the intensity of shock that is imposed on the drive.

Load magnitude The load magnitude is the physical size of the tensile force being applied to a chain at a particular time. The load magnitude range is the difference between the largest and the smallest tensile forces applied during a complete loading cycle.

Loading frequency (time) Loading frequency is the number of times per unit of time that the chain in a conveyor or a bucket elevator is exposed to a complete cycle of loading. A complete loading cycle normally occurs when a particular link moves completely around the system and returns to its starting point.

Longitudinal stresses Stresses that act along the span of a bushing or cylinder.

Measuring load The specified standard load under which a chain is to be measured for length.

Mohs hardness A ranking of minerals' hardnesses from 1 to 10. A higher-numbered mineral can scratch a lower-numbered mineral, but not vice versa. The Mohs hardness of a diamond is 10.

Multiple Strand Chain A Roller Chain (or other chain) made up of two or more strands assembled as a single structure on pins extending through the entire assembly.

Multiple strand factor A factor by which the horsepower rating of a single-strand chain is multiplied to obtain the horsepower capacity of a chain with two or more strands.

Offset link A special offset link, as termed in Roller Chain descriptions, made for use in a straight link chain when an uneven number of links in the total strand is required. Also applies, as a specific term, to Silent Chains.

Offset section For a Roller Chain, a factory-assembled section, made up of a roller link and an offset link; for a Silent Chain, a factory-assembled section, made up of three or more links, one or more of which consists of offset plates. Offset sections are used to connect strands of chain having an odd number of pitches.

Overchaining A drive is overchained when it incorporates a chain of substantially higher rating than that indicated by normal selection procedures to have been necessary.

Over-gauge diameter The measurement over gauge pins of specified diameter inserted in opposite tooth gaps of a sprocket. For an odd number of teeth, the pins are placed in the nearest opposite gaps.

Pitch *See* Chain pitch; Pitch diameter.

Pitch circle The circle passing through the chain joint centers if the chain were wrapped around a sprocket of correct theoretical dimensions.

Pitch diameter The diameter of the sprocket pitch circle.

Pocket curve The curve or configuration of a sprocket at the bottom of the tooth gap, tangent to the bottom diameter circle.

Pulsations Fluctuations of a cyclic nature in load or speed.

Roller flex path The path taken by the roller (or barrel) as it flexes onto or off of the sprocket.

Root diameter The theoretical bottom diameter of a sprocket, equal to the pitch diameter minus the chain roller or barrel diameter.

Run-in The initial period of operation of any mechanism, during which the component parts seat themselves.

Scoring Marring or scratching of pin or bushing caused by metallic debris being picked up in the contact surfaces of one of the parts.

Seating curve A specific term for the pocket curve of a Roller Chain sprocket.

Seizing Stiffening (or "freezing") of a chain joint as a result of roughness and high friction caused by galling.

Service factor A factor by which the specified horsepower or working load of a chain is multiplied to compensate for operating conditions.

Sidebar The tension member of an Engineering Steel Chain.

Single-cut sprocket For double-pitch Roller Chains, a sprocket having one set of effective teeth.

Staggered Pitch Chain A chain with alternate links of differing pitches, one usually being considerably greater than the other.

Steel Thimble Roller Chain Specific manufacturer's term for steel roller chain of the Engineering Steel Type.

Takeup *See* Chain takeup.

Taper-lock sprocket Manufacturers' term for sprockets equipped with a split tapered bushing for rigid mounting on a shaft.

Template An accurate scale outline of a sprocket tooth contour which is followed by a photoelectric cell, cam, or other follower, on special flame-cutting equipment used to cut sprocket teeth.

Tension linkage A chain application primarily transmitting motion (tensile force rather than horsepower) at low speeds.

Thermoplastic A plastic material that can be softened and resoftened without significant changes in chemical composition.

Thimble Specific manufacturers' term for the bushing of an Engineering Steel Chain.

Tooth flange thickness The width, in profile, of a Roller Chain sprocket tooth.

Tooth form The shape of the working surface of a sprocket tooth.

Tooth gap The space between two sprocket teeth.

Tooth profile The outline of a sprocket tooth as projected on a plane through the sprocket axis and the center of the tooth.

Topping curve The curve of the outer portion of the tooth form. It is shaped to guide the roller (if any) smoothly into mesh with the sprocket.

Torque Torsional moment, expressed in inch-pounds in chain calculations, which is the product of chain pull and one-half the sprocket pitch diameter.

Transverse pitch The lateral distance between the centerlines of each strand of a multiple strand chain, or between the tooth profiles on a sprocket for a multiple strand chain.

Underchaining A drive is underchained when it incorporates a chain of substantially lower rating than that indicated to be needed from normal selection procedures.

Working face That portion of the face of the sprocket tooth contacted by the barrel, roller or bushing, or equivalent.

Working load An allowable recommended tensile load for chains used on conveyors, drives with nonstandard chains, or other applications of low relative speeds.

Index